A Laboratory for General, Organic, and Biochemistry

Third Edition

Charles H. Henrickson
Larry C. Byrd
Norman W. Hunter
of
Western Kentucky University

Boston Burr Ridge, IL Dubuque, IA Madison, WI New York San Francisco St. Louis
Bangkok Bogotá Caracas Lisbon London Madrid
Mexico City Milan New Delhi Seoul Singapore Sydney Taipei Toronto

McGraw-Hill Higher Education 

*A Division of The **McGraw-Hill** Companies*

A LABORATORY FOR GENERAL, ORGANIC, AND BIOCHEMISTRY
THIRD EDITION

 This book is printed on recycled, acid-free paper containing 10% postconsumer waste.

1 2 3 4 5 6 7 8 9 0 QPD/QPD 0 9 8 7 6 5 4 3 2 1 0

ISBN 0–07–231785–X

Vice president and editor-in-chief: *Kevin T. Kane*
Publisher: *James M. Smith*
Sponsoring editor: *Kent A. Peterson*
Developmental editor: *Shirley R. Oberbroeckling*
Editorial assistant: *Jennifer L. Bensink*
Senior marketing manager: *Martin J. Lange*
Senior marketing assistant: *Tami Petsche*
Senior project manager: *Jayne Klein*
Media technology project manager: *Phillip Meek*
Production supervisor: *Sandy Ludovissy*
Design manager: *Stuart D. Paterson*
Cover designer: *Joshua Van Drake*
Cover image: *"Brazilian Amazon Rainforest" PhotoDisc*
Supplement coordinator: *Brenda A. Ernzen*
Printer: *Quebecor Printing Book Group/Dubuque, IA*

Some of the laboratory experiments included in this text may be hazardous if materials are handled improperly or if procedures are conducted incorrectly. Safety precautions are necessary when you are working with chemicals, glass test tubes, hot water baths, sharp instruments, and the like, or for any procedures that generally require caution. Your school may have set regulations regarding safety procedures that your instructor will explain to you. Should you have any problems with materials or procedures, please ask your instructor for help.

www.mhhe.com

One must learn by doing the thing;
for though you think you know it,
you have no certainty until you try.

Sophocles

CONTENTS

To the Instructor:

A principal difference between science courses and nonscience courses is the inclusion of the laboratory. It is obvious to us who teach chemistry why this is so, but too often our students fail to realize the extensive experimental work behind the facts presented so freely in lecture. The laboratory can bring abstract words and concepts to life. Something as elementary as a boiling point becomes real when observed firsthand.

For more than twenty-five years we have been concerned with the laboratory experience, and have developed these laboratory exercises to support the training of students in health-related areas, agriculture and general education. Each exercise has been modified, rewritten and modified again until we were confident that it demonstrated the concepts we wanted to show. To use a worn term, these experiments "work" the way we describe them. Detailed, comprehensive instructions for preparing each experiment are in the *Instructor's Manual*. In addition, the *Instructor's Manual* contains many hints that we have learned over the years to ensure student success.

The transition from the second to the third edition of **Laboratory Experiments in General, Organic and Biochemistry** brought about some changes. The first two experiments of the previous edition were combined into one. This allowed the inclusion of an exercise in the preparation and interpretation of straight-line graphs. An experiment focusing on aldehydes and ketones was also added. Two experiments were deleted: Distillation and Qualitative Analysis of Cations. This gives us twenty-seven experiments that begin with the basics of measurement and conclude with the chemistry of digestion. Some experiments require more than one laboratory period. Each is written as a self-contained teaching and learning unit. Though each experiment benefits greatly from supporting material presented in lecture, sufficient information is provided in the discussion section to understand the fundamentals of the laboratory activity. There are several features that we think, when taken together, make this manual unique:

- Laboratory safety is presented first thing, requiring a signed verification that each student is aware of the safety regulations in the laboratory.

- Each experiment begins with a stated purpose.

- A comprehensive discussion follows that introduces the experiment and provides much of the background needed to understand the activity.

- Preliminary Exercises are designed to get the student into the investigation before the laboratory begins. They can be collected as part of the experiment or used as a guide for short quizzes.

- Each experimental procedure and its accompanying Report Sheet are tightly linked with identical section titles and organization. Each Report Sheet has clearly identified areas for observations, calculations and responses to questions raised in the experimental procedure.

- Ample use of photographs and line drawings make equipment setups and concepts easier for the student to understand.

- Care has been taken to minimize use of expensive equipment and reagents on a large scale.

- Caution warnings appear in boxes when important safety concerns merit repeating.

- Environmental alerts appear in boxes, alerting students to follow proper chemical waste collection procedures.

- Hazardous reagents are avoided wherever possible, but if concentrated acids or corrosive substances are required, adequate warning is given before they are encountered in the procedures.

- Calculations are clearly described in a step-by-step sequence with examples for each experiment that requires calculations.

- Several special sections appear throughout the manual addressing concepts or computations that are important in the experiment at hand. The language of acid-base chemistry appears in Experiment 15. A discussion of molarity and calculations using molarity appears in Experiment 14. The concept of percent yield is examined in Experiment 18, and formula writing and Lewis structure writing are presented in Experiments 7 and 9 respectively. The step-by-step method of performing calculations based on balanced equations is presented in Experiment 8. In addition, brief biographical sketches of several famous scientists appear in Experiments 10 and 13.

- An introduction to chemical nomenclature appropriate to this level appears in Appendix F with two sets of student exercises.

- A brief discussion of significant figures and rounding numbers is presented in Appendix G and is accompanied by calculation exercises.

We wish to express our gratitude to the chemistry faculty and our students at Western Kentucky University for their helpful suggestions over the years. Special thanks goes to Shirley Oberbroeckling, our developmental editor at McGraw-Hill for assistance in preparing the manuscript. The manuscript was prepared camera-ready by C.H.H. using Word Perfect®, Draw Perfect® and Bitstream® fonts.

Charles H. Henrickson
Larry C. Byrd
Norman W. Hunter

Bowling Green, Kentucky

SECTION I:

General Laboratory Procedures and Laboratory Safety

Welcome to chemistry lab. Before you can begin to study chemistry in the laboratory you need to learn about:

1. The general operating procedures of a chemistry laboratory
2. The proper handling of chemical reagents
3. Laboratory safety

Please read the following sections carefully since they specifically address these topics. During the first laboratory period you will be assigned a work space and a set of equipment. This equipment will be exclusively yours, so you will need to ensure it is all there and in good condition through the check-in procedure.

General Laboratory Procedures:

A. Working in the Laboratory

1. *Keep your assigned working area reasonably clean at all times.* Clean your work space with a damp towel before leaving the lab for the day.

2. *Keep your lab equipment clean.* Do not store dirty glassware or chemical reagents in your locker between laboratory periods.

3. *You are responsible for the equipment assigned to you.* If you fail to lock your desk and equipment is stolen, you will have to pay to replace it.

4. *Coats, books and other apparel should be stored in the assigned areas* while laboratory experiments are being performed.

5. *If you make a mess, clean it up right away.* Be especially careful not to spill reagents around the balances, reagent shelves and fume hoods. If you are not certain how to clean up a spill, ask the instructor.

6. *Follow proper procedures for the disposal of liquid and solid waste.* To reduce the impact on the environment, certain materials must be collected in specific containers for proper disposal. Your instructor will inform you when this is necessary. At other times it is satisfactory to dispose of waste liquids by washing them down the drain with plenty of water. *If you have any questions about proper disposal, ask your instructor.*

7. *Do not attempt to dispose of solid waste in the sink.* Place solid waste in the designated disposal cans. If you litter a sink, *you* will be required to clean it.

8. *Be certain that the water and gas jets are shut off* before leaving the lab for the day.

B. Handling Laboratory Reagents

1. *Read the labels on reagent bottles carefully.* Serious accidents may result if the wrong reagent is used in an experiment. There is a considerable difference between:

    a. magnesium chloride ($MgCl_2$) and manganese(II) chloride ($MnCl_2$)

    b. 6 M H_2SO_4 (dilute acid) and 18 M H_2SO_4 (concentrated acid)

    c. iron(II) sulfate, $FeSO_4$, and iron(III) sulfate, $Fe_2(SO_4)_3$

    d. charcoal and manganese dioxide (both black solids)

2. *Please do not move a reagent bottle far from the reagent shelf* and *always return the bottle to the same spot from which it came.* If these simple rules are followed, everyone will know where a reagent is when it is needed.

3. *Help keep reagents pure by not contaminating them with dirty utensils.* Scoopulas and spatulas should be wiped clean before obtaining solid reagents from bottles. If you need a few drops of a liquid, pour a small amount into a clean beaker and use a dropper to obtain the desired volume as shown in Figure 1. Never insert a dropper into the reagent bottle.

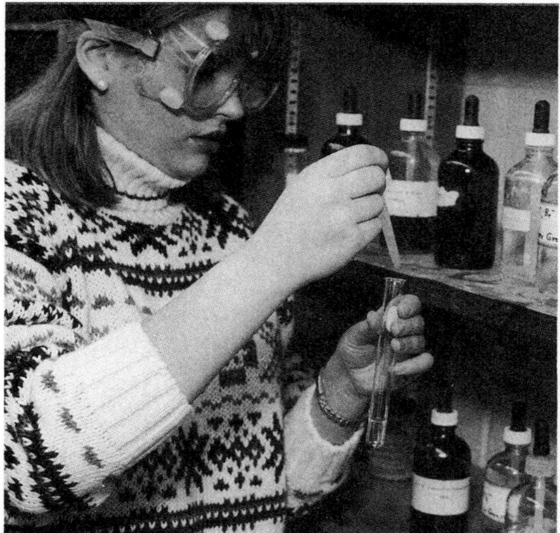

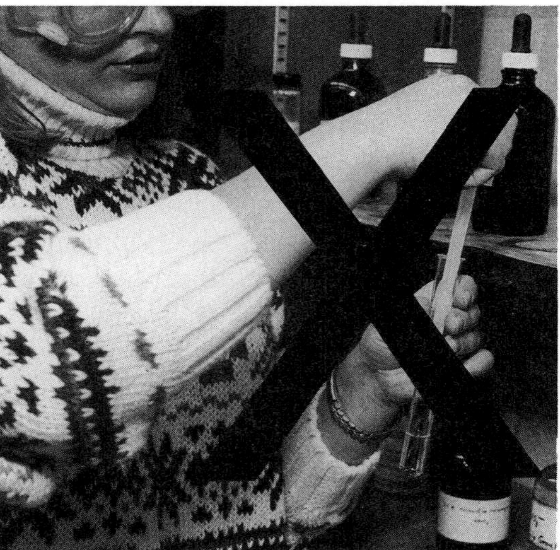

Figure 1. The correct way to obtain a sample with a dropper: Hold the loaded dropper *above* the mouth of the test tube and release the reagent as needed. *Never* place the dropper in the test tube. This will contaminate the dropper and, in turn, the solution it came from.

4. *Unused reagent should never be returned to the reagent bottle.* If you get more than you need, either give the excess to a fellow student or discard it in the proper waste container. This will eliminate contamination of the reagents.

5. *Never place a reagent directly on a balance pan.* Use weighing paper (Figure 2), a plastic weighing tray or a small beaker to hold the reagents. Some balances allow you to automatically compensate for the weight of the container.

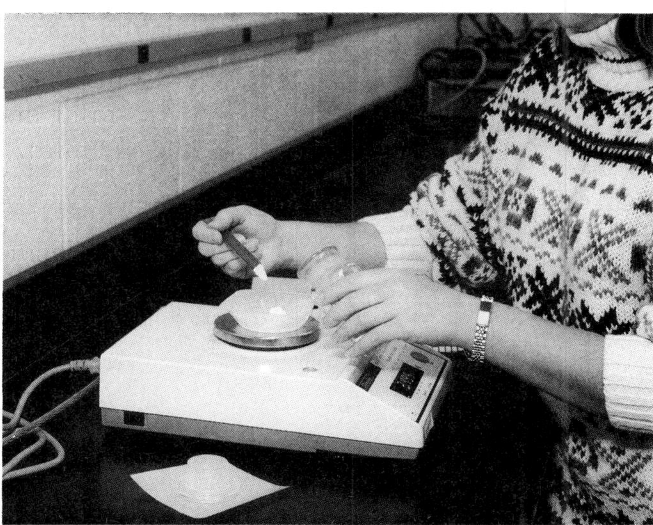

Figure 2. Always use weighing paper or a plastic weighing cup when weighing out reagents. Never place the reagent directly on the balance pan.

6. *If a corrosive liquid gets on your skin, quickly rinse the affected area with running water and inform the instructor.* The instructor will know how best to treat the problem.

C. The Experiments

> **CAUTIONS, ENVIRONMENTAL ALERTS and INSTRUCTIONAL NOTES will be set off in boxes in the experiments where necessary. Pay close attention to these. They provide safety instruction, information about disposal of hazardous waste and important notes so you can correctly prepare for the experiments.**

1. *Attend laboratory every time it is scheduled.* A missed laboratory is difficult to make up, and special reagents that may have been prepared might not be available the following week. *If you must miss lab, inform your instructor beforehand* so arrangements can be made for you to make it up. It is always best to make up the lab during the *same* week you have the absence.

2. *Prepare for the laboratory by reading the experiment and completing the Preliminary Exercises before coming to lab.* If you have a general understanding of the experiment, procedures and calculations, you will save time and reduce the chances of making serious errors. Your instructor may collect the Preliminary Exercises first thing each day, or they may be used to prepare for lab quizzes. Your instructor will let you know.

3. *You must do your own work in lab.* You are encouraged to discuss any aspect of any experiment with your assistants, fellow students and the instructor, but the work you turn in for grading must be your own.

4. *Complete the Report Sheet and questions that accompany each experiment with complete sentences and use proper English.* All calculations must be clearly shown and all measured values must be accompanied by the correct units (grams, centimeters, seconds, etc.). Calculated values must be written with the proper number of significant figures.

Laboratory Safety:

The laboratory instructor and assistants will not be able to watch every student at all times to see if things are being done correctly and safely. Because of this, *you* must look out for your own well-being and that of your neighbors. To point out areas of greatest concern, a list of 15 safety rules has been compiled:

1. *Think about what you are doing at all times.* Study the experiments before coming to lab, know the hazards before you begin, and be prepared for the unexpected. Think not only of your own safety but that of others working around you (Figure 3).

Figure 3. Charlie is thinking more about Mozart than mixtures and splashes Paula with acid. But Paula, being a bright student, is wearing her safety goggles. It could have been worse.

2. Safety glasses, safety goggles or prescription safety glasses must be worn at all times when laboratory work is being done. Shoes with tops (not sandals) are also required.

3. Report *at once* any accident, however minor, to your instructor.

4. Do not do unauthorized experiments in the lab. Do only the experiments specified by your instructor.

5. Know the location of the fire extinguishers, safety showers and eye-wash equipment in the laboratory room. Also, know where the nearest first-aid kit and telephone are located.

6. At no time will a student be permitted to work in a laboratory alone.

7. Never aim the opening of a test tube or flask in which a reaction is taking place at yourself or anyone else. Avoid the temptation to peer into the test tube to see what's up. It may come up to meet you.

When heating a liquid in a test tube, hold the tube at an angle and apply the flame near the surface of the liquid, as shown in Figure 4. Do not look directly into the test tube, and do not point the mouth of the test tube at yourself or a neighbor.

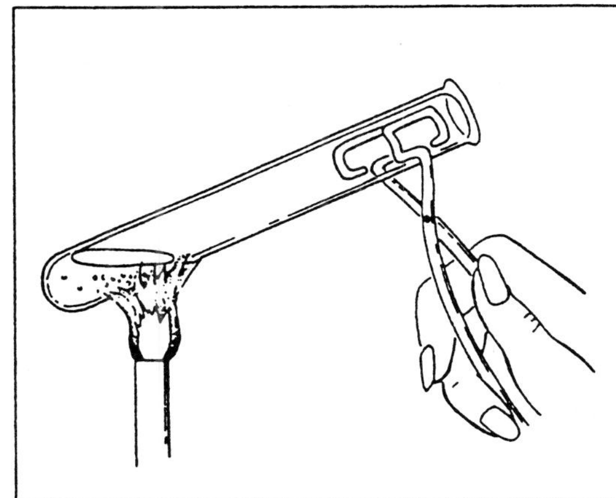

Figure 4

8. Never leave an experiment unattended while it is being heated or reacting rapidly.

9. Be very cautious when checking the odor of a substance.

When observing the odor of a substance, fill your lungs half-full of air, then cautiously inhale the vapors as you wave them toward your nose, as shown in Figure 5. If the vapor is irritating, the air you inhaled at first will let you exhaust it quickly from your nostrils.

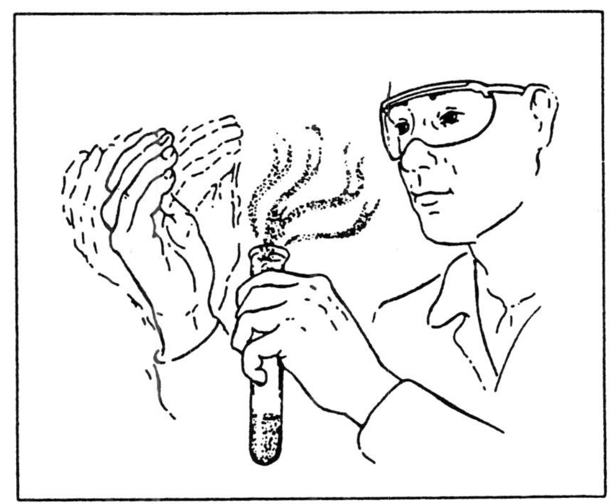

Figure 5

10. Never put anything in your mouth while working in the laboratory. Eating, drinking and smoking are not allowed. Do not pipet by mouth. Always use a pipet bulb to draw liquid up into the pipet.

11. Carry out experiments in the fume hoods when asked to do so to allow irritating or toxic fumes to be vented away from you and others in the lab.

12. Clean up any spilled reagents immediately. Ask your instructor for assistance if you are not sure how to clean up a particular chemical.

13. When inserting glass tubing into rubber stoppers, always lubricate the glass with glycerine (glycerol) first and wrap the tubing and stopper in a towel, as shown in Figure 6. The tubing should be fire polished and inserted with a gentle twisting motion. Avoid using too much force or the glass may snap and an injury occur. Your hands should be closer together than is shown in the figure below. Here, the hands are kept apart to allow the tubing to be seen.

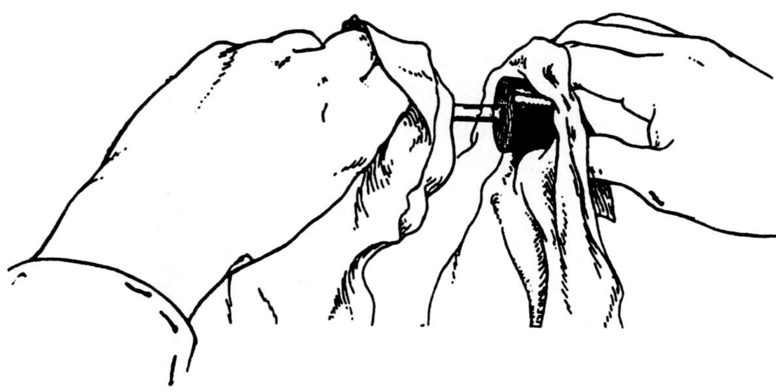

Figure 6. The correct way to insert tubing in a stopper: *Lubricate* the stopper and glass with glycerine. Protect your hands with a towel and insert the tubing with a gentle twisting motion.

14. Bundle long, loose hair so it will not become fuel for a nearby burner.

15. Think about what is hot and what is not. A common injury in laboratory occurs when someone tries to hurry things along and grabs a ring stand, glass or burner before it has had time to cool off. As the first safety rule says, *think about what you are doing at all times.*

Figure 7. If you need a small volume of reagent, place a sample in a small beaker and then transfer what is needed using a dropper. This avoids contamination of the stock reagent. Note: When pouring liquids, the bottle label should face the palm of your hand and the bottle cap should be held between your fingers. If the cap must be set down, place it on a clean piece of paper, watch glass or glass plate.

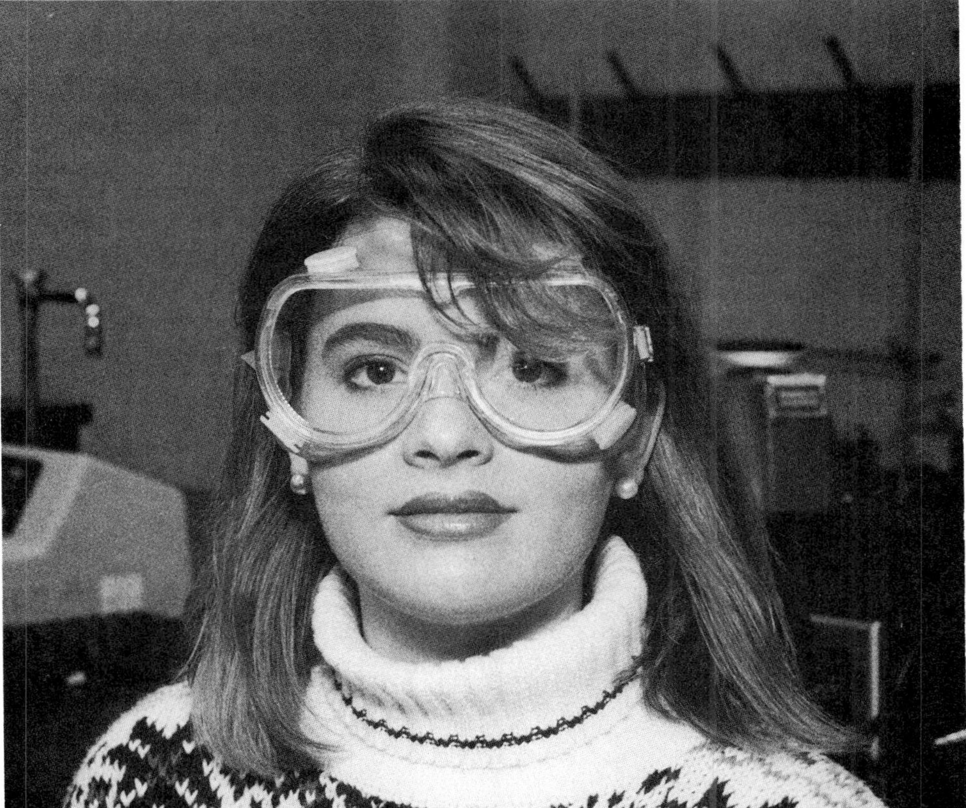

Figure 8. Eye protection must be worn at all times in the laboratory. Goggles provide the best protection from the front and sides.

LABORATORY SAFETY RULES VERIFICATION FORM

After you have carefully read the following safety rules, sign this form and return it to your instructor.

1. *Think about what you are doing at all times.* Know the hazards before you begin an experiment and be prepared for the unexpected. Think not only of your own safety but that of others working around you.

2. Safety glasses, safety goggles or prescription safety glasses must be worn at all times when laboratory work is being done. Shoes with tops (not sandals) are also required.

3. Report at once any accident, however minor, to your instructor.

4. Do not do unauthorized experiments in the lab. Do only the experiments specified by your instructor.

5. Know the location of the fire extinguishers, safety showers and eye-wash equipment in the laboratory room. Also, know the location of the nearest first-aid kit and telephone.

6. At no time will you be permitted work in the laboratory alone.

7. Never aim the opening of a test tube or flask in which a reaction is taking place at yourself or anyone else.

8. Never leave an experiment unattended while it is being heated or reacting rapidly.

9. Be very cautious when checking the odor of a substance.

10. Never put anything in your mouth while working in the laboratory. Eating, drinking and smoking are not allowed. Do not pipet by mouth.

11. Carry out experiments in the fume hoods when asked to do so to allow irritating fumes to be vented away from you and others in the lab.

12. Clean up any spilled reagents immediately. Ask your instructor for assistance if you are unsure about handling a spilled chemical.

13. When inserting glass tubing into rubber stoppers, always lubricate the glass with glycerine first and protect your hands with a towel. The tubing should be fire polished and inserted with a gentle twisting motion. Avoid using too much force.

14. Bundle long, loose hair.

15. Think about what is hot and what is not. A common injury in laboratory occurs when someone tries to hurry things along and grabs a ring stand or burner before it has had time to cool off. As the first safety rule says, *think about what you are doing at all times.*

I have read and understood these laboratory safety rules in their entirety.

______________________________ (signature) _______________ (date)

Checking into the Laboratory:

Your instructor will have specific check-in instructions that are appropriate for your laboratory design. Several common items of laboratory glassware and hardware are shown below to help you identify them during the check-in procedure.

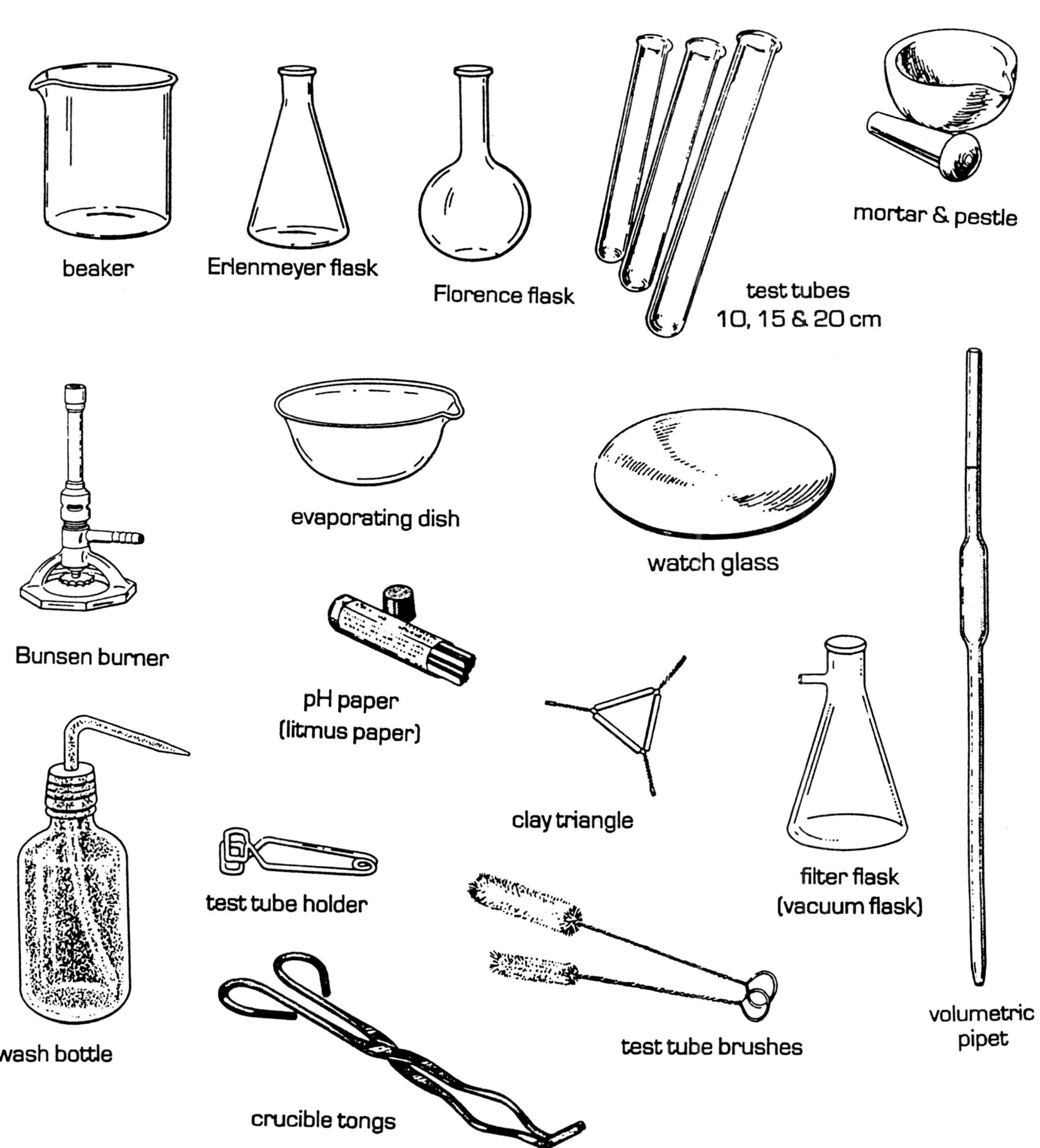

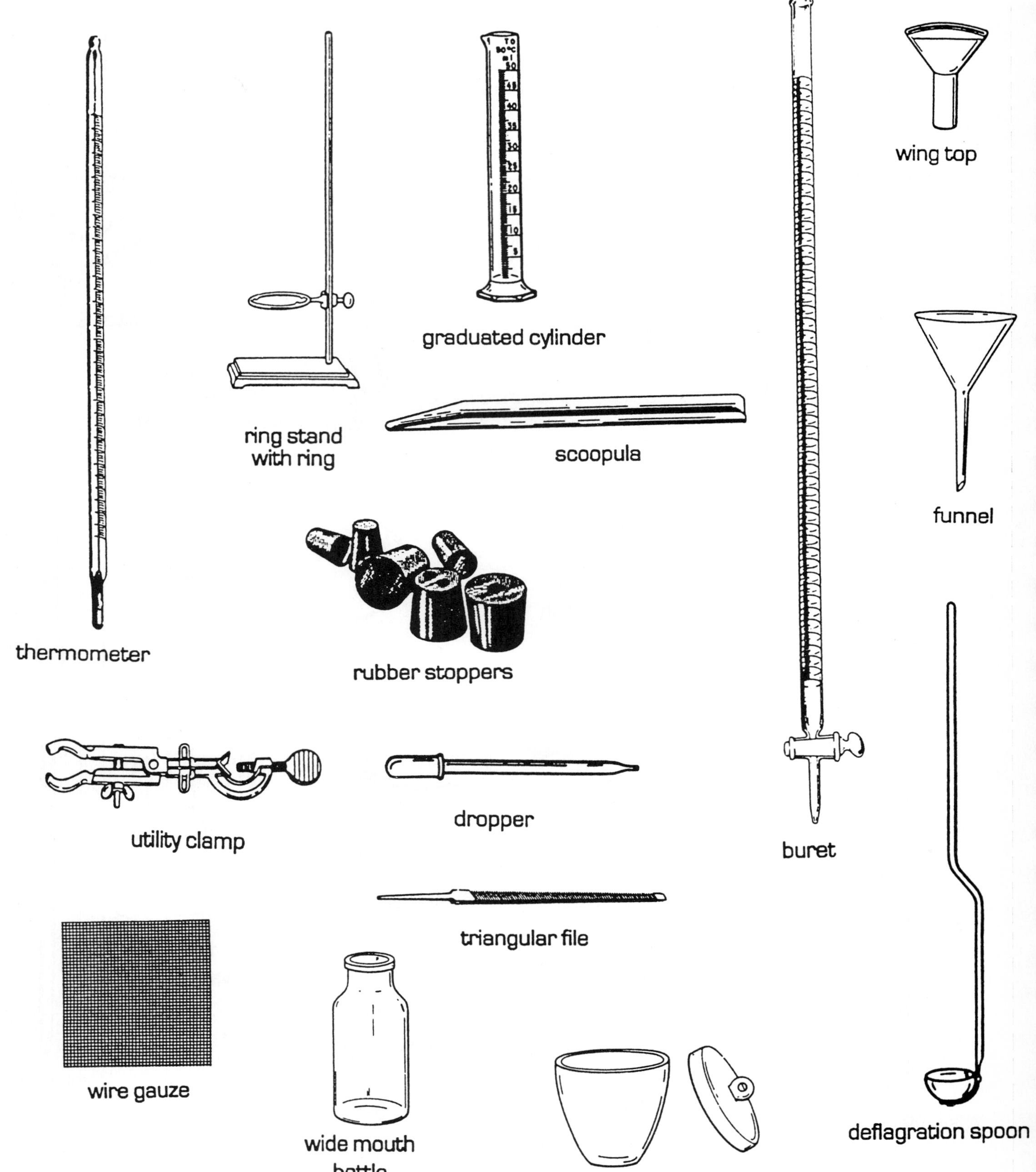
thermometer
ring stand
with ring
graduated cylinder
scoopula
buret
wing top
funnel
rubber stoppers
utility clamp
dropper
triangular file
wire gauze
wide mouth
bottle
crucible & lid
deflagration spoon

SECTION II:

Basic Concepts, Measurement, and Properties

Experiment 1

Measurement and Density[1]

Purpose:

This activity will teach you about measurements, their accuracy, precision and uncertainty and how each of these can be stated. You will learn techniques for measuring solids and liquids in the laboratory and you will learn how to evaluate the quality of a balance. Then you will use these skills to determine the density of a solid.

Discussion:

Many experiments involve the measurement of mass, volume or length. No matter how carefully you may carry out a measurement, no matter how carefully you read the measuring instruments, there will always be some uncertainty in every measured value. The uncertainty can be minimized but never completely eliminated. The quality of a set of measurements is stated in terms of accuracy and precision.

A. Accuracy of a Measurement

The **accuracy** of a measured value concerns the agreement of that value with the "true" value, or with what is generally accepted as the "true" value. Suppose a bar of silver has a known mass (true mass) of exactly 500 grams, and two students weighed the bar and reported 498 g and 512 g as the mass. The value of 498 g has greater accuracy than the other. The accuracy of a value can only be known, then, if the true value is also known. The accuracy of a measurement is quantitatively expressed in terms of either **absolute error** or **relative error**. If the error has a positive sign, the experimental value is greater than the true value, and vice versa if the error is negative. But frequently relative error is expressed without the + or − sign. Relative error is frequently stated as percent error.

$$\text{absolute error} = (\text{experimental value} - \text{true value})$$

$$\%\ \text{relative error} = \frac{\text{absolute error}}{\text{true value}} \times 100\%$$

[1] Appendix VII: **Significant Figures and Rounding Numbers** complements this exercise.

Most often the "true" value is not known, so to increase one's confidence in a reported value, several measurements are made and averaged, or alternate methods of measurement may be used to insure an accurate value is obtained. The United States Bureau of Standards provides samples of known "true values" for technical and scientific use.

B. Precision of Measured Values

Precision is a measure of the agreement between two or more measurements that are averaged to give the reported value (each measurement being made under the same conditions). Whereas accuracy is a measure of the agreement between a reported value (which may be the average of several trials) and the true value, precision is a measure of agreement between the individual measurements, i.e., the values used to obtain the average. Errors that affect precision occur randomly and have an equal probability of being on the high side or the low side of the average of a set of measurements.

A set of measured values can be of high precision and high accuracy, or of low precision and low accuracy or of any other high-low combination. Figure 1 shows the results of three weighings done by four different students on four different balances. The true mass of the weighed sample is indicated by the vertical line.

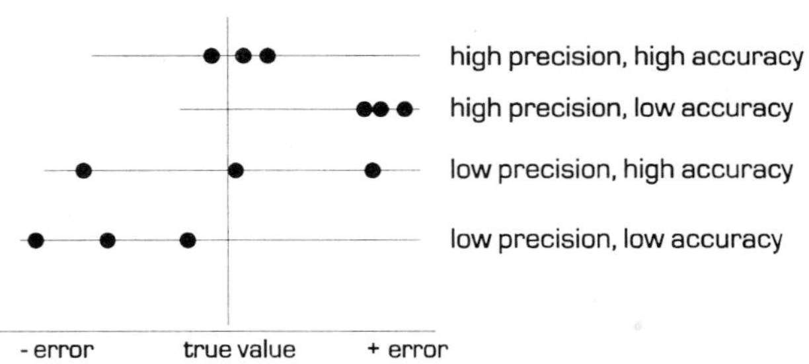

Figure 1

The most common method used to express the precision in a set of measured values is the **standard deviation**, a statistical measure of the scatter of data about the average, or mean, of a set of measurements. *The smaller the standard deviation the higher the precision of the set of measurements.* Standard deviation (s.d.) is defined by the following equation:

$$\text{s.d.} = \sqrt{\frac{\Sigma\,(X_i - X_{mean})^2}{N - 1}}$$

X_i = value of one measurement
X_{mean} = mean (average) of all N measurements
N = number of measurements in the data set

An approximate value for s.d. can be obtained by dividing the *range* of the data (highest measured value minus lowest measured value) by the square root of the number of measured values, N. As N gets larger, the reliability of this approximation increases.

$$\text{standard deviation} = \text{s.d.} \approx \frac{\text{range}}{\sqrt{N}} = \frac{(\text{highest value} - \text{lowest value})}{\sqrt{N}}$$

$\sqrt{3} = 1.732$ $\quad\sqrt{4} = 2.000$ $\quad\sqrt{5} = 2.236$ $\quad\sqrt{6} = 2.449$

This approximate relationship is used below to find the standard deviation of a set of six measurements.

measurement	mass
1.	6.93 g
2.	6.88 g
3.	6.99 g
4.	7.02 g
5.	7.12 g
6.	6.94 g

$$\text{range} = (7.12\text{ g} - 6.88\text{ g}) = 0.24\text{ g}$$

$$N = 6 \quad \text{and} \quad \sqrt{N} = 2.449$$

$$\text{s.d.} = \frac{(\text{range})}{\sqrt{N}} = \frac{0.24\text{ g}}{2.449} = 0.098 = 0.10\text{ g}$$

$$\text{mass} = \text{mean} \pm \text{s.d.} = 6.98 \pm 0.10\text{ g}$$

C. Uncertainty in Measurements

Suppose you want to measure the mass of a coin. The mass you measure will have a degree of uncertainty. The size of this uncertainty will depend on the quality of the balance and your skill in using it. If the smallest division of mass shown on the balance is 0.01 g, you can measure the mass of the coin to within $\pm$ 0.01 g. If the balance indicates the coin has a mass of 22.87 g, its mass is then 22.87 $\pm$ .01 g. A better balance might show the mass to be 22.868 g, with uncertainty of $\pm$ 1 in the third place after the decimal, 22.868 $\pm$0.001 g.

In general chemistry the tools used to measure volume, mass or length vary in quality. If we only need "about" 10 mL of water, a graduated cylinder will do the job. If we need to measure out exactly 10.75 mL of water, a buret would be a better choice. Both graduated cylinders and burets are scaled in units of volume, but the markings on a buret are in units of a tenth of a milliliter (0.1 mL) while those on a 50 mL graduate are in milliliters, a division ten times larger than that on a buret. The smallest division on a meter stick is 1 mm, but if more accuracy is needed a caliper can be used to measure length to the hundredths of a millimeter.

When using precision equipment like a buret, it is common practice to estimate the value of the digit beyond the closest scaled subdivision. The volume scale on a buret is marked to the 0.1 mL as shown in Figure 5, and the volume can be read with certainty to the tenth of a milliliter. But, the next place after the decimal, the hundredths place, can be estimated. There will be some uncertainty in the second place, but if a volume is stated as 10.57 mL, it means the volume is closer to 10.57 mL than to 10.56 or 10.58 mL.

The degree of uncertainty in a measured value must be reflected in the way you record the value on paper. Only those digits that are considered significant should appear in the recorded number. The number of **significant digits** or **significant figures** is equal to those digits you know with certainty plus the estimated digit. The volume measured in the buret, 10.57 mL is stated to 4 significant figures. The "1", the "0" and the "5" are known exactly, the fourth digit, the "7" is the estimated digit. The better the quality of the tools used to make measurements, the greater the number of significant figures the reported values can have. A more complete discussion of significant figures is presented in Appendix VII, *Significant Figures and Rounding Numbers.*

When you record a measured value, it must be stated to the proper number of significant digits. If the bottom of the meniscus (the concave surface of water in a graduate or buret) rests exactly on the 8.0 mL mark, record the volume as 8.0 mL, not simply as 8 mL. A number written as 8 mL implies a volume of 8 ± 1 mL, a rather crude measurement. When stated as 8.0 mL, the greater accuracy of the measurement is shown, 8.0 ± 0.1 mL. The usual limits of certainty in values measured for several common laboratory tools are given below:

10 mL graduate: ±0.1 mL
50 mL graduate: ±1 mL
multiple beam balance: ±0.01 g
electronic balance: ±0.001 g
buret: ±0.01 mL
volumetric pipet: ±0.02 mL
meter stick: ±1 mm
(Check with your instructor, not all balances are created equal.)

One problem with hand-held calculators is that they present answers to 8 or 10 digits. *You* need to be aware of how many digits are significant and round off the displayed number accordingly. The calculator will not know. That is *your* job.

D. Balances – Measuring Mass

Two types of balance are used in chemistry laboratories: the multiple-beam balance (MBB) and the single-pan automatic balance (SAB).

The Multiple-Beam Balance

Most multiple-beam balances used in chemistry laboratories are accurate to the second place after the decimal, that is, to the hundredths of a gram. A skeleton drawing of a MMB is shown in Figure 2. The *beam* of the balance rests on a *fulcrum* or knife edge. A *pan*, which holds the object to be weighed, is suspended from one end of the beam, and on the opposite end movable weights, or *poise*, can move along the beam to bring the pointer on the end of the beam in alignment with the *zero mark*. Before the mass of an object can be measured, the balance must be "*zeroed*" by moving the *counterweight adjustment knob* on one end of the beam until the *beam pointer* indicates zero. The balance pan must be empty and all poise set to their zero positions while this is being done. If you will be using a multiple-beam balance, try to identify each of these important balance parts before proceeding. The following steps outline the correct use of the multiple-beam balance:

1. The balance must be resting on a vibration-free, level, horizontal surface.

2. The balance must be *zeroed* before it can be used for accurate work. With the pan empty and all poise seated in their zero positions, adjust the counterweight until the beam pointer rests on the zero or balance mark. *This is a very important adjustment and should not be overlooked.*

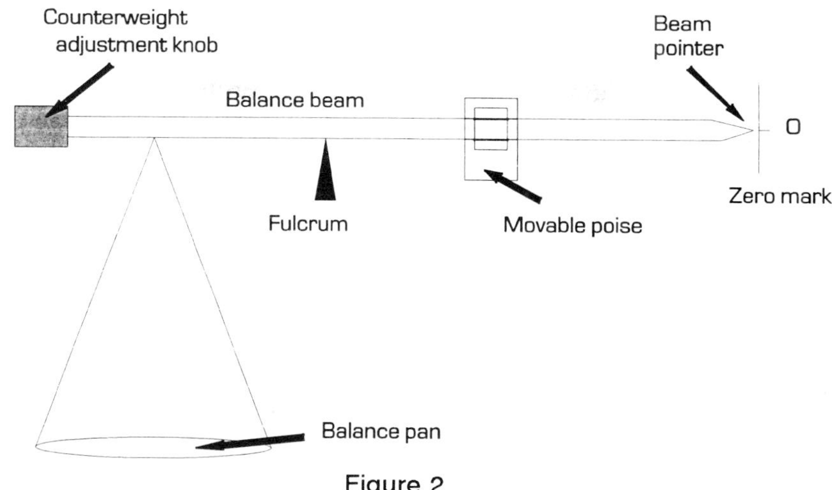

Figure 2

3. Place the object you wish to weigh in the *center* of the pan. Use weighing paper or a weighing boat if you are weighing solid reagents.

4. Add mass to the beam, starting with the largest poise and working down until the pointer is aligned with the zero mark. Move the heavier poise to the right until the beam falls below the zero mark, then back it up one unit and seat it firmly in the notch in the beam. Then do the same with the next heavier poise. The smallest poise (0 to 9.99 g) does not ride on a notched beam and can be placed in the exact location to achieve balance.

5. When finished, remove the weighed object and return all poise to their zero positions except for the largest, which should be left at the 100 gram mark to keep the beam from oscillating due to air currents in the lab.

The Single Pan Automatic Balance

Figure 3

The single pan automatic balance is more expensive and easier to use than the multiple-beam balance. A typical student-grade automatic balance is shown in Figure 3. The mass of the object being weighed is displayed in digital format in a window on the front of the balance. The balance is automatically zeroed when it is turned on (of course, the pan must be empty). Depending on the particular balance, an automatic balance can be capable of accuracy to the second, third or fourth place after the decimal. If you will be using an automatic balance during your laboratory work, your instructor will demonstrate its correct use.

Observe the following precautions no matter which type of balance you use:

1. The balance must be level and placed on a vibration-free surface.
2. Do not move the balance around; keep the balance in its place.
3. Do not place reagents or containers with wet exteriors directly on the balance pan. Dry exterior surfaces and use weighing paper or weighing boats.
4. Record a measured mass directly into the laboratory notebook, not on a scrap of paper which might be lost.
5. Clean up spilled reagents from the balance and the surrounding area immediately.

E. Measuring Volumes of Liquids

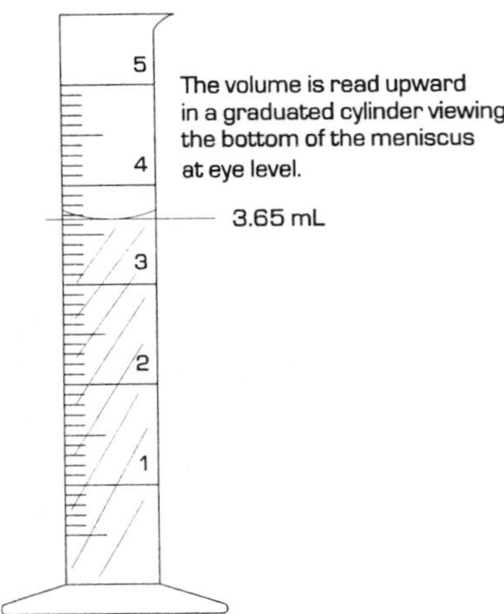

Figure 4 Graduated Cylinder

The volume of a liquid can be measured with a graduated cylinder or a buret. Which ever is used, the volume is measured viewing the bottom of the meniscus at eye level as shown in Figures 4 and 5. Notice in Figure 4 that the numbers on the graduated cylinder increase moving upward from bottom to top. It is just the opposite for a buret, a portion of which is shown in Figure 5. Here the numbers increase moving downward. It is possible to read either to the accuracy of one-tenth of the smallest marked division, which would be the second place after the decimal. The bottom of the meniscus in the graduated cylinder lies almost exactly between 3.6 and 3.7 mL so it would be reasonable to state the volume as 3.65 mL (it is estimated to be closer to 3.65 than 3.64 or 3.66 mL). In practice, graduated cylinders are not often used to this degree of accuracy as stated earlier, though it is possible to read them this accurately.

The drawing of a section of a buret is shown to the right and the volumes are marked in the tenths of a milliliter. Reading downward, you can see the bottom of the meniscus marks a volume just a little larger than 5.6 mL. The next decimal place, the hundredths place, is then estimated to be about one-tenth of the span between 5.6 and 5.7 mL, so the volume is stated as 5.61 mL. The number in the hundredths place is the estimated digit and the volume of 5.61 mL has three significant digits.

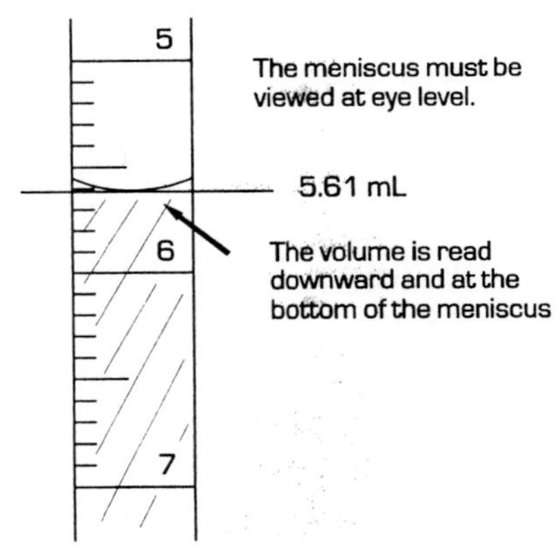

Figure 5 A Section of A Buret

F. Measuring The Density of a Solid

Once you are confident that you can measure mass and volume well, you will apply those skills as you determine the density of a solid. The **density** of a sample of matter is equal to its mass divided by its volume. Density is usually stated as grams per milliliter, g/mL, for liquids and grams per cubic centimeter, g/cc or g/cm^3, for solids.

$$\text{density} = \frac{\text{mass of the sample}}{\text{volume of the sample}} = \frac{m}{v}$$

Density is a physical property of matter, and is useful in identifying substances.

Experimental Procedure:

A. Measuring the Mass of a Solid and Determining the Quality of a Balance

1. Obtain a metal tag or other object of unknown mass from the instructor and record its identification number in the proper space on the Report Sheet. If the balance you are using is numbered or identified with a particular lab space, record the number of the balance or your desk number on the Report Sheet also.

2. Zero the balance and obtain the mass of the object to the maximum accuracy of the balance. Record this mass on the Report Sheet as the "first weighing" to the allowed number of significant digits. Be certain to accompany the measured mass with the correct unit of measurement. Remove the object from the balance, zero the balance again, and measure the mass of the object a second time. Record the mass as the "second weighing" on the Report Sheet. Repeat the zeroing and weighing sequence four more times, recording all data on the Report Sheet as you read it from the balance.

3. Calculate the average, or mean, of the masses you measured. Then, calculate the standard deviation using the approximate method described earlier. The smaller the standard deviation, the higher the quality of the balance *and* your technique.

B. Measuring the Volume of a Liquid

1. Find the 25 mL or 50 mL graduated cylinder in your equipment locker and observe the volume markings along its surface. Determine what each mark represents. On the Report Sheet you will find a figure that represents the bottom portion of a graduate. The line drawn across the figure represents the 10.0 mL division. Draw in the remaining division marks *as you see them on your graduated cylinder.* Make the drawing as accurate as possible. Label the 5.0 mL mark in the same manner as shown for the 10.0 mL division.

2. Fill the graduate with water until the bottom of the meniscus just rests on the 15.0 mL mark (for 25 mL cylinders) or the 30.0 mL mark (for 50 mL cylinders) when viewed at eye level. Using a medicine dropper, held vertically, count the number of drops of water required to raise the meniscus to the point where it just rests on the division mark for either 16.0 or 31.0 mL, that is, 1.0 mL greater than where you started. Record the number of drops on the Report Sheet.

Continue adding water, counting each drop, to raise the meniscus to the 17.0 or 32.0 mL mark. Record the number of drops needed to increase the volume by one more milliliter.

Again add water with the dropper, counting the drops needed to raise the meniscus from the 17.0 mL mark to the 18.0 mL mark (or from the 32.0 mL mark to the 33.0 mL mark), recording the number of drops on the Report Sheet.

3. From these data, calculate the average number of drops in one milliliter and record this value on the Report Sheet.

C. Measuring the Length of a Line

A common device for measuring distance in the beginning laboratory is the meter stick (a bit cumbersome for shorter distances) or a small plastic scale calibrated in centimeters and millimeters. Frequently, these devices also have scales calibrated in inches and, if long enough, in feet.

1. Obtain a meter stick or other measuring device suggested by your instructor from the stock in the laboratory. It should have both centimeter and inch scales. Examine both scales and determine what each scale mark represents.

2. From the instructor, obtain a card with a line drawn on it of unknown length. Record the number of the card on the Report Sheet in the space provided. As an alternative, your instructor may have you measure an object near you such as the length of the cover of the lab manual, or the line used for your name on the Report Sheet.

3. Using the inch scale on the measuring stick, measure the length of the line as accurately as you can to the nearest one-tenth (0.1) of an inch. (Each ⅛ inch equals 0.125 inch.) You will need to carefully estimate the tenths of an inch. Record your measurement on the Report Sheet. Then, measure the length of the same line using the centimeter scale and record the length you measured on the Report Sheet to the nearest one-tenth of a centimeter, 0.1 cm, the millimeter. Be certain to include the units with each recorded measured value.

4. Divide the measured length in centimeters by the measured length in inches. Express this ratio, the number of centimeters per inch, to the first place after the decimal and record the value on the Report Sheet using the correct units.

D. Determining the Density of A Solid

1. Obtain a density unknown from the instructor and record its number and label color on the Report Sheet.

2. On a zeroed balance, determine the mass of the solid and record the value to the second place after the decimal on the Report Sheet. Use proper units.

3. The volume of an irregularly shaped object can be determined by measuring the volume of water it displaces when completely submerged. Obtain a buret from the stock in the laboratory, and mount it vertically in a buret clamp attached to a ring stand. The buret is designed to deliver accurately measured volumes of liquids, but here it will be used simply as a container calibrated in units of volume.

   Fill the buret about half full with water, and after all air bubbles have escaped, read the exact volume of water contained in the buret to the hundredths of a milliliter, the second place after the decimal. Because the calibration marks increase from 0.00 mL at the top to 50.00 mL at the bottom, *a buret must always be read downward.* Seeing the bottom of the meniscus can be made easier by using a **buret card** as shown in Figure 2. You can make a buret card by drawing a heavy black line on a sheet of white paper. When you hold the card behind the buret with the black line a little below the meniscus, the black line will be reflected in the curvature of the meniscus. Record the volume of water in the buret on the Report Sheet.

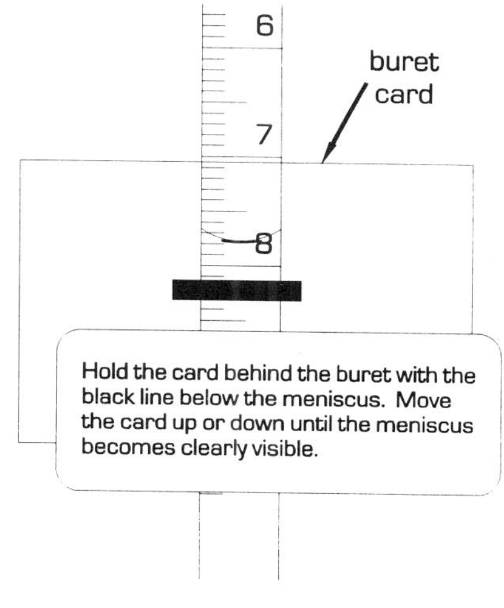

Figure 2

4. Tilt the buret and slowly slide the density unknown down the tube into the water. After all air bubbles have escaped (be certain of this) read and record this second volume reading to the hundredths of a milliliter. Remember, read *down* the buret.

   The volume of the density unknown equals the difference between the two volume readings.

5. Calculate the density of your unknown, and record this value on the Report Sheet. Be certain the correct units accompany the number.

   After the instructor approves your data, dry off the density unknown and return it to the instructor.

Name ______________________________ Locker Number__________ Date____________
Please print; last name first

PRELIMINARY EXERCISES: *Experiment 1*

Measurement and Density

1. Why should weighing data, or any other measurement data, be recorded immediately in your laboratory manual and not on a scrap of paper?

2. An object was carefully weighed five times on a multiple-beam balance. The balance was zeroed each time. The masses were: 11.36 g; 11.37 g; 11.40 g; 11.38 g and 11.39 g.
   a. Calculate the average value of these measurements. (Please show all work.)

   average = ___________

   b. Calculate the standard deviation, **s.d.**, of this set of measurements using the range of the data and the square root of the number of measurements averaged. (Please show all work.)

   s.d. = ___________

3. Calculate the density of a solid if it has a mass of 8.47 g and a volume of 3.24 cm^3. Please show your calculations, with units, below.

   density = ___________

4. The following illustrations show water levels in a buret, one *before* an object was submerged in the water and one *after*. The volume of the object is equal to the difference in the two volume readings. What is the volume of the submerged object? (Remember, read *down* the buret.)

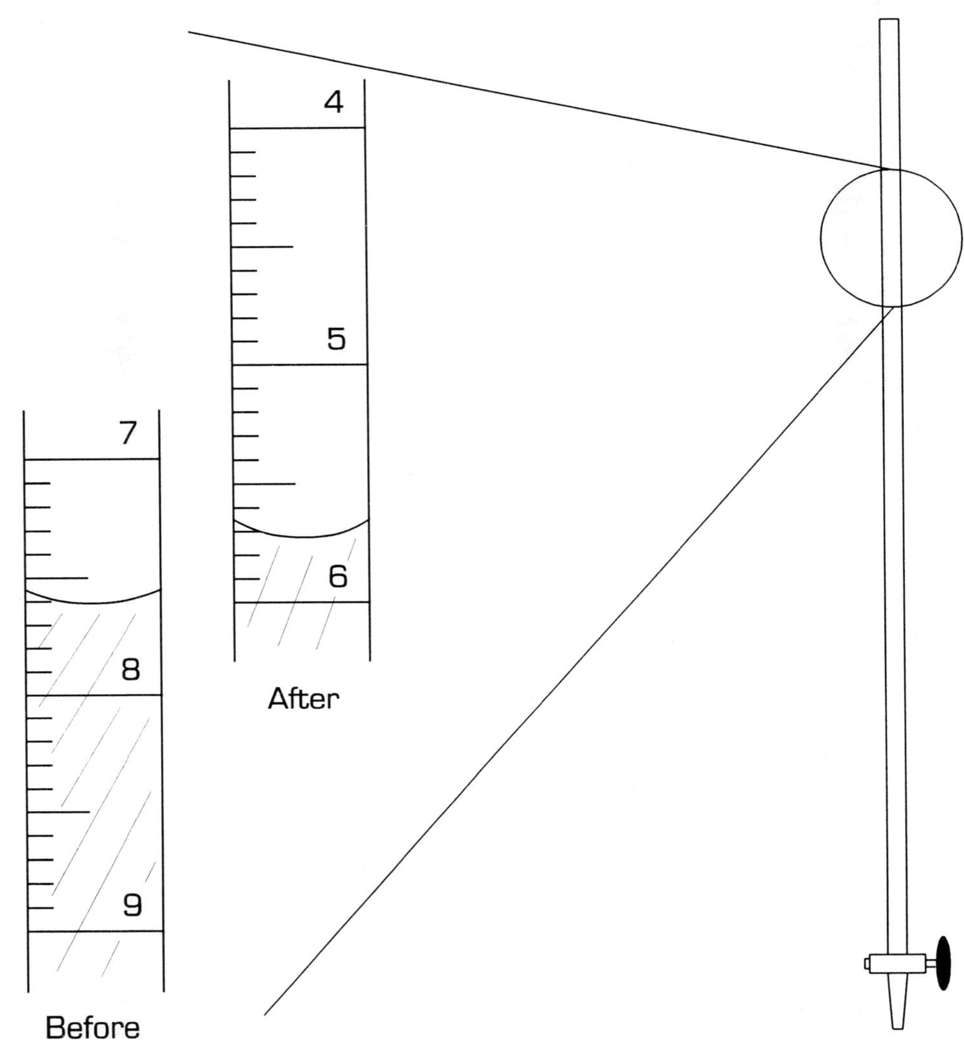

volume before submerging density unknown: ____________ (be certain to include the unit.)

volume after submerging density unknown: ____________

volume of density unknown: ____________

Name ______________________ Locker Number__________ Date____________
Please print; last name first

REPORT SHEET: *Experiment 1*

Measurement and Density

Make certain you record all measured values to the proper number of significant figures and state the unit of each measured value.

A. Measuring the Mass of a Solid and Determining the Quality of a Balance

Unknown Number: __________ Balance/Desk Number: __________

Mass of unknown, first weighing: ______________
(Reminder: Significant figures and units.)

Mass of unknown, second weighing: ______________

Mass of unknown, third weighing: ______________

Mass of unknown, fourth weighing: ______________

Mass of unknown, fifth weighing: ______________

Mass of unknown, sixth weighing: ______________

Average mass of the unknown: ______________

(Please show calculations here.)

Highest measured mass: ______________

Lowest measured mass: ______________

Standard deviation of the averaged values: ______________

(Please show calculations here.)

B. Measuring the Volume of a Liquid

1. Briefly describe the graduated cylinder you are using:

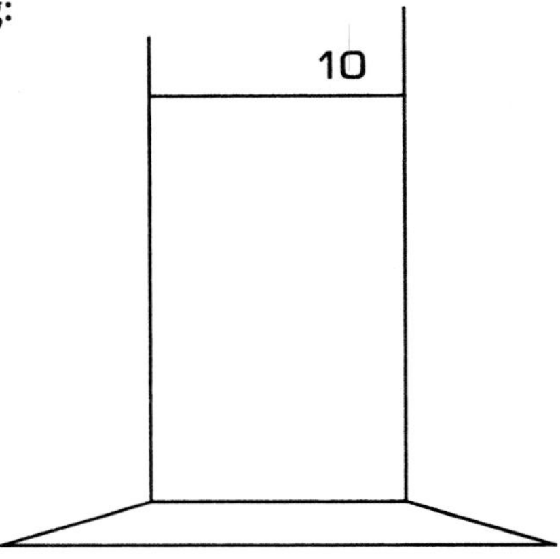

Complete the drawing on the right as instructed in the Experimental Procedure:

What volume is represented by the nearest (closest spaced) divisions on your graduated cylinder? ______________________ (include units here and below)

2. Number of drops to increase volume 1.0 mL: ____________
(from 15.0 mL to 16.0 mL or from 30.0 mL to 31.0 mL)

Number of drops to increase volume 1.0 mL: ____________
(from 16.0 mL to 17.0 mL or from 31.0 mL to 32.0 mL)

Number of drops to increase volume 1.0 mL: ____________
(from 17.0 mL to 18.0 mL or from 32.0 mL to 33.0 mL)

3. Average number of drops per milliliter: ____________
(Please show calculations below)

C. Measuring the Length of a Line

Line Number (or object measured): ____________

Length of line or object in inches: ____________

Length of line or object in centimeters: ____________

Number of centimeters in one inch $= \dfrac{\text{length in } cm}{\text{length in } in} =$ ________ $=$ ____________ cm/in

D. Determining the Density of a Solid

Density Unknown Number ____________

Trial II is to be carried out if the results of Trial I are unsatisfactory. Accompany each measured and calculated value with the correct unit.

	Trial I	Trial II
1. Mass of density unknown:	____________	____________
2. Buret reading – before adding density unknown:	____________	____________
3. Buret reading – after adding density unknown:	____________	____________
4. Volume of density unknown:	____________	____________
5. Density of unknown:	____________	____________

(Show all calculations below with units and state the answer to the correct number of significant figures.)

Experiment 2

Preparing Graphs

Purpose:

The purpose of this exercise is to learn about graphs, their use, preparation and interpretation. Emphasis will be placed on straight-line graphs. You will also become familiar with the terminology associated with graphs.

Discussion:

Graphs help us understand information more easily. Often we can see quickly the relationship between two measured values when they are displayed as a graph as opposed to the same values presented in a table. The volume occupied by a sample of air at six temperatures is shown below in Figure 1. On the left the temperature and volume are presented in a table and on the right in a graph. The straight-line (or linear) relationship between volume and temperature is clearly shown in the graph, but not so in the tabulated data. The graph reveals more to us than just numbers.

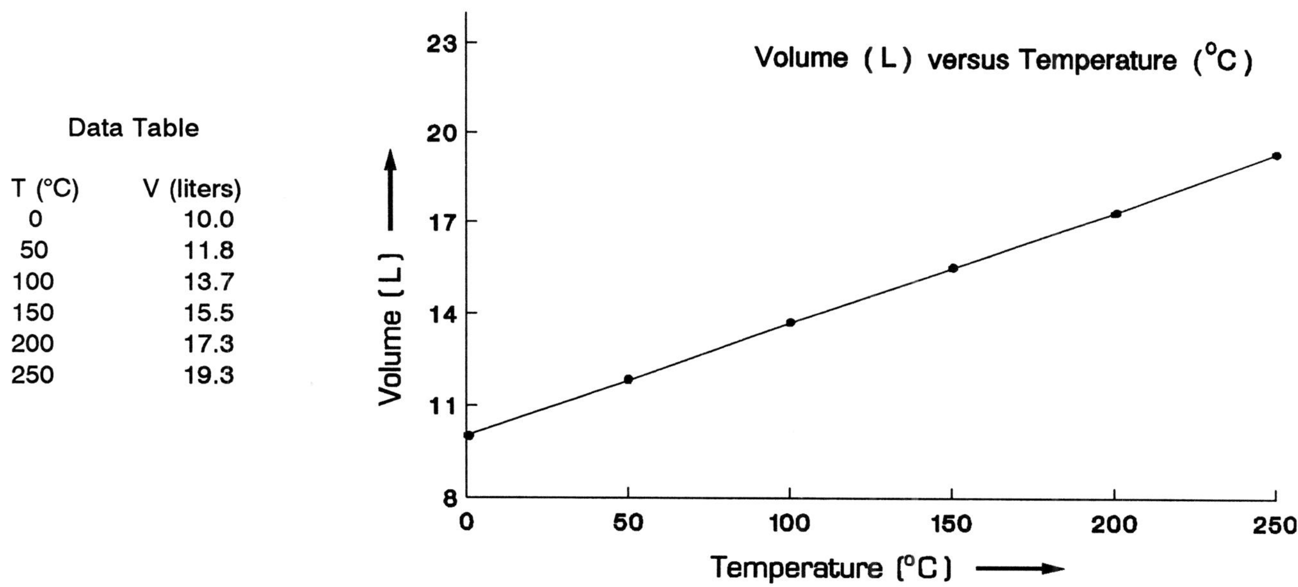

Data Table

T (°C)	V (liters)
0	10.0
50	11.8
100	13.7
150	15.5
200	17.3
250	19.3

Figure 1

There are many kinds of graphs used in science and business to present information. A **line** graph, a **bar** graph and a **pie** graph are used in Figure 2 to show the distribution of grades on an exam. All three graphs are prepared using the same data, though in this case you may find the bar or pie graph easier to interpret.

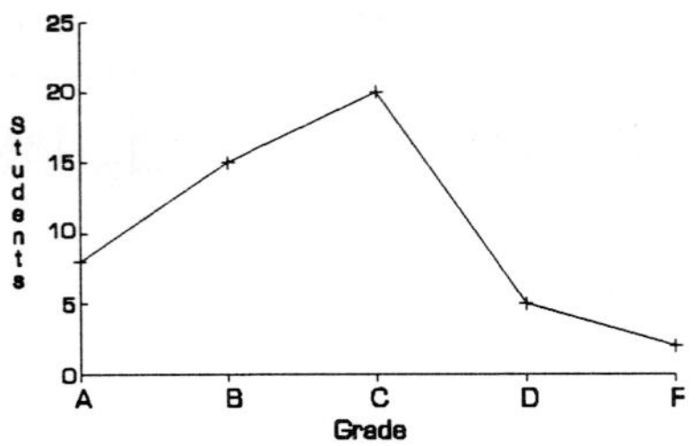

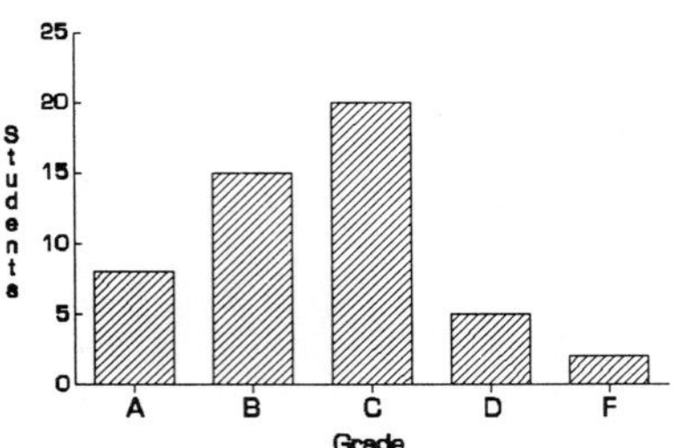

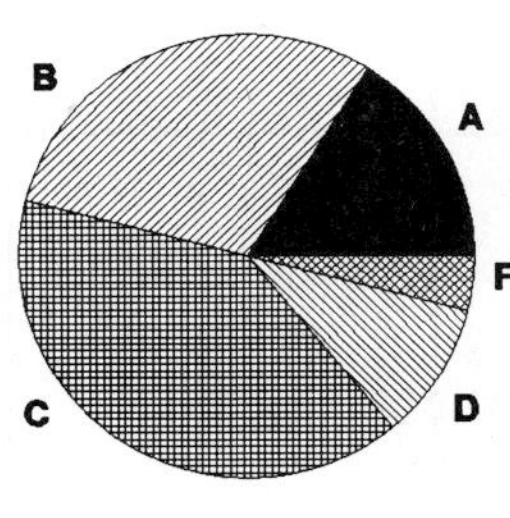

Figure 2 A Line Graph, A Bar Graph and A Pie Graph

In science graphs are used most often to show the relationship between two measured values (as opposed to distribution of grades as shown above), and line graphs are better for that purpose. For that reason the preparation and use of line graphs will be emphasized in this exercise.

Important Terms:

Line graphs are drawn on **axes**, two straight lines drawn at right angles to one another. The vertical line is customarily taken as the **y-axis** (the **ordinate**) and the horizontal axis as the **x-axis** (the **abscissa**). The axes pictured in Figure 3 cross at the **origin**, the point at which the values of x and y equal 0. In actual practice it is not necessary that the axes cross at x = 0, y = 0, though only the 0,0 point is considered the origin of a **Cartesian coordinate system**. Each axis is divided into **scale units** set off by **tic marks**. The x-axis and y-axis need not have scale units of the same size, though all scale units along an axis must be of the same size. The signs of x and y in each quadrant in a Cartesian system are shown in Figure 3. Data are plotted by placing a "dot" or other mark at the point corresponding to the values of x and y. In Figure 3, the point marked **A** corresponds to the values x = 9, y = 7, and point **B** to x = −6, y = −5. Each dot or mark represents a **data point**, and a complete graph will consist of a series of data points which, when connected, will show the relationship between the two variables.

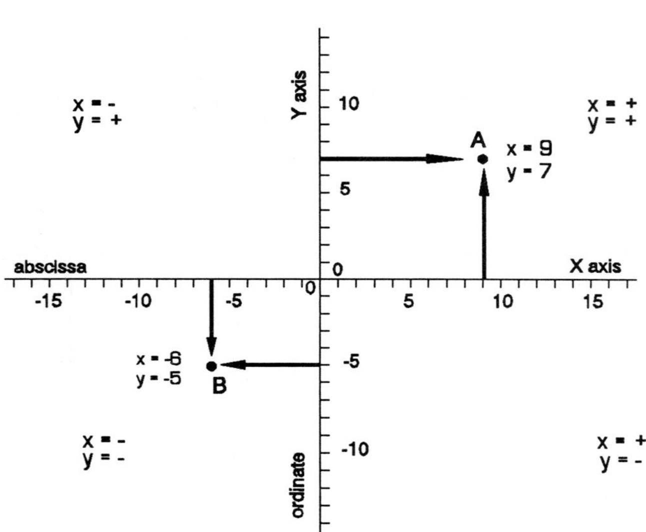

Figure 3 The Cartesian Coordinate System

It is customary to plot the **independent variable** along the x-axis and the **dependent variable** along the y-axis. The independent variable is the quantity under control of the person doing the experiment (temperature, time, etc.) and the dependent variable is the quantity that changes in response to changes in the independent variable. Sometimes the assignment of dependent and independent variables is arbitrary. At other times, because of custom or the nature of the experiment, assignment is without question.

Many times only positive values of x and y numbers are being plotted, and it is unnecessary to draw all four quadrants of the Cartesian system. You only need to draw one quadrant, the one in which both x and y are positive as shown in Figure 4. The other three quadrants are omitted.

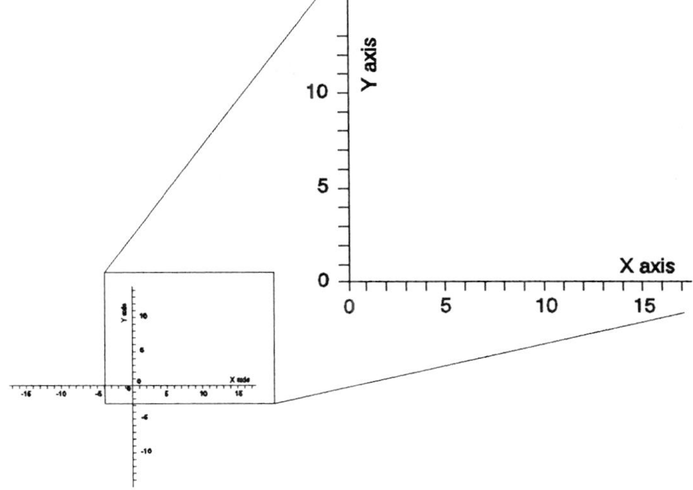

Figure 4 Using One Quadrant of the Cartesian System

Guidelines for Preparing A Graph:

1. *Always use graph paper.* Do not try to hand draw a grid of lines or estimate point placement on a blank sheet of paper or theme paper.

2. *Make the graph as large as possible.* The larger the plot, the greater the accuracy in placing data points and the greater the accuracy of values read from the graph.

3. *Draw axes with a straight edge and label each axis.* The label should indicate the variable being plotted *and* the unit for that variable written in parentheses, such as, "temperature (°C)" or "mass (g)". A small arrow pointing in the direction of increasing value may accompany the label.

4. *Identify the variable to be plotted on the x-axis.* This should be the **independent variable**, that is, the variable you control. For example, if you were measuring the volume of a gas at different temperatures you should choose temperature as the independent variable. The volume of the gas *depends* on the temperature you set. To prepare a graph of volume and temperature, plot temperature on the horizontal axis as the independent variable and volume on the vertical axis as the dependent variable, as is done in Figure 1.

5. *Scale each axis so the data span all or most all of its length.* The axes do not have to begin at zero. In Figure 1 the y-axis begins at 8 liters while the x-axis begins at 0°C. This reduces unused length on the y-axis. You will have to calculate the amount of change between tic marks. If you had to plot numbers from 0 to 100 along an axis with 12 tic marks, you could set the first tic mark to be 0 (the origin), the second to be 10 and so forth. You will leave one tic mark unused, but your numbers span nearly the entire length of the axis which is satisfactory.

6. *Label the graph.* In an unused area of the sheet of graph paper indicate the terms being plotted in this order: "y-axis variable" versus "x-axis variable." You can see how this is done in Figure 1 which has the title "Volume (L) versus Temperature (°C)."

Numbers must be plotted with the algebraically increasing values going to the right on the x-axis and upward on the y-axis.

Straight-Line Graphs:

Straight-line graphs have two advantages. The first is that a straight-line is easy to draw, only a straight edge is required. The other, and perhaps the more important advantage, is of a mathematical nature. Any straight-line can be described by an equation that relates the values of x and y used to generate the line. The general straight-line equation is

$$\mathbf{y} = m\,\mathbf{x} + \mathrm{b}$$

where:

- **y** = the dependent variable (plotted on y-axis, ↕)
- **x** = the independent variable (plotted on x-axis, ↔)
- m = the slope of the straight line (see below)
- b = the intercept (the value of **y** when **x** = 0)

All linear (straight-line) relationships fit this equation. The relationship between temperatures on the Celsius and Fahrenheit scales is linear and is shown in Figure 5. Here the Celsius temperature is arbitrarily chosen as the independent variable and so is plotted on the x-axis.

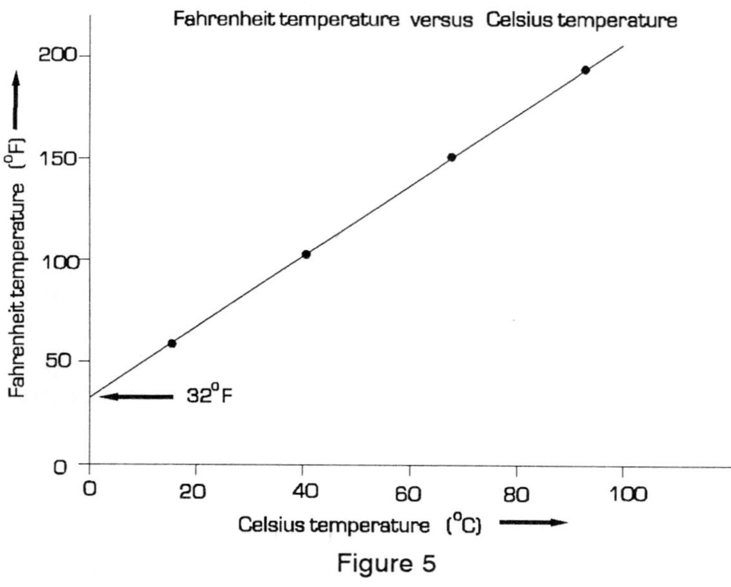

Figure 5

The equation that relates the two temperature scales is:

$$\begin{array}{cccc} {}^\circ\mathrm{F} = & (1.8\ {}^\circ\mathrm{F}/{}^\circ\mathrm{C}) & {}^\circ\mathrm{C}\ + & 32^\circ\mathrm{F} \\ \uparrow & \uparrow & \uparrow & \uparrow \\ \mathbf{y} = & (\mathit{slope}) & \mathbf{x} + & \text{intercept} \end{array}$$

This is a straight-line equation. The **slope** of the line equals 1.8 °F/°C and the **intercept** is 32°F; the value of the Fahrenheit temperature (on the y-axis) when the Celsius temperature (on the x-axis) equals 0°.

The value of the **slope** tells us that the Fahrenheit temperature changes 1.8° for each 1.0° change in the Celsius temperature. So, a 10 degree increase in the Celsius temperature would be the same as an 18 degree increase on the Fahrenheit scale. The values of the slope and intercept are taken from the graph. Remember, the intercept (32°F) is the value of the y-variable (the Fahrenheit temperature) when the value of the x-variable (the Celsius temperature) is zero. It takes a little more effort to calculate the slope. You may wonder how the value of 1.8 °F/°C came about? The slope of any straight line can be calculated using any two points on that line with their corresponding coordinate values (x_1, y_1) and (x_2, y_2).

$$\text{slope} = \frac{(y_2 - y_1)}{(x_2 - x_1)}$$

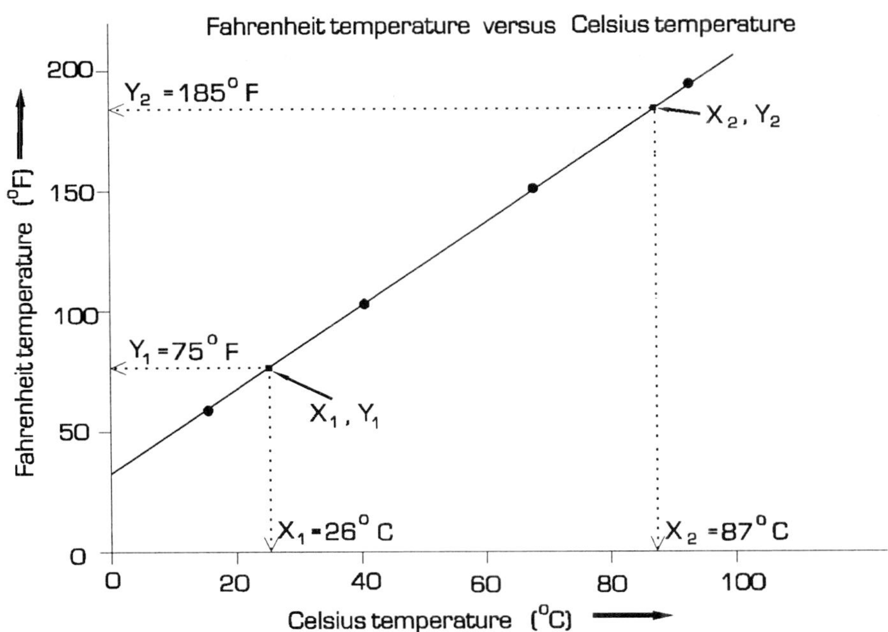

Figure 6 Calculating the Slope of A Straight Line

Two points are chosen on the straight line, and the coordinates of these points are:

$$x_1 = 26°C,\ y_1 = 75°F \qquad \text{and} \qquad x_2 = 87°C,\ y_2 = 185°F$$

$$\text{slope} = \frac{(y_2 - y_1)}{(x_2 - x_1)} = \frac{(185°\,F - 75°\,F)}{(87°\,C - 26°\,C)} = \frac{110°\,F}{61°\,C} = 1.8\,\frac{°\,F}{°\,C}$$

Notice that the two points chosen to determine the slope (symbolized ▪) were not data points (symbolized •), they are two points placed *on the line*. The reason data points were not used is to minimize errors that might arise using data points. There will most always be some experimental scatter in any set of data, and the points may not make a perfectly straight line. One has to use judgement to draw the "best" straight line through the points with some of them being a little above the line and some a little below the line. Drawing this "best" straight line will average out the scatter in the data, and the line will represent a more accurate relationship between the two variables.

The slope of a line can be *positive* or *negative*. The slope of a line is *positive* if the "y" term gets larger as the "x" term gets larger. The slope is positive in Figure 6. But, if the value of "y" decreases as "x" increases, then the slope is *negative*. All the graphs you will prepare in this experiment will have positive slopes.

Also, be aware that the slope can have units; the unit of "y" over the unit of "x." The slope of the line in Figure 6 has units of °F over °C. The units are an important part of the slope.

Data gathered in the study of three different metals are shown in Figure 7. The mass of several different volumes are plotted for each metal showing the straight-line relationship between mass and volume. Notice that the slope of the line is different for each metal. The line with the steepest rise is that for lead. The lead line has the largest slope, while the line for calcium has the smallest. In a plot of mass versus volume, the slope of the line equals the **density** of the metal, mass divided by volume (the y-term divided by the x-term). Each line has the same intercept of zero, because as the volume of the sample goes to zero so does its mass.

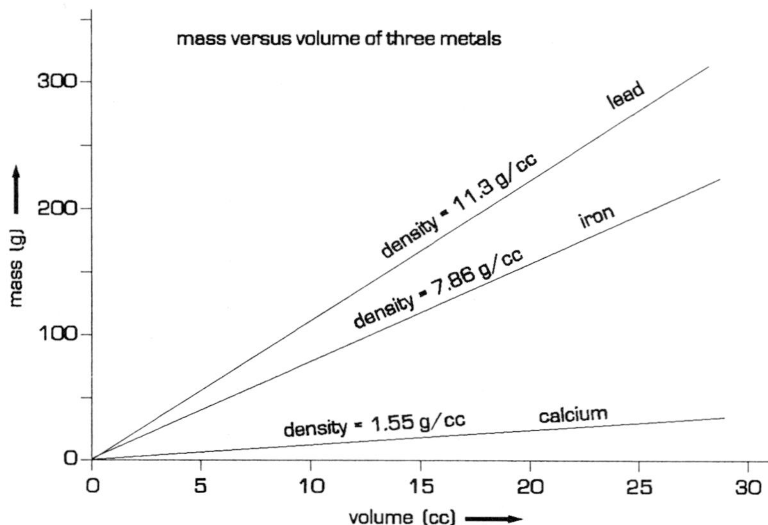

Figure 7 Mass Versus Volume for Three Metals

The slope has units of mass over volume, g/cc, the unit of density.

y	=	slope	**x**
↓		↓	↓
mass	=	(density)	volume

Remember, even though we speak in terms of "x" and "y" we must interpret the graph and write the equation using the actual variables, in this case, *volume* for the "x" term and *mass* for the "y" term. The straight line equation derived from this graph for each metal is:

$$\text{mass (g)} = \text{density} \left(\frac{\text{g}}{\text{cc}}\right) \text{volume (cc)} + 0$$

For lead, the equation would be:

$$\text{mass} = \left(\frac{11.3 \text{ g}}{\text{cc}}\right) \text{x volume}$$

Rearranging;

$$\text{density} = \frac{\text{mass}}{\text{volume}} = \frac{11.3 \text{ g}}{\text{cc}}$$

In the following exercises you will be asked to prepare several straight-line graphs. Follow the guidelines presented in this discussion as you prepare these graphs. In two exercises you will be asked to determine the slope of a straight line, and in exercise 5 to use the slope to derive a straight line equation.

Name ______________________________ **Locker Number**__________ **Date**__________
Please print; last name first

PRELIMINARY EXERCISES: *Experiment 2*

Preparing Graphs

1. Consulting the graph to the right:

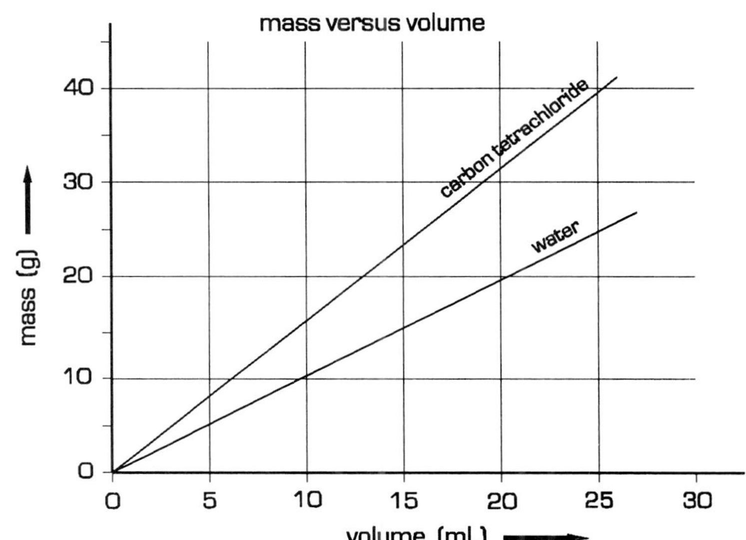

(a) Which has the larger mass, 10 mL of carbon tetrachloride or 15 mL of water?

____

(b) What is the mass of 25 mL of carbon tetrachloride?

____

(c) Which line has the greater slope?

____

(d) What unit would accompany the value of the slope of either line?

____

2. On the graph shown to the right:

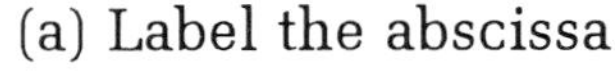

(a) Label the abscissa.

(b) Label the ordinate.

(c) Draw a circle around the dependent variable.

(d) draw a small arrow pointing at a tic mark.

(e) Circle the origin.

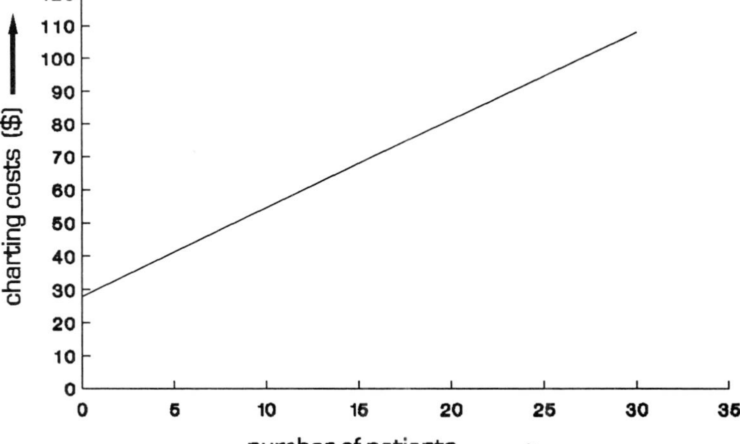

Name ______________________ Locker Number__________ Date___________
Please print; last name first

REPORT SHEET: ***Experiment 2***

Preparing Graphs

1. A series of data points are plotted on the graph to the right.
(a) Draw the "best" straight line through the points while not passing directly through any data point, and
(b) determine the value of the intercept.

Intercept = ______________

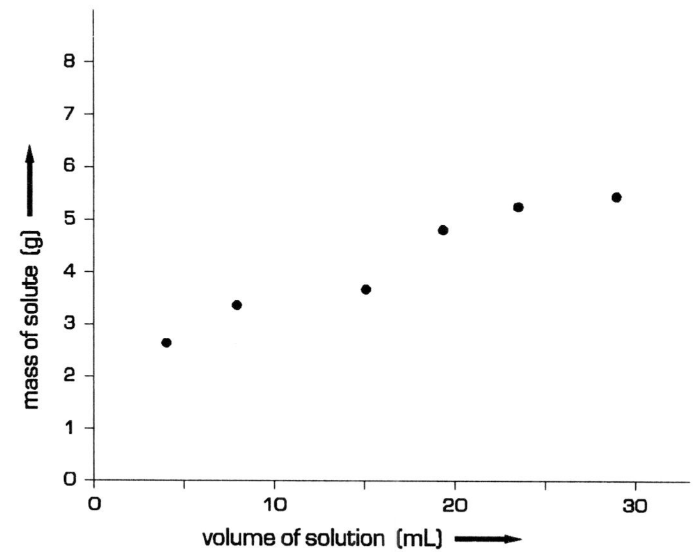

2. To the right is a plot of the mass of solute in various volumes of a concentrated acid solution.

(a) Calculate the slope of the line and report the value with the correct units.

slope = ____________

(b) Write the mathematical equation for the straight line that relates mass of solute and volume of solution.

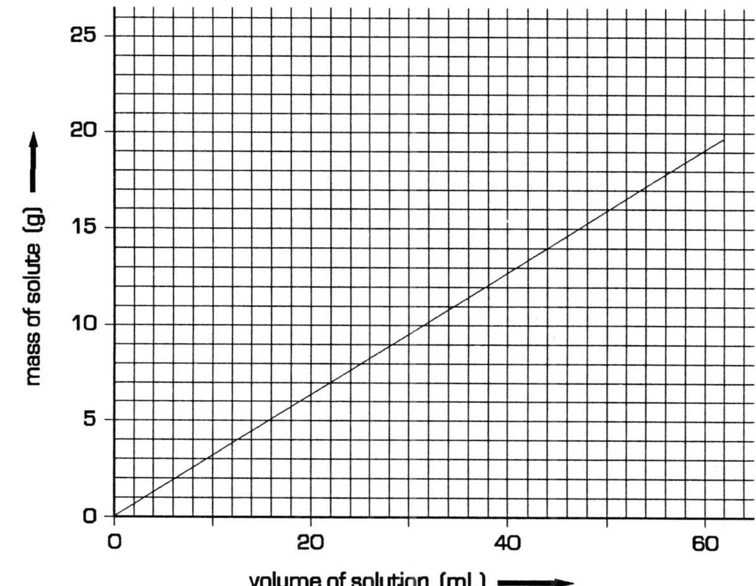

____

3. Gases expand when heated. An experiment was carried out in which the volume of a quantity of helium gas was measured at several temperatures (while the pressure applied to the gas remained constant). Temperature is recorded in Kelvins and volume in liters. Construct a plot of these data to show the linear relationship between absolute temperature and volume. This relationship is known as Charles' Law. Temperature is the independent variable. Scale the axes using 0, 0 at the origin. Your instructor may ask you to use a larger sheet of graph paper which can be found at the end of the Appendix.

Temperature (K)	Volume (L)
200	8.0
300	12
400	16
500	20
600	24

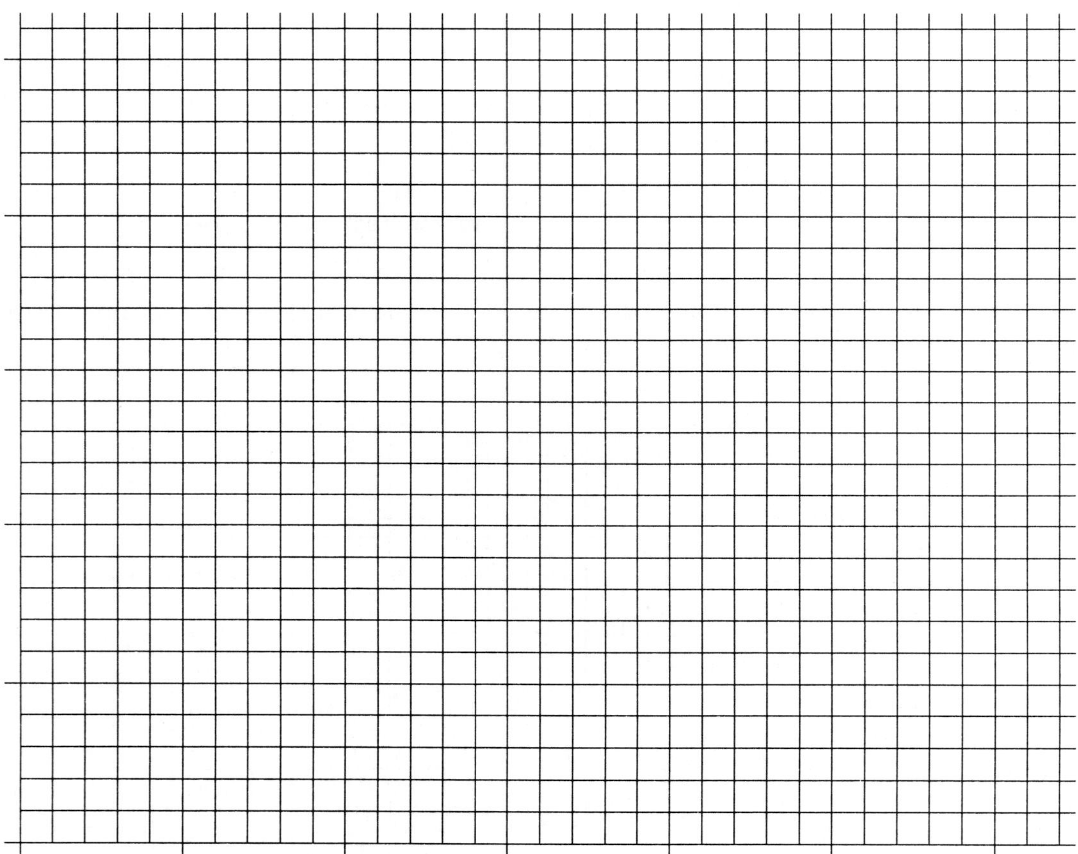

Name ______________________ Locker Number__________ Date__________
Please print; last name first

REPORT SHEET (continued): *Experiment 2*

Preparing Graphs

4. As an exercise, a student weighed several bunches of nails on a balance. The nails were not all the same size. Here are the student's data:

number of nails	mass in grams
5	82
8	122
12	208
20	324
26	435

Prepare a plot of these data on the grid below or on the graph paper at the end of the Appendix. Use "number of nails" as the independent variable and be certain to scale and label the axes correctly. Scale the axes starting at the origin; 0, 0.

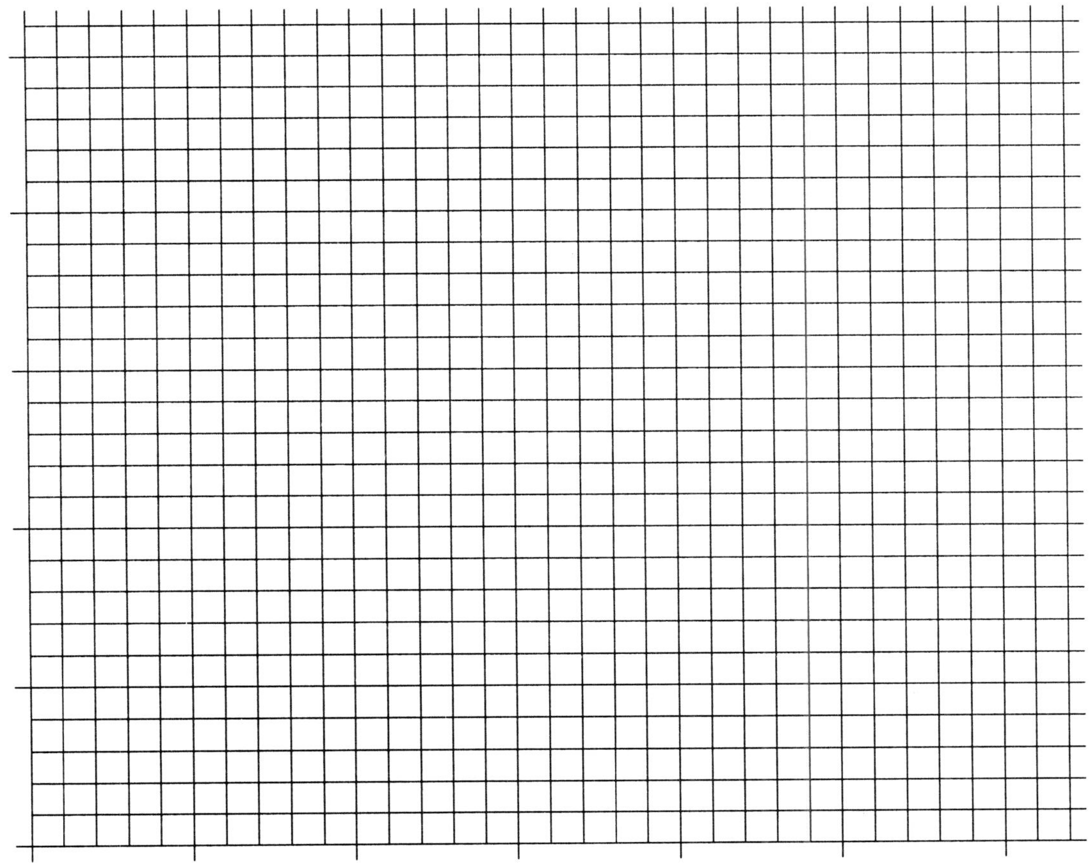

5. In this exercise you are to develop the equation that relates temperature on the Fahrenheit scale (°F), to temperature on the Kelvin scale (K). You may want to prepare this plot on a full sheet of graph paper which can be found at the end of the Appendix. Use Fahrenheit temperature as the independent variable.
(a) Prepare a plot of these data.
(b) Determine the value of the intercept.
(c) Calculate the slope of the straight line.
(d) Using you slope and intercept, derive the straight line equation that relates temperature in degrees Fahrenheit and temperature in Kelvins.

Temperature (°F)	Temperature (K)
0	255
50	283
100	311
150	339
200	366
250	394
300	422

intercept = ______________

slope = ______________

The equation: ______________________

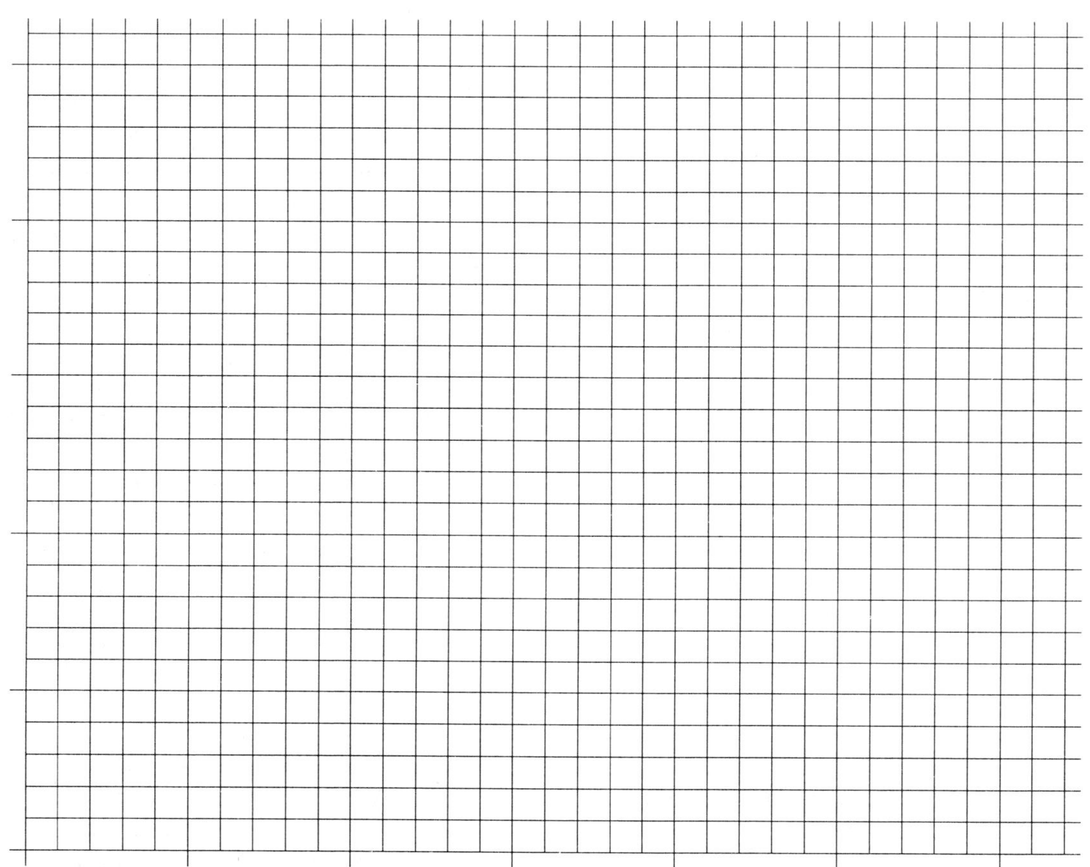

Experiment 3

The Identification of an Unknown Liquid

Purpose:

In this experiment you are to identify an unknown liquid by comparing its physical properties with those of known liquids. In addition, you will learn how to use a volumetric pipet, measure the density and boiling temperature of a liquid and judge its solubility in water.

Discussion:

Properties are those characteristics of a substance that allow it to be identified and distinguished from other substances. There are two classes of properties: physical and chemical. **Physical properties** can be determined without changing the identity of the substance. Some physical properties are color, crystal shape (if a solid), melting temperature, boiling temperature, density, solubility, odor and index of refraction. **Chemical properties** are determined by the chemical changes the substance can undergo. Some examples of chemical changes are reaction with oxygen or water, reaction with acids or bases, decomposition if heated or exposed to light, and instability when subjected to mechanical shock.

Every chemist has handbooks that are essentially tabulations of the properties of thousands of compounds. Chemists involved in the synthesis of new compounds use this information constantly in their work. When a chemist performs a synthesis for the first time, he or she carefully analyzes the products of the reaction to see if a new compound has been formed, or if a new route to an already known compound has been discovered. The chemist separates the mixture of products into pure substances and examines each one to determine its identity. This requires measuring properties of each substance. Every substance has its own set of properties that acts as its fingerprint. No matter how sophisticated a chemist's techniques may be when attempting to identify a compound, he or she is doing so by determining the properties of the compound.

Usually a chemist has a good idea what the products of a chemical reaction will be and can, by consulting handbooks, prepare a list of likely compounds with some of their chemical and physical properties. If the properties of one of the compounds isolated from the product mixture match those of one of the listed possibilities, then the identity of that product is known. If no matches can be made, perhaps a new substance has been synthesized; but only further work can prove this is so.

There are many properties a chemist can determine routinely in a well-equipped laboratory, but a chemist cannot determine *every* physical or chemical property. Only those properties deemed most revealing in a particular situation will be measured. Sometimes two determinations are sufficient; other times, as many as a dozen are necessary to peg the identification with certainty. Some of the new techniques available to chemists today are so powerful and reveal so much information about a compound, that often a single determination can tell a chemist the identity of a compound with near certainty.

In this experiment, you will face a problem not unlike that of a synthetic chemist. You are to determine the identity of a liquid by measuring three of its physical properties: density, solubility in water and boiling temperature. Comparison of your measurements with those tabulated for several liquids will allow you to learn its identity.

Experimental Procedure:

CAUTION: As always, eye protection must be worn at all times.

Obtain a sample of unknown liquid from the laboratory instructor. Record the number of the unknown on the Report Sheet. Perform the investigation in the order the tests are presented so you do not run out of liquid before your task is complete. After each determination consult the properties listed in Table 1 on page 40 and eliminate those that are obviously not close to your observations. By doing so, you will identify the unknown by a process of elimination.

ENVIRONMENTAL ALERT: Your instructor will tell you how to dispose of leftover samples of unknown liquid. The liquids should not go down the drain. Safe disposal of waste is an important part of each experiment.

A. Determining a Liquid's Solubility in Water

If two liquids are mixed and do not separate upon standing, they are considered **soluble** in each other. Water and ethyl alcohol are soluble in each other. If two liquids separate into distinct layers after they are mixed, then they are **insoluble** in one another. Water and oil are insoluble in one another. Even after being shaken together, water and oil left standing will form two layers with a **visible boundary** separating the more dense water layer on the bottom from the oil layer on top. Though almost all liquids are soluble in one another to at least a slight degree, using equal volumes of both liquids greatly lessens the need to make judgments about partial solubility.

Method: Place 1 mL (about 20 drops) of the unknown liquid in a small test tube. Add an *equal volume* of water to the liquid. Seal the test tube with a small cork or plastic film[1] and mix the two liquids by shaking the tube for 15 seconds. Set the test tube aside for one minute, then check to see if the two liquids separate. Record your observations and conclusion on the Report Sheet. You should be able to eliminate some liquids in Table 1 even now. Your instructor will tell you how to dispose of the mixture in the test tube.

---

1 Parafilm® is satisfactory, though it may partially dissolve if exposed to certain organic solvents for extended periods.

B. Determining the Density of a Liquid

Density is a physical property of matter and is equal to the mass of a sample divided by its volume. The density of a solid or liquid is usually expressed in units of grams per milliliter (g/mL) or grams per cubic centimeter (g/cc). The element mercury has a density of 13.6 g/mL. This means that 1.00 mL of mercury has a mass of 13.6 g.

You will determine the density of the unknown liquid by measuring the mass of 10.00 mL of the liquid. The 10.00-mL volume will be measured using a volumetric pipet. A **volumetric pipet** is a precision piece of glassware designed to deliver a known volume of liquid. A volumetric pipet is shown in Figure 1. Note the location of the etched line. The pipet must be clean so that as liquid drains from it droplets of liquid do not adhere to the inner wall. Also, the tip must be open, free-flowing and not chipped.

Follow these instructions when using a volumetric pipet:

1. Place the liquid to be measured in a small beaker. The volume of the sample must be greater that the volume of the pipet.
2. Holding the pipet vertically, immerse the tip into the liquid and, using a **pipet bulb**, draw the liquid into the pipet until the meniscus is several centimeters above the etched line. (For obvious reasons of safety, never draw liquid into a pipet by mouth. Always use a pipet bulb.)
3. Quickly slip the pipet bulb off the pipet as you cover the end of the pipet with the soft tip of your index finger. (You may need to practice this bulb-finger switch until you can do it smoothly.) Once your finger is in place, allow liquid to drain from the pipet until the *meniscus just rests on the etched line when viewed at eye level.* Wipe any excess liquid from the tip with a tissue. The pipet is now ready to deliver 10.00 mL of liquid.
4. Position the pipet vertically above the container that is to receive the liquid, remove your finger tip, and allow the pipet to drain by gravity alone. When the flow ceases, wait 10 seconds, touch the tip to the inside wall of the container to remove the adhering droplet, and remove the pipet. You will notice a small amount of liquid remaining in the pipet tip. This is normal, and you should *not* attempt to add it to the volume already delivered. The pipet is designed to deliver a known volume of liquid by the action of gravity alone. A small amount of liquid should remain in the tip when the pipet is used properly.

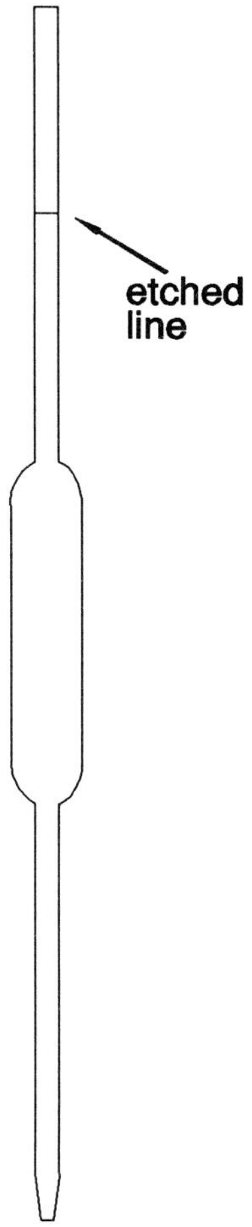

Figure 1

It would be good to practice using the pipet to obtain 10.00-mL volumes of deionized water. Seek help if you have problems.

Method: Obtain a pipet from the supply in the laboratory. If the tip is chipped, turn it over to the instructor. Any moisture in the pipet may be removed with a small amount of ethyl alcohol or acetone. Rinse the pipet with a small portion of the unknown liquid before measuring out the sample for weighing.

Weigh a clean, dry, 50-mL beaker. Record the mass on the Report Sheet to the maximum accuracy allowed by the balance. Using the pipet, place 10.00 mL of the unknown liquid in this beaker and measure the mass of the beaker containing the unknown liquid on the balance. You may want to cover the beaker with a clean watch glass to prevent evaporation if you must wait to use a balance. Record this second mass on the Report Sheet, again to the maximum accuracy of the balance. The mass of 10.00 mL of liquid is the difference between the two measured values.

Knowing the mass of 10.00 mL of liquid, calculate the density of the unknown and record the value, with units, on the Report Sheet. At this point you should be able to eliminate most of the liquids in Table 1.

Save the volume of liquid used in the density determination for the boiling temperature determination. Rinse the pipet thoroughly with deionized water, then with alcohol or acetone, and return it to the lab storage area.

C. Measuring the Boiling Point of a Liquid

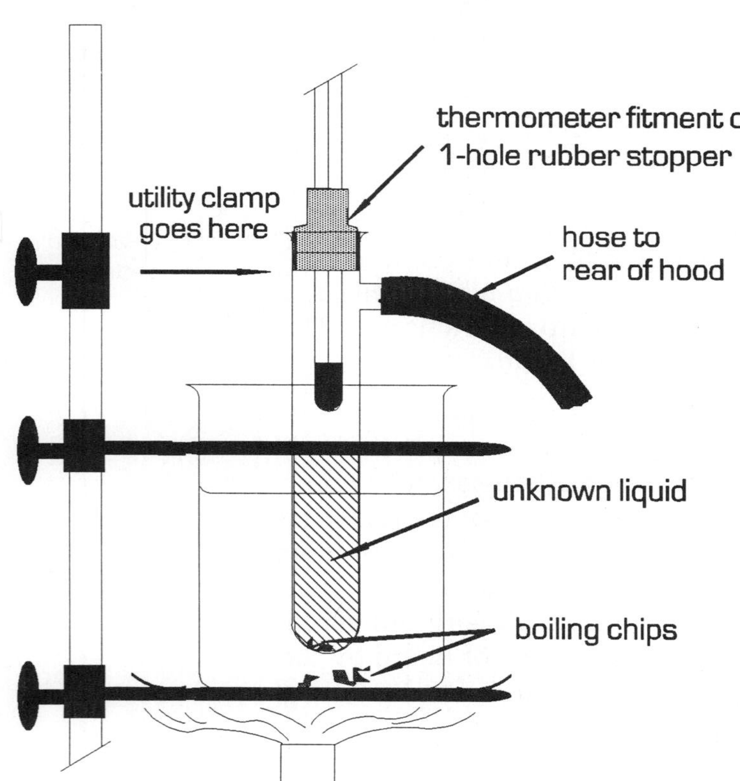

Figure 2

The temperature at which a liquid boils is a physical property of the liquid. As long as the external pressure (atmospheric pressure) remains constant, the boiling temperature will remain constant until the liquid is completely boiled away.

You will measure the boiling temperature of your liquid in an apparatus like that shown to the left in Figure 2. The side-arm test tube and other equipment will be made available to you. The bottom of the thermometer is about 1 cm above the level of the liquid in the side-arm test tube and not immersed in the liquid.

Method: In the fume hood, set up a ring stand supporting a 400 mL beaker two-thirds full of tap water. Use a utility clamp to support the side-arm test tube above the beaker as shown in Figure 2. Once you are certain the side-arm test tube is clean and dry, pour 10 to 15 mL of your liquid in the test tube and fit it in the clamp on the ring stand.

Place two or three boiling chips (pieces of porcelain or brick) in the test tube and two or three boiling chips into the water bath. (Boiling chips are employed to prevent the liquid from "bumping," or boiling in such a way that hot liquid suddenly shoots out of the test tube or beaker.) The rubber hose leading from the arm of the test tube should extend to the rear of the fume hood to ensure the vapors are vented properly. If necessary, use two lengths of rubber hose connected by a short length of glass tubing.

The water level in the 400 mL beaker should be below the level of the liquid in the side-arm test tube, and the thermometer bulb should be about 1 cm *above* the surface of the liquid. Heating the unknown liquid in the water bath will allow you to control heating better and obtain an accurate measure of the boiling temperature.

Heat the water bath until the liquid in the test tube begins to boil at a steady rate. Remove the burner and observe the temperature while the liquid is boiling steadily. Once a constant temperature is reached, *record it to the tenth of a degree on the Report Sheet.* This should be the highest temperature reached; if not, you are observing a false value and it will be necessary to heat the liquid directly with the burner flame.

For those liquids that boil above 80°C, it is better to measure the boiling temperature using direct heating. Remove the side-arm test tube from the water bath (caution: it's going to be hot!) and clamp it to the ring stand. Holding the burner in your hand, wipe the flame back and forth across the base of the test tube. Control the heating carefully as you observe the temperature. *Record on the Report Sheet the highest temperature that is reached and holds as the liquid boils.*

Eliminating all liquids in Table 1 with boiling temperatures different from the temperature you measured should reduce the number of possible liquids to a precious few; in fact, if you have made accurate observations, you should be able to identify your unknown liquid.

D. Identifying the Unknown Liquid

Compare your measurements with the data in Table 1. Your liquid is the one with properties that are identical, or nearly so, to those you measured. Be aware that the boiling points in the table are normal boiling points, that is, the boiling temperatures measured at sea level, where the average atmospheric pressure is 760 mmHg (1.00 atmosphere). Since the pressure in the laboratory will most likely be a bit lower than 1.00 atmosphere, the boiling point you measure may be slightly lower than that shown in the table. This should not create a problem for you if you have determined the density and solubility correctly.

Just as an example, suppose you determined your liquid to be *insoluble* in water, with a *density of 0.72 g/mL*, and a *boiling point of 70.2°C*. No liquid exactly matches these values, but methylbutyl ether agrees best and would be the best choice.

Table 1. Several Liquids and Some of Their Properties*

CAUTION: These liquids are flammable and must be used with care around open flames.

ENVIRONMENTAL ALERT: Dispose of leftover liquids properly. Your instructor will inform you of the safest disposal method.

Liquid (in order of increasing boiling point)	**Boiling Point (°C)**	**Density (g/mL)**	**Solubility in Equal Volume of Water**
ethyl bromide	38.0	1.43	insoluble (I)
methylene chloride	39.5	1.32	I
cyclopentane	49.3	0.74	I
ethyl formate	54.0	0.92	I
ethylisopropyl ether	54.0	0.74	I
acetone	56.6	0.79	soluble (S)
1,1-dichloroethane	57.3	1.17	I
isobutyraldehyde	63.0	0.79	I
methyl alcohol	64.3	0.79	S
tetrahydrofuran	65.5	0.90	S
isopropyl ether	67.5	0.72	I
hexane	69.0	0.66	I
methylbutyl ether	71.0	0.74	I
1,1,1-trichloroethane	74.0	1.32	I
ethyl acetate	77.1	0.89	I
ethyl alcohol	78.5	0.79	S
2,2-dimethylpentane	79.2	0.67	I
cyclohexane	81.4	0.78	I
isopropyl alcohol	82.3	0.79	S
ethylene chloride	83.5	1.26	I
1,2-dichloropropane	95.0	1.16	I
propyl alcohol	97.2	0.80	S
heptane	98.5	0.68	I
ethyl propionate	99.1	0.88	I
propyl acetate	101.8	0.89	I
isobutyl alcohol	108.4	0.80	I
toluene	110.8	0.86	I

* Due to environmental and safety concerns, not every liquid listed here will be available as a possible unknown. Several liquids are placed in the table simply as foils to test your skill.

Name ______________________________ Locker Number__________ Date____________
Please print; last name first

PRELIMINARY EXERCISES: *Experiment 3*

The Identification of an Unknown Liquid

1. Indicate which of the following would be considered physical properties (P) and which would be considered chemical properties (C) of iron.

   _____ a. Iron reacts with oxygen to form iron oxides.

   _____ b. Iron is a good conductor of heat and electricity.

   _____ c. Iron melts at 1535°C.

   _____ d. When iron is immersed in a solution of hydrochloric acid, hydrogen gas is produced.

   _____ e. Iron becomes red-hot when heated.

   _____ f. A 500-cc volume of iron has a mass of 3930 g.

   _____ g. Iron can be polished to a high luster.

2. The mass of 10.00 mL of carbon tetrachloride was measured to be 15.94 g. What is the density of carbon tetrachloride?

   Density = __________

3. (a) The density of water is 1.00 g/mL. Carbon tetrachloride is insoluble in water. Would carbon tetrachloride float on water or sink beneath water when they are combined?

   (b) Hexane has a density of 0.66 g/mL and is also insoluble in water. Would hexane float on water or sink beneath water when mixed with water?

4. A student determined that an unknown liquid was soluble in an equal volume of water, had a density of approximately 0.80 g/mL and a boiling temperature of 55.5°C. Consult Table 1 and identify the liquid.

   The liquid is: ______________________________

Name ______________________ Locker Number__________ Date__________
Please print; last name first

REPORT SHEET: *Experiment 3*

The Identification of an Unknown Liquid

Unknown number: __________

Trial 2 is to be carried out if Trial 1 is unsatisfactory.

	Trial 1	Trial 2
A. Determining a Liquid's Solubility in Water		
Observation	________	________
Conclusion (soluble/insoluble)	________	________
B. Determining the Density of a Liquid		
Mass of beaker + 10.00 mL liquid	________	________
Mass of beaker	________	________
Mass of 10.00 mL unknown liquid	________	________
Density	________	________

C. Measuring the Boiling Point of a Liquid

Determine which liquid in Table 1 has a set of physical properties similar to those you determined. If your work was done carefully, only one liquid will have data that match yours. List your values and those from the table for the liquid that best agrees with your values.

	Trial 1		Trial 2	
Property	**Your Values**	**Table Values**	**Your Values**	**Table Values**
Solubility	________	________	________	________
Density	________	________	________	________
Boiling point	________	________	________	________

D. Identifying the unknown liquid

Trial 1 ______________________________

Trial 2 ______________________________

SECTION III:
Chemical Separations

Experiment 4

Elements, Compounds and Mixtures

Purpose:

In this experiment you will see some of the differences between elements, compounds and mixtures. You will also learn several methods used to separate one species from another in the laboratory.

Discussion:

A. Classification of Matter

Properties are those characteristics of matter that allow one substance to be distinguished from another. The properties of copper are different from those of gold and allow us to tell the metals apart. All properties can be classified as being intensive or extensive. **Intensive properties** are those that do not depend on the amount of matter being examined; color, odor, hardness, electrical conductivity and density are intensive properties. **Extensive properties** depend on the amount of matter in the sample being examined; mass and volume are extensive properties.

As an experimental science, chemistry relies on observations and measurements to gain an understanding of matter. An early goal in chemistry was the classification of matter into different groups based on similarities and differences in observed properties. One such organizational scheme is shown below:

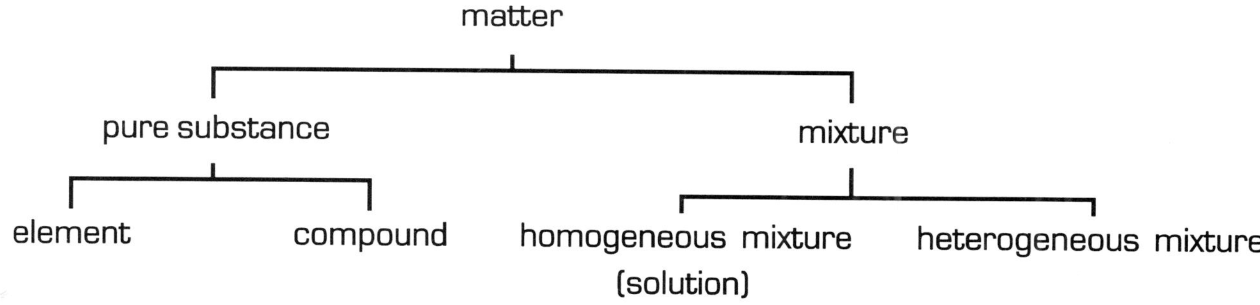

As shown above, each substance may be classified as a pure substance or a mixture. A **pure substance** (1) has the same intensive properties throughout and (2) cannot be separated into two or more other substances except by means of a chemical reaction. Water is a pure substance. All samples of water have the same intensive properties (color, density, etc.), and physical changes, such as freezing and boiling, cannot change its identity.

Pure substances may be further subdivided into elements and compounds. An **element** is a pure substance that cannot be decomposed by ordinary chemical methods into simpler substances. There are more than 105 known elements. They are considered the building blocks of all matter. A **compound** is composed of two or more elements chemically combined in a definite weight ratio. A compound can be decomposed into simpler substances (elements or other compounds) by chemical means. **Mixtures** are composed of two or more substances (elements or compounds), but not in any particular fixed weight ratio. A soft drink is a mixture. More coloring, sugar or flavor can be added if desired, and it still remains a soft drink. There are many more compounds than elements, and there are many more mixtures than pure substances.

A **heterogeneous mixture** is not uniform in its composition, and one can see individual particles within it with the naked eye. Sand, wood, soil and rock are all examples of heterogeneous mixtures. A **homogeneous mixture** has a uniform composition, and the individual particles that make it up (atoms, ions or molecules) cannot be seen with the naked eye. **Solutions** are homogeneous mixtures. One cannot see molecules of sugar in a soft drink; the solution looks identical throughout.

Substances that compose mixtures can be separated and isolated. **Physical methods** of separation do not change the identity of any substance during the procedure, while **chemical methods** of separation do. Separating a metal from its ore by consuming it with acid represents a chemical separation. Using a magnet to remove iron filings from a mixture of sand and iron filings would be a physical separation. The iron is not changed by the magnet. Several physical methods of separation commonly used in general chemistry are described below.

B. Methods Used to Separate Mixtures

Laboratory work often involves separating mixtures into their component parts. The purification of the products of a reaction is nothing more than a separation process, removing small amounts of impurities from larger amounts of product.

Six common physical methods used to separate the components of a mixture are (1) filtration, (2) distillation, (3) solubility extraction, (4) fractional crystallization, (5) decantation and (6) evaporation. These are described below.

1. **Filtration** is used to separate insoluble solids from a liquid. The mixture is passed through a porous paper held in a funnel. The liquid passes through and the solid is held back by the paper. The liquid that passes through is called the **filtrate**. The solid remaining on the porous paper is the **residue** or **precipitate**.

2. **Distillation** is used to separate volatile substances, especially liquids, from one another or from nonvolatile substances. A liquid can be purified by converting it to a gas (by heating it to its boiling point), then cooling the vapors to condense it back

to the liquid and collecting the liquid in another container. A mixture of liquids can be separated using distillation if they boil at different temperatures. The lower boiling liquids will distill off first, leaving the higher boiling liquid(s) behind. A liquid can also be separated from nonvolatile impurities (dissolved minerals) using distillation. Pure water can be obtained from sea water by distillation. The nonvolatile salts in sea water remain behind as the water boils off and condenses into another container.

3. **Solubility extraction** removes the soluble components of a mixture from the insoluble components. A mixture of powdered iron and sulfur can be separated by dissolving the sulfur in carbon disulfide, a liquid that does not dissolve iron. The liquid phase can be separated from the iron by filtration. Evaporation of the liquid will leave the solid sulfur behind.

4. **Fractional crystallization** separates solids based on their different degrees of solubility. In a sense, fractional crystallization is the reverse of solubility extraction. A saturated solution of the mixture is prepared at an elevated temperature. As this solution cools, the substance of lowest solubility crystallizes out of the solution, leaving small quantities of impurities in solution. For most substances, solubility is lower at low temperatures than at high temperatures.

5. **Decantation** is a method to remove a liquid from an insoluble solid. The insoluble substance is allowed to settle, then the liquid is poured off into another container without disturbing the solid. This means of separation seldom affords a complete separation of the solid from the liquid.

6. **Evaporation** – In **evaporation**, the volatile liquid is allowed to evaporate, leaving nonvolatile impurities behind. Heat may be used to increase the rate of evaporation.

C. Common Laboratory Techniques for Evaporation and Filtration

1. Evaporation

Rapid Evaporation: The solution to be evaporated is placed in an evaporating dish and covered with a watch glass to prevent loss of materials due to spattering as the solution boils, as shown in Figure 1. The solution is heated gently until most of the liquid has boiled away and is then heated even more gently. As the residue dries, the burner should be held in your hand and the gentle flame waved back and forth under the evaporating dish.

Slow Evaporation: If it is important not to risk loss of any material by spattering, the evaporation may be carried out more slowly on a steam bath. A beaker of boiling water serves as a steam bath. Simply place the evaporating dish on the beaker and boil the water until all of the liquid in the dish has evaporated. A simple steam bath setup is shown in Figure 2.

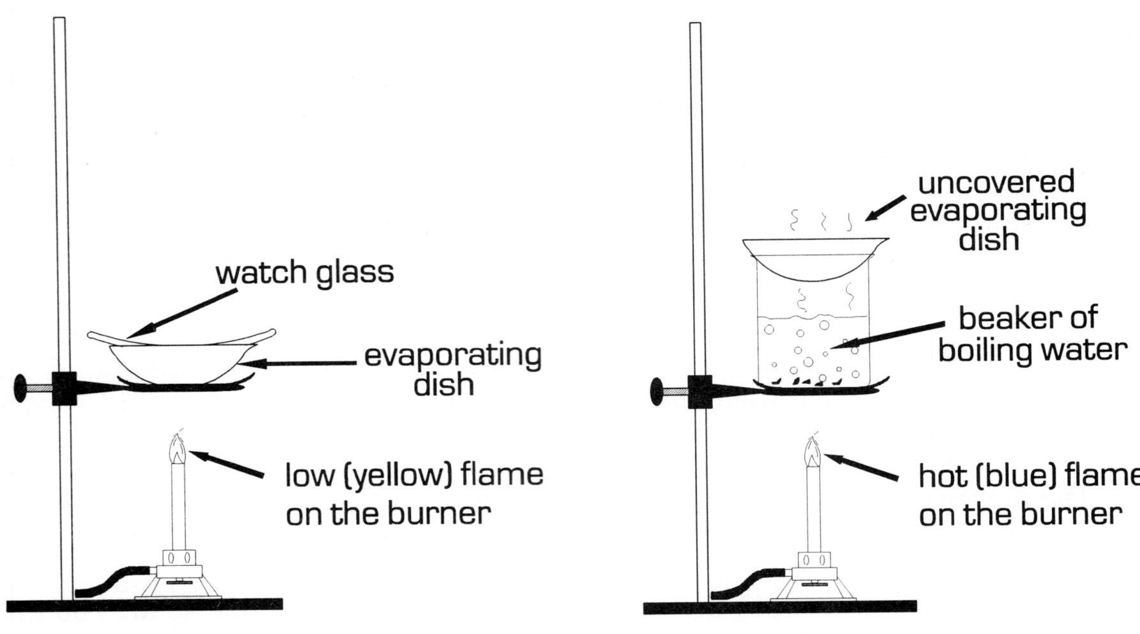

Figure 1. Rapid evaporation

Figure 2. Slow evaporation

2. Filtration

Porous filtering materials include cloth, paper, a mat of fiberglass, sintered glass or other material. The most often used material is paper.

A disk of filter paper must be folded in a special way, as shown in Figure 3, before it is placed in a funnel. Tearing off a corner of the folded paper allows it to make a better seal to the glass when it is wet. The filter paper is opened so that three thicknesses of paper are on one side and one on the other. Place the folded filter paper in a funnel, moisten the paper with water and press the edge of the paper to the side of the funnel. If this is done correctly, the first liquid that passes through the filter will fill the funnel stem and "pull" the remaining liquid through the paper.

Use a stirring rod to guide the liquid and suspended matter into the funnel as shown in Figure 4. The majority of solid should stay in the beaker until most of the liquid has passed through the paper. Then transfer the remaining solid to the funnel using the stream of water from a wash bottle.

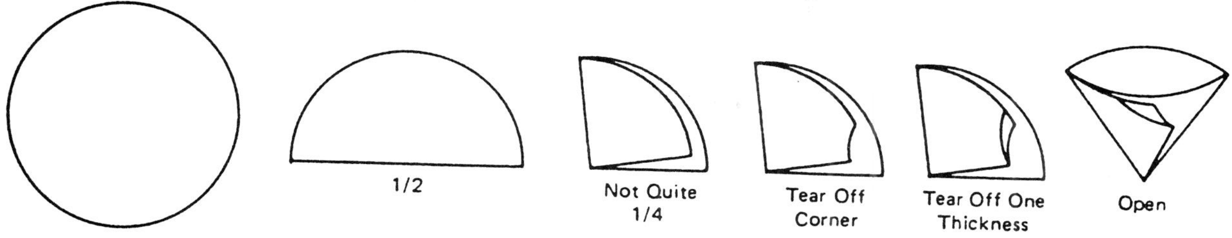

Figure 3. The correct way to fold a disk of filter paper

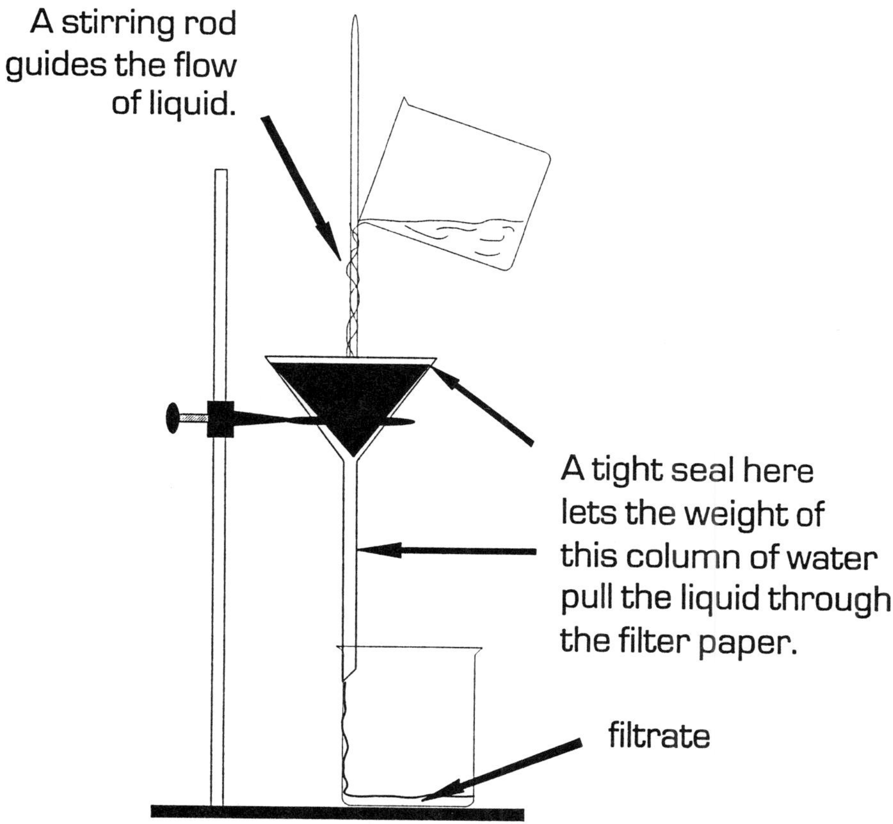

Figure 4. Use a stirring rod to guide the flow of liquid into the cone of filter paper. Sealing the moist filter paper around the rim of the funnel will speed the filtering process.

Experimental Procedure:

A. Characteristics of Elements, Compounds and Mixtures

1. Samples of four elements – iron (Fe), sulfur (S), mercury (Hg) and magnesium (Mg) – are on display in the laboratory. Examine each one carefully and record your observations on the Report Sheet.

2. Examine the mixture of sulfur and iron filings on display in the laboratory. Describe the appearance of the mixture on the Report Sheet. Test the mixture with a bar magnet and record your observations on the Report Sheet.

3. Examine the sample of the compound iron(II) sulfide, FeS, which is on display. This compound is formed when iron and sulfur chemically react with one another. Describe the appearance of the compound on the Report Sheet. Test the material with a bar magnet and record your observations.

B. A Study of a Mixture and Its Separation

To separate a mixture into its components, it is necessary to know some of the properties of these components. The mixture you are to separate later on in this experiment will be composed of two or three of the following: an inert material (sand and/or powdered carbon), powdered magnesium, and sodium chloride. The powered carbon is used so that you will not automatically assume that a black or gray sample must contain magnesium power, which also looks black or gray.

1. Use a spatula to place small amounts (about the size of a match head) of each of these four substances into separate 15-cm test tubes. Add 2 mL (40 drops) of tap water to each sample. Seal each test tube with a cork or plastic film and shake the contents for a minute or two, then note which has dissolved. If you cannot readily decide if a substance has dissolved, pour the mixture through filter paper and look for a residue on the paper. Record those that dissolved in water on the Report Sheet.

> **CAUTION: Hydrochloric acid is used in steps 2 and 4. If you spill acid on your skin, flood the area with water for several minutes and notify your instructor.**
>
> **CAUTION: As always, eye protection must be worn at all times.**
>
> **ENVIRONMENTAL ALERT: Dispose of all chemical waste as directed by your instructor.**

2. Decant (pour off) the water from the substances that did not dissolve in water. Add 2 mL (40 drops) of 6 M hydrochloric acid, HCl, one drop at a time, to each and note on the Report Sheet which is soluble in the acid. At this point it is safe to assume

that evidence of a chemical reaction indicates that the substance is being converted into a water-soluble species. As the reaction proceeds, you will see the undissolved substance "disappearing" into the solution. Collect all waste in a 150-mL beaker for disposal at the end of the experiment.

Using the facts learned in the previous tests, you will now separate an unknown mixture into it component parts. The prepared mixture will be composed of some combination of inert materials (either sand *or* powered carbon), powdered magnesium and sodium chloride. There may be one, two or all three components in the unknown.

3. Obtain your unknown mixture and record its number on the Report Sheet. Weigh out a 2.0-g sample of the prepared unknown mixture on weighing paper and transfer it to a 15-cm test tube. Add about 5 mL of tap water and heat the mixture to dissolve any water-soluble material.

CAUTION: To heat a substance in a test tube, hold the tube at an angle in the flame using a wire test-tube holder. Use a gentle flame. *Do not* point the mouth of the test tube at yourself or another person. Move the tube back and forth constantly for even heating. If you feel the test tube begin to vibrate, let it cool until the vibration stops. The vibration is a warning that the liquid is about to boil violently or "bump."

After heating the mixture, separate the liquid from any undissolved solid by filtration. Pour the hot mixture, including as much of the undissolved solid as possible, onto a folded filter paper fitted in a funnel. Collect the filtrate in an evaporating dish.

Using the procedure for rapid evaporation shown in Figure 1, evaporate the filtrate to dryness to determine if a water-soluble component is present.

4. Slowly add 6 M HCl in small amounts with a dropper to the residue that remains on the filter paper until any chemical reaction ceases. If the substance reacts with HCl, the reaction will convert it to a water-soluble compound and the solution of the compound will pass through the filter paper. Collect the acidic filtrate in the 150-mL waste collection beaker. The observation of a reaction with acid indicates the presence of a component that reacts with HCl. Of course, you may have a component in your mixture that does not react with acid and remains unchanged on the filter paper. Your observations should tell you which water-insoluble component is in your mixture.

5. Once the component that reacts with HCl (if present) is removed, any residue remaining on the filter paper is an inert material in the mixture.

6. Dispose of the acidic solution you collected in the 150-mL beaker as instructed.

Name ______________________________ Locker Number__________ Date____________

Please print: Last name first

PRELIMINARY EXERCISES: *Experiment 4*

Elements, Compounds and Mixtures

1. Define the following terms:

a. element
b. compound
c. mixture
d. intensive property

2. Indicate whether each sample below is a mixture, M, a compound, C, or an element, E.

a. air	___________	e. distilled water	___________
b. blood	___________	f. wood	___________
c. milk	___________	g. sodium chloride	___________
d. nitrogen	___________	h. oxygen	___________

3. Sulfur does not dissolve in water, but it does dissolve in carbon disulfide; potassium nitrate does not dissolve in carbon disulfide, but does in water; and carbon does not dissolve in either of these liquids. Using the above data, outline a process for separating these three components of gunpowder.

Name ______________________ Locker Number__________ Date__________
Please print: Last name first

REPORT SHEET: *Experiment 4*

Elements, Compounds and Mixtures

A. Characteristics of Elements, Compounds and Mixtures

1. Complete the table:

Element	Physical State	Color	Luster
iron			
sulfur			
mercury			
magnesium			

2. a. Describe the appearance of the mixture of iron and sulfur.

   b. What effect did the bar magnet have on this mixture?

3. a. Describe the appearance of the compound FeS.

   b. What effect did the bar magnet have on this compound?

B. A Study of a Mixture and Its Separation

1. Indicate whether each substance below is soluble (S) or insoluble (I) in water and 6 M HCl.

Solvent	White Sand	Powdered Carbon	Powdered Magnesium	Sodium Chloride
water				
6 M HCl				

2. Components of an unknown mixture:

Unknown mixture number: ___________

a. Observations upon evaporation of the water filtrate.

b. Observations upon adding 6 M HCl to the residue on the filter paper.

c. Conclusions: List component(s) *present* in the mixture.

List component(s) *not present* in the mixture.

Experiment 5

Separation Using Chromatographic Techniques

Purpose:

In this experiment you will become familiar with chromatography, a technique used to separate small quantities of mixtures into their component parts. You will use paper chromatography to analyze mixtures of food colors, inks and ions and thin-layer chromatography to separate the colored components of plant chloroplast.

Discussion:

In 1906, Michael Tswett, a Russian botanist, described a method to separate the chloroplast extract obtained from plant tissue into its components. Because many of the components he observed were highly colored, he called the separation technique **chromatography** (*chroma* means "color" in Greek) and the visible result of the separation a **chromatograph**. Over the next several decades, the chromatographic technique became increasingly refined and highly varied. Today chromatography is considered one of the most powerful physical methods of separation we have.

All forms of chromatography employ a stationary phase and a moving phase. The **stationary phase** may be very small pieces of clay, porcelain, cellulose or other material. The **moving phase** (also called the developing phase) is a solvent or mixture of solvents that flows over the stationary phase. The separation of substances is based on the difference in attraction each has for the stationary phase *and* for the moving phase.

Perhaps it will be easier to understand chromatography if we describe the operation of column chromatography, the same method used by Tswett in his original work. The column is a long glass tube with a stopcock on one end. The column is packed with a finely divided a*d*sorbent solid which acts as the stationary phase. Tswett used powdered limestone. The tube is mounted vertically on a stand with the stopcock at the bottom, and the solid stationary phase, (held in place with a cotton plug), is just covered with the developing solvent. A small amount of the mixture to be separated is poured onto the top of the solid. Solvent (the moving phase) is then slowly added and passed down the column and drained out the bottom. If the stationary and moving phases are properly chosen, the mixture will separate because:

1. Each substance in the mixture will be a*d*sorbed to a different degree on the surface of the stationary phase.

2. Each substance in the mixture will have a different affinity for the moving phase.

A compound that has a high affinity for the moving phase and a low affinity for the stationary phase will move down the column rapidly. A compound with a low affinity for the solvent and a high affinity for the stationary phase will move down the column quite slowly.

The rates at which other components of the mixture move down the column between these extremes will be controlled by the relative degree of attraction for the stationary and moving phases. Tswett found sharply separated colored bands along the length of the column. Each band was one component of chloroplast.

Notice that substances are *ad*sorbed onto the surface of the stationary phase and not *ab*sorbed into the body of the solid. Adsorption and absorption are not the same.

The principles that apply in column chromatography form the basis of more than a dozen other types of chromatography used today. Each has certain advantages and works best with specific kinds of mixtures. Three types of chromatography are described below.

In **thin-layer chromatography**, the adsorbent is spread in a thin layer on a plastic or glass plate. A small amount of the mixture to be separated is applied to the adsorbent layer near one end of the plate. This procedure is called **spotting**. The plate is then placed vertically in a container with the lower end resting in a shallow pool of developing solvent. The spot must be above the solvent surface. As the developing solvent ascends the thin layer by capillary action, the components of the mixture are carried with it at varying rates of ascension, undergoing separation as shown in Figure 1.

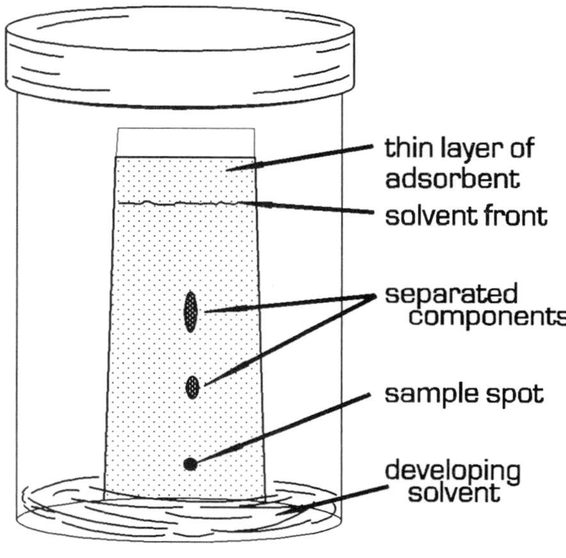

Figure 1

Gas chromatography uses a gas, such as helium, as the moving phase that carries the gaseous mixture. Because everything must be in the gas phase, this technique is usually done at elevated temperatures. As the gaseous mixture flows over the adsorbent, which is packed in a long, coiled tube, separation of the mixture takes place. Separation is based on the varying degrees of attraction that the components of the mixture have for the stationary phase. Those with low degrees of attraction pass through the column faster than those with high degrees of attraction.

Paper chromatography is one of the simplest variations on this theme. The stationary phase is paper, very much like filter paper. A small sample of the mixture is "spotted" near one end of a strip of paper and dried. The paper is then suspended in a vessel with its lower edge immersed in the developing solvent. As the solvent moves up the paper, the difference between each component's affinity for the paper and for the solvent effects the separation. In general, paper chromatography is not as effective as the other methods described, but it is inexpensive, fast and capable of producing good results.

In paper and thin-layer chromatography, the mobility of a substance separated in the analysis is stated in terms of its R_f value. The R_f value equals the ratio of the distance traveled by the substance divided by the distance traveled by the solvent. Each distance is measured from the point at which the mixture is applied to the adsorbent, as shown in Figure 2.

$$R_f = \frac{\text{distance traveled by the substance}}{\text{distance traveled by the solvent}}$$

The R_f value is a physical property of the substance under the conditions of the analysis. R_f values for the two components separated in Figure 2 are calculated below.

$$R_{f(1)} = \frac{100.\ \text{mm}}{120.\ \text{mm}} = 0.833$$

$$R_{f(2)} = \frac{68.0\ \text{mm}}{120.\ \text{mm}} = 0.567$$

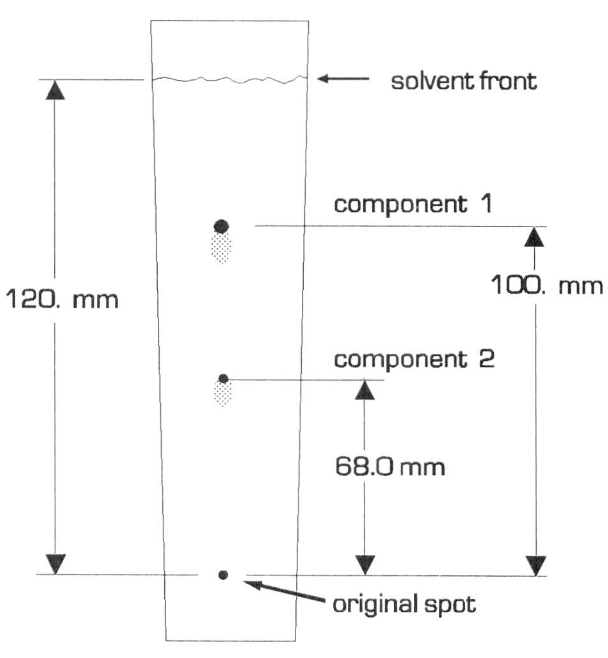

Figure 2

Whether the component spot travels 10 mm or 30 mm from the starting point, the ratio of that distance to the distance traveled by the solvent will remain constant. Also, the same compound under the same conditions will have the same R_f value each time it is analyzed. This means that R_f values can be used to identify compounds. For example, vitamin C has the same R_f value whether it is part of a mixture of vitamins or is present by itself. If a spot produced from a mixture has the same R_f value as pure vitamin C developed under identical conditions, then vitamin C is in the mixture.

Experimental Procedure:

CAUTION: As always, eye protection must be worn at all times.

A. Paper Chromatography of Food Coloring Components

In this section of the experiment, you will learn what colors are blended to obtain the red, blue, green and yellow commercial food colors.

1. In a 50-mL graduated cylinder, collect 15 mL of deionized water, 15 mL of 1-propanol and 15 mL of 1-butanol. Pour the mixture into a clean 600-mL beaker and cover the top tightly with Saran® wrap. Hold the plastic sheet tightly in place with a rubberband, if necessary. Gently swirl the liquid to mix the components of the developing solvent. The covered beaker will serve as the development chamber.

2. Obtain a 21.5-cm x 11.5-cm rectangle of chromatography paper from the supply in the lab. Handle the paper by the edges at all times. *Using a pencil* and a ruler, draw a line across the long dimension of the paper 1.5 cm in from one edge, as shown in Figure 3. Do not use a pen to draw this line. Starting from the left end of the line, mark off ten points 2 cm apart from one another. Label each point with a letter, writing the letters beneath the points. Write your initials in the upper left-hand corner of the paper.

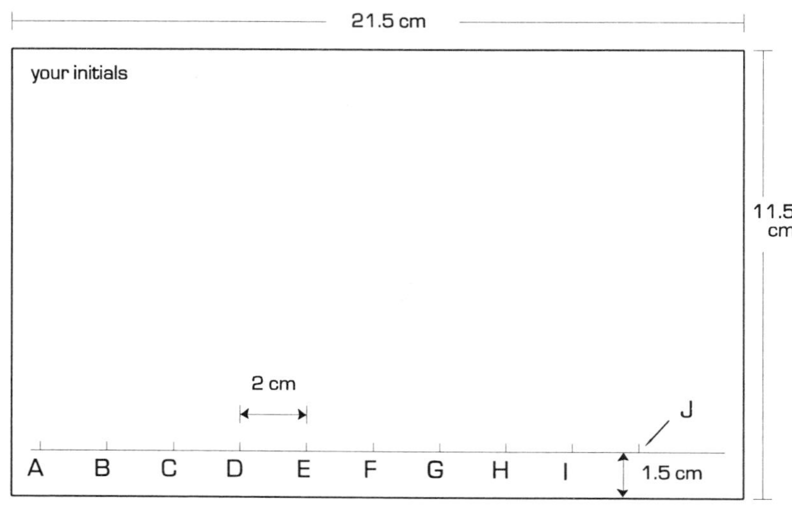

Figure 3

3. The food colors to be analyzed will be "spotted" at each point along the line. Each spot is made by drawing up a small volume of the liquid into a thin capillary tube, then lightly touching the tube to the paper. The spot must not be more than 3 mm in diameter, the size of an "O"; otherwise separation will be difficult. Before you attempt to spot food colors on the large paper, practice making spots on scrap pieces of chromatographic paper until you can make them the correct size.

4. Obtain one unknown color sample to analyze from your instructor and record the sample number on the Report Sheet. The unknown will be analyzed at the same time you examine four FD&C food colors and two commercial food colors.

   Spot the food colors on the chromatographic paper at the points prescribed:

   A. FD&C blue No. 1
   B. FD&C blue No. 2
   C. FD&C red No. 3
   D. FD&C red No. 40
   E. FD&C yellow No. 5
   F. FD&C yellow No. 6
   G. mixture of A through F
   H. commercial red coloring
   I. commercial green coloring
   J. your unknown

   Once all the points have been spotted, set the paper aside for a few minutes to allow the spots to dry.

5. Once all the spots are dry, roll the paper into a cylinder and carefully staple the ends together about 2 cm from the top and 1.5 cm from the bottom, as shown in Figure 4. *The edges of the paper should not touch.*

   Handling the paper by the edges, place the paper cylinder in the development chamber. The spots should be just above the layer of developing solvent, and the paper should not touch the walls of the beaker. Tightly seal the development chamber with Saran® wrap and let it stand undisturbed as the solvent ascends the paper. A properly assembled development chamber is shown in Figure 5.

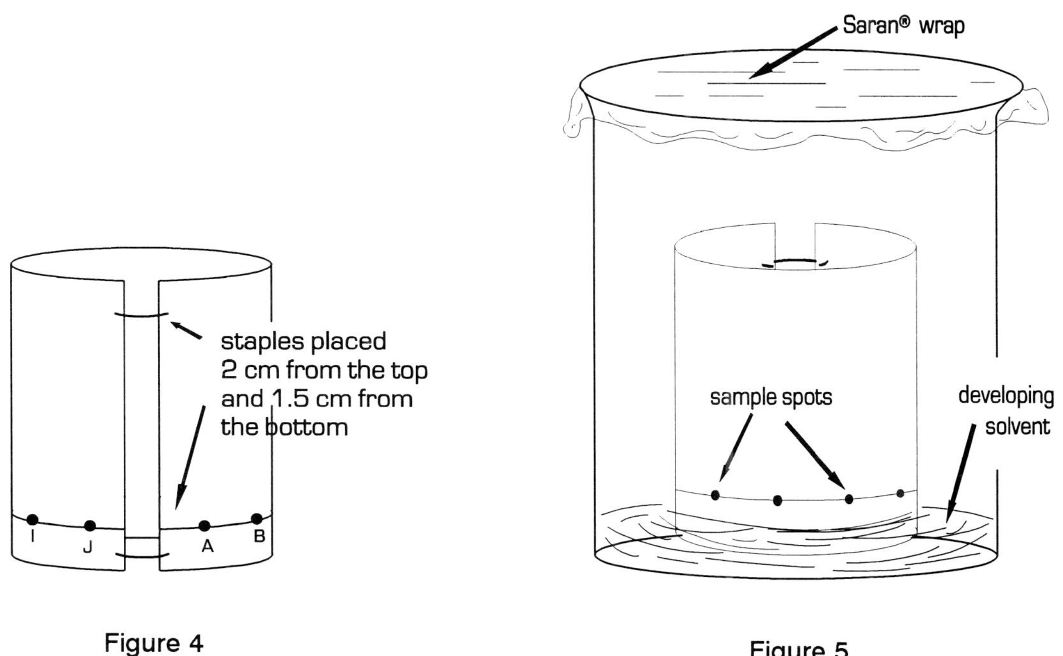

Figure 4

Figure 5

6. After the developing solution has moved at least two-thirds of the way up the strip, (or after one hour), remove the paper cylinder, undo the staples, and lay it flat on a paper towel to dry. *Quickly mark the position of the solvent front* with a pencil. After the paper has dried, outline the colored spots with a pencil and make a dot in what you judge to be the center of each spot. Because of the sample size used here, the components of food colors often form elongated spots with a definite head and an extended tail. The dot should be in the center of the head.

7. Measure the distance in millimeters from the starting line to the dot in the center of each separated component and measure the distance from the starting line to the solvent front, as shown on the chromatogram in Figure 2. Write these distances in pencil on the chromatogram.

   If a particular food color produces only one spot, then it is not a mixture of dyes. Those that produce two or more spots are mixtures. Identify the FD&C food colors present in the red and green commercial brands. Also, identify which of the known samples (A through F) was given to you as an unknown. Record your findings on the Report Sheet.

ENVIRONMENTAL ALERT: The developing solvent mixture is composed of alcohols that are considered biodegradable. Still, your instructor will advise you how to dispose of the developing solvent.

8. Tape the chromatogram to the Report Sheet. Calculate the R_f value for each FD&C color and each component of the two commercial food colors, samples H and I. Calculate the R_f values of the component(s) of your unknown. Show all calculations on the Report Sheet. Use the back side if necessary.

B. Paper Chromatography of Inks

Repeat the procedure described in Section A, but this time use inks suggested by your instructor. Fountain pen ink and felt pen ink work well. When finished, tape the chromatogram to the Report Sheet and show, if requested to do so, the calculation of each R_f value.

C. Paper Chromatography of Cations

ENVIRONMENTAL ALERT: The three metal ions Ag^+, Pb^{2+} and Hg_2^{2+} must not go down the drain. A disposal procedure will be given by your instructor.

Chromatography can be used to separate ions as well as neutral molecules. In this experiment you will separate silver ion, Ag^+, lead ion, Pb^{2+}, and mercury(I) ion, Hg_2^{2+}. You will also analyze an unknown that will contain one, two or all three of these ions. Since these ions are colorless in solution, each chromatogram will be chemically treated to allow the colorless spots to be seen on the paper. Strips of chromatographic paper will be used instead of the cylinder of paper used in Sections A and B.

1. In each of eight clean 20-cm test tubes, place 0.5 mL (10 drops) of 6 M acetic acid ($HC_2H_3O_2$), 2.5 mL of deionized water (H_2O), and 2.0 mL of butyl alcohol (C_4H_9OH), also called 1-butanol.

   Stopper each test tube with a No. 4 rubber stopper that has a slit cut across the bottom (obtained from the supply in the lab). Swirl the test tubes to thoroughly mix the three liquids. *Do not allow the liquids to touch the stoppers.* Let the test tubes stand in a test tube rack until the liquids separate into two distinct layers. Keep the stoppers in place.

2. Obtain eight 18-cm x 1.5-cm strips of chromatography paper. Using the centimeter scale on the Report Sheet or a plastic scale available in the laboratory, draw a faint line *with a pencil* across the width of each strip 2.0 cm from one end. Number each strip on the opposite end from 1 to 8, again using a pencil, and place your initials next to the number.

The metal ion solution to be analyzed will be "spotted" in the center of each line. The spots are made by drawing up a small volume of the liquid into a thin capillary tube, then lightly touching the tube to the paper. The spot must not be more than 0.4 cm in diameter or separation will be difficult. Before you attempt to spot the ion solutions, practice making spots on a piece of chromatographic paper until you can make them the correct size.

3. Obtain an unknown from the instructor and record its number on the Report Sheet.

4. Spot the *silver ion solution* on strips 1 and 2. Spot the *lead ion solution* on strips 3 and 4. Spot the *mercury(I) ion solution* on strips 5 and 6. Spot the *unknown ion solution* on strips 7 and 8. Allow the spots to dry and repeat the same spotting procedure *two* more times, so that each spot has three applications. *Allow the spots to dry completely before proceeding to the next step.*

5. Fix the top edge of each strip into the slit cut in the bottom of a rubber stopper. Check to be sure that when each strip is placed in the test tube, its lower end will be immersed in the *top layer* of the developing solution. Once the length is correct, carefully lower each strip into one of the 20-cm test tubes until its lower end is immersed in the *top layer* of the developing solution. The spot must be *above* the developing solvent, as shown in Figure 6.

   The paper strips should hang nearly straight down and *should not touch the inner wall of the test tube*. If, during the development stage, a paper strip touches the wall of the test tube, rotate the stopper to pull the strip away from the wall. The end of the paper strip should remain in the developing solution. Allow the tubes to stand undisturbed in a test tube rack for one hour.

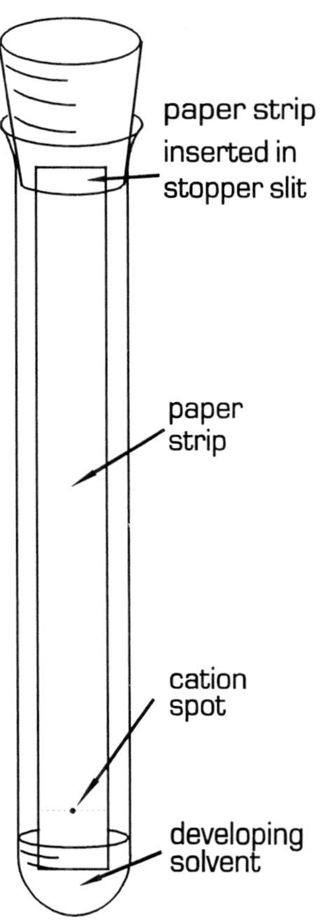

Figure 6

6. After the developing solution has moved at least two-thirds of the way up the strip (or after one hour), remove each strip and lay it on a paper towel to dry. *Mark the location of the solvent front* on each strip with a pencil. After the strips have dried, *quickly* dip each one up to the solvent line into a solution of potassium chromate, K_2CrO_4, that is set out in the lab for this purpose. This should be done quickly. Dry the paper strips in an 80°C drying oven or set them out in the air to dry. A *red spot* indicates *silver ion*. An *orange spot* indicates *mercury(I) ion* and a *yellow spot* indicates *lead ion*. Outline the location of each ion with a pencil and determine the composition of the unknown.

7. Tape the better chromatogram of each pair to the Report Sheet and, if requested to do so, calculate the R_f value for each ion. Show all calculations on the Report Sheet.

D. Extraction and Thin-Layer Chromatography of Plant Chloroplast

Though commercial thin-layer chromatographic sheets are available, your instructor may ask you to prepare your own plates for this experiment. If so, the instructions for doing so are given in Section E on the next page.

1. Extraction of Plant Chloroplast

The colored substances in chloroplast can be extracted by grinding plant material to a paste in organic solvents. Good vegetable sources for this experiment are spinach, broccoli and turnip greens. Your instructor may suggest others.

a. Fill a clean mortar about half-full of chopped spinach or other vegetable. Add enough clean, white sand to cover the bottom of the mortar. Grind the plant material with the pestle until it is reduced to a coarse paste.

> **CAUTION: The solvent mixture used to extract chloroplast in the next step is flammable. Make certain all flames are extinguished in your work area before working with the solvent mixture.**

b. Obtain about 10 mL of the petroleum ether-acetone solvent mixture (the chloroplast extraction solution) from the supply in the lab and add it to the mortar. Grind the mixture for five minutes, then pour the chloroplast extract into a 15-cm test tube. (If you are using spinach, the extract should be dark emerald green. The extract can stain your skin, so take care.) If asked to do so, repeat the extraction with a second 10-mL volume of solvent. You should collect enough extract to cover the bottom of a 15-cm test tube to a depth of about 0.5 cm.

c. After the water (which comes from the vegetable) and petroleum layers have separated, use a capillary dropper to transfer the colored layer (this should be the upper layer) to a clean 10-cm test tube. If the extract will not be used within a few hours, seal and wrap the test tube in aluminum foil and store it in a refrigerator. Be sure to label your sample. If instructed, repeat the extraction using another vegetable.

2. Thin-Layer Chromatography of Plant Chloroplast

a. Obtain two thin-layer chromatographic slides for each plant extract you are to analyze. The slides may be either those you prepare or commercial slides provided by the instructor. Place them on a clean paper towel, coated side up.

b. Make a small scratch mark on opposite edges of each slide, 1.0 cm from the coated end. This will mark where you spot the chloroplast extract. Use the scale printed on the Report Sheet or a small plastic scale from the supply in the lab.

c. Use a thin capillary tube to apply a small volume of the chloroplast extract to the thin-layer slide. Dip the capillary tube into the extract to pick up a small volume of liquid. Then, without actually touching the thin-layer coating, place a droplet of the extract on the coating between the two scratch marks on the slide. *Do not touch the coating.* Allow the spot to dry for several seconds. Continue to place droplets of extract on the slide, allowing each to dry before adding the next droplet, until a *thin band* of extract extends across the width of the slide. Three spottings should be enough. Prepare two slides for each vegetable.

CAUTION: The developing solution used in the next step is flammable. Make certain all flames are extinguished in your work area before working with this solution.

d. Prepare one development chamber for each vegetable extract you are analyzing. Use a wide-mouth jar fitted with a screw cap as the development chamber. Add 3.0 mL of developing solution to each jar. The developing solution is available in the laboratory and is a mixture of petroleum ether and acetone.

e. Place the two slides for the vegetable to be analyzed back to back (coated sides facing away from each other) in the development chamber. Arrange the plates to form a small tepee. The band of spots must be above the level of the developing solvent. Place the screw cap on tightly. The solvent will quickly rise up the thin layer separating the components of chloroplast. If a second vegetable is being analyzed, set up the development chamber for it at this time.

f. When the developing solvent nearly reaches the top of the thin-layer coating, remove the slides and place them face up on a paper towel to dry. Quickly mark the location of the solvent front if you are to calculate R_f values.

g. If you used commercially prepared chromatographic plates, tape them to the Report Sheet. Label the colors you observe next to each plate.

If you prepared your own plates on microscope slides, make a sketch of what you observe on each plate and label all colors.

If requested, calculate the R_f values for each component. In addition, observe each slide under ultraviolet light. Draw any spots that appear *only* under ultraviolet light on your sketches. Identify the extract used for each chromatogram.

E. Preparation of Thin-Layer Chromatographic Plates (optional)

1. Wash two 25-mm x 75-mm glass microscope slides with soap and water for each vegetable extract you are to analyze. Rinse each one thoroughly with tap water, then with acetone (from the squeeze bottle in the lab). Acetone vapors are flammable, so stay away from open flames. Allow the slides to air-dry on a clean paper towel.

2. Obtain a jar of silica gel coating suspension. Thoroughly stir the silica gel-acetone slurry. Then, holding a glass slide by the narrow end, dip it into the suspension up to your finger tips. Swish the plate back and forth several times, then withdraw it

vertically from the mixture in one smooth, continuous motion. Hold the slide vertically over a paper towel to drain off any excess slurry.

3. When the solvent has evaporated from the silica gel coating, decide which side has the better coating. You instructor may help you make this decision. Scrape the poorer coating off, letting the silica gel powder fall into a waste can.

4. A properly coated slide will have a thin, even coating without scratches, chunks, or ridges. If you can see through the coating, it is too thin and will not be satisfactory. Place the slides, coated side up, on a clean paper towel for future use. From this point, handle each slide only by its uncoated end.

Name ______________________________ Locker Number__________ Date____________
Please print; last name first

PRELIMINARY EXERCISES: *Experiment 5*

Separation Using Chromatographic Techniques

1. Considering the attraction a substance may have for the stationary phase as opposed to the moving phase, why do some substances move up the stationary phase faster than others in a chromatography experiment?

2. Describe how the rate of movement of a substance compared to the rate of movement of the solvent can be used to identify a substance in a chromatography experiment.

3. Define the following terms:

a. moving phase
b. adsorption

(The Preliminary Exercises continue on the next page.)

4. Substances A, B and C have R_f values of 0.90, 0.35 and 0.60 respectively. Draw the spots and the solvent front that would be consistent with these values on the chromatogram below.

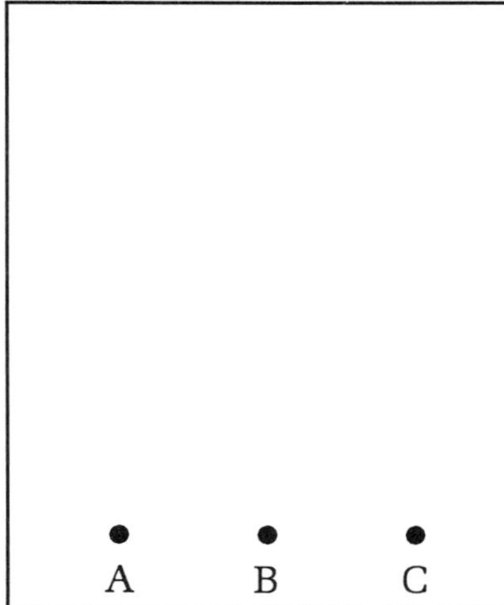

5. An experiment was performed in lab that was supposed to produce aspirin. How could chromatography be used to tell if aspirin was actually the product of the reaction?

Name ______________________________ Locker Number__________ Date____________
Please print; last name first

REPORT SHEET: *Experiment 5*

Separation Using Chromatographic Techniques

At the bottom of the page, attach the chromatograms for the paper chromatography experiment(s) you performed.

A. Paper Chromatography of Food Coloring Components

Food coloring unknown number: __________

Indicate all distances in millimeters on the chromatogram and calculate the R_f value for each component. Show how you calculated the R_f values.

B. Paper Chromatography of Inks

Outline each component in pencil found on the chromatogram. If R_f values are required, indicate those values on the chromatogram. Show all calculations.

C. Paper Chromatography of Cations

Cation unknown number _________

Identify each ion alongside the chromatogram and in the space below. If R_f values are required, indicate those in the same place. Show how you calculated each value.

Ions in the unknown: ______________

METRIC 1 2 3 4 5 6 7 8 9 10 11 12 13 14 15

D. Extraction and Thin-Layer Chromatography of Plant Chloroplast

If you used commercially prepared plastic chromatographic plates, tape them in the rectangles drawn below. Beside each plate, indicate each band you detect under visible light. Also, indicate any band that appears only under ultraviolet light.

If you prepared glass chromatographic plates (Section E), you will not be able to attach them to the Report Sheet. Instead, for each plate, sketch the bands you observe and the location of the solvent front in the rectangles below. State the color of each band. Also, draw any bands that are made visible under ultraviolet light.

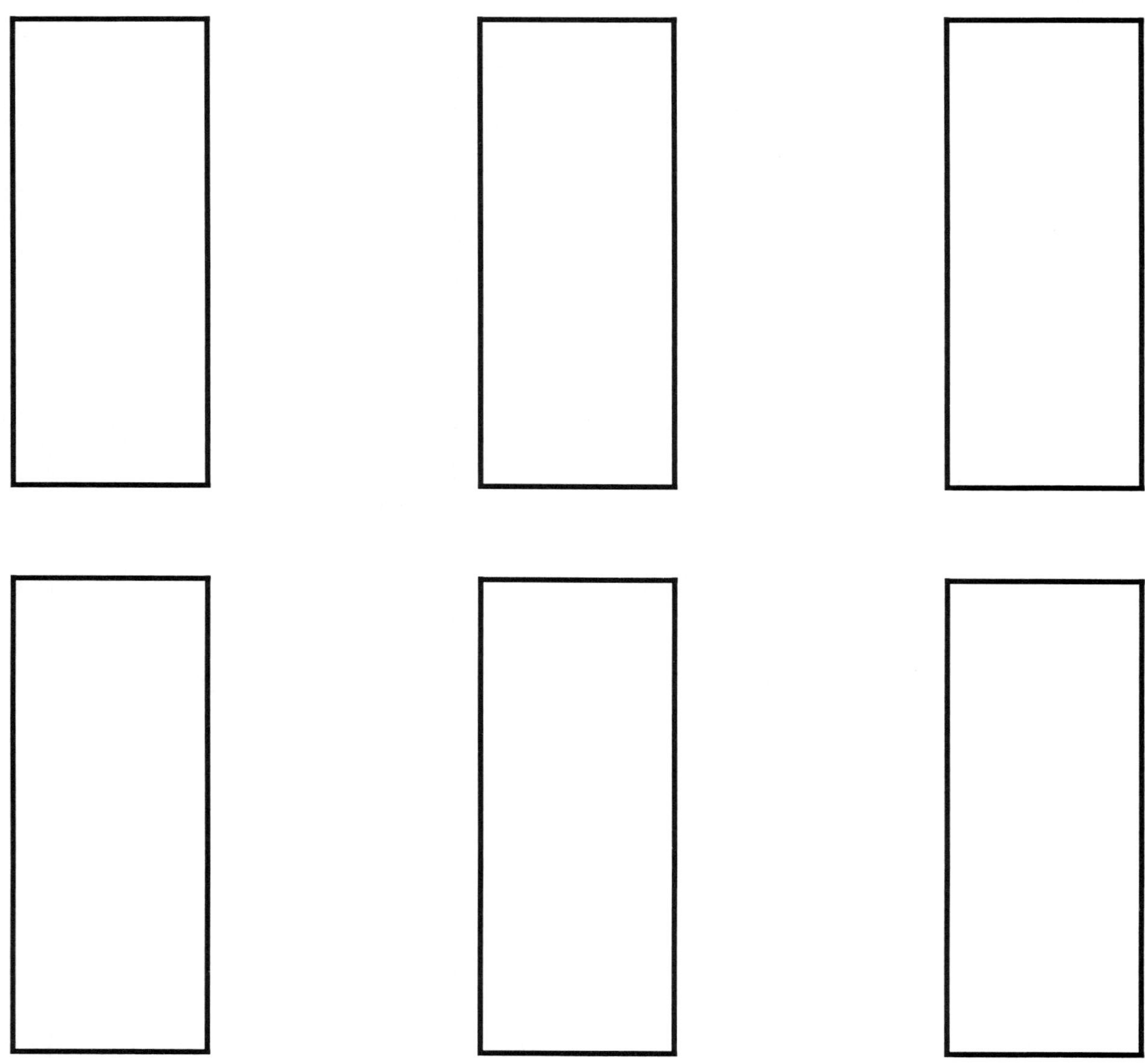

SECTION IV:

Chemical Formulas, Equations, Mass Relationships and Structure

Experiment 6

Hydrates and the Determination of the Formula of a Hydrate

Purpose:

This experiment will introduce you to a class of compounds known as hydrates, and you will determine the percent water in a hydrate as well as its formula by measuring the mass of water driven from a hydrate sample with heat.

Discussion:

Many salts, when exposed to moist air, gain a small amount of weight due to the a*d*sorption of water *onto* the surface of the crystals. This adsorbed water can be easily removed with gentle heating. But some salts can gain larger amounts of water by a*b*sorption, actually incorporating the water molecules *into* their crystalline structure. In these salts, a definite number of water molecules becomes associated with the positive and negative ions, and these water molecules are firmly held within the crystal. Salts of this kind are called **hydrates**, or **hydrate salts**, and their formulas show the fixed water content in their crystals. The formula of a hydrate of calcium chloride, $CaCl_2 \bullet 4\ H_2O$, clearly shows that four molecules of water are associated with each formula unit of calcium chloride. Hydrates can also form when water evaporates from a salt solution. The crystals that remain frequently contain water bound within them (water of hydration) so the crystals are hydrates. When an aqueous solution of copper(II) sulfate is evaporated to dryness, for example, the hydrate of copper(II) sulfate, $CuSO_4 \bullet 5\ H_2O$, is obtained as beautiful blue crystals.

You can see that the formula of a hydrate may be written showing the number of moles of water set off from the formula of the salt and separated by a dot. The formula of the hydrate of copper(II) sulfate shows that 5 mol[1] of water are present for each mole of copper(II) sulfate, and it is named copper(II) sulfate pentahydrate. Some salts can form more than one hydrate. Cobalt(II) chloride, $CoCl_2$, forms a red hydrate with the formula $CoCl_2 \bullet 6\ H_2O$ and a violet hydrate, which is $CoCl_2 \bullet 2\ H_2O$. The mole ratio of water to salt is revealed in their names:

$CoCl_2 \bullet 6\ H_2O$ cobalt(II) chloride hexahydrate
$CoCl_2 \bullet 2\ H_2O$ cobalt(II) chloride dihydrate

[1] Mol is the abbreviation for mole when used as a unit on a number.

Most hydrates will lose their water of hydration when heated strongly, leaving the **anhydrous** (water-free) salt behind. Some hydrates display a color change as the water is lost.

$$\underset{\text{red}}{CoCl_2 \bullet 6\ H_2O} \ \underset{\Delta}{\overset{-\ 4\ H_2O}{\longrightarrow}} \ \underset{\text{violet}}{CoCl_2 \bullet 2\ H_2O} \ \underset{\Delta}{\overset{-\ 2\ H_2O}{\longrightarrow}} \ \underset{\text{blue}}{CoCl_2}$$

The loss of water by a hydrate is reversible. If the anhydrous salt is exposed to moist air or some other source of water, it will regain the water of hydration. The property of reversibility can be used to distinguish true hydrates from other compounds that produce water when heated. The hydrate can be reformed from the anhydrous salt by addition of water. Other substances that produce water when heated strongly may do so because of extensive decomposition, with water simply a product of the decomposition. Adding water to the decomposition products will not reform the original substances; thus, they are not hydrates.

The absorption of water by an anhydrous salt may occur slowly, but their great affinity for water allows these substances to be used as scavengers of small amounts of water in storage containers, gases and organic liquids (in which they are insoluble). When used to absorb water, anhydrous salts are called **desiccants** (to dry). Desiccants are **hygroscopic** (water-attracting) substances. Some desiccants have such a high affinity for water that they can dissolve in their own water of hydration. Salts displaying this property are termed **deliquescent** (to melt from). Calcium chloride is deliquescent and if left in contact with moist air, will absorb enough water to dissolve itself in time.

In this experiment, you are to determine the percent water in a hydrate obtained from your instructor. If you do this well, you will be given the general formula of the hydrate as "salt $\bullet$ X H_2O." With this formula and the percent water, you can then calculate the value of X and complete the formula of the compound. The following equations summarize the calculations required in this experiment:

1. **Calculating the Percent Water in a Hydrate**
   The percent water (by mass) in a hydrate is determined by dividing the mass of water lost by the hydrate with heating by the mass of the original hydrate sample. Multiplying this fraction by 100% will give the percent water in the hydrate.

$$\text{mass of hydrate sample} = (\text{mass of crucible} + \text{lid} + \text{unknown}) - (\text{mass of crucible} + \text{lid})$$

$$\text{mass of water driven off} = (\text{mass of crucible} + \text{lid} + \text{unknown}) - (\text{mass of crucible} + \text{lid} + \text{residue after final heating})$$

$$\text{percent } H_2O = \frac{\text{mass of water driven off}}{\text{mass of hydrate sample}} \times 100\%$$

2. **Calculating the Formula of a Hydrate**
Here you are to calculate the value of X in the formula of the hydrate, salt • X H_2O. X is determined by dividing the number of moles of water lost by the hydrate sample by the number of moles of salt in the sample. Since this is the moles of salt without the water of hydration, it will be called the **anhydrous salt**. Your instructor will give you the formula of the anhydrous salt, and from this you can calculate its formula weight by summing the atomic weights of all the atoms in the formula. Knowing the formula weight, you can convert the mass of the anhydrous salt to moles of the anhydrous salt. The formula weight of water, H_2O, is 18.0.

$$\text{mass of anhydrous salt} = (\text{mass of crucible} + \text{lid} + \text{residue after final heating}) - (\text{mass of crucible} + \text{lid})$$

$$\text{moles of anhydrous salt in sample} = \frac{\text{mass of anhydrous salt}}{\text{formula weight of anhydrous salt}}$$

$$\text{moles of water in hydrate sample} = \frac{\text{mass of water driven off}}{\text{formula weight of water}}$$

$$X = \frac{\text{moles of water in hydrate sample}}{\text{moles of anhydrous salt in sample}}$$

The formula of the hydrate is then salt • X H_2O

Experimental Procedure:

During those times in Section A when the crucible is cooling perform the tests in Section B. Do not wait until Section A is complete before beginning Section B.

ENVIRONMENTAL ALERT: Dispose of all solid residues as directed by your instructor. In Section B, nickel(II) chloride and potassium dichromate must not go down the drain.

CAUTION: As always, eye protection must be worn at all times.

A. Determining the Percent Water and the Formula of a Hydrate

Obtain an unknown hydrate from the instructor. Record the number of the unknown on the Report Sheet.

The unknown hydrate is to be heated in a crucible to drive off the water of hydration. But before you start, check your crucible for cracks by carefully dropping it on its side onto the lab bench from a height of about 2 in. A bell-like ring indicates the crucible is not cracked. A dull thud signals a problem, and a new crucible should be obtained.

During the course of the experiment, handle the crucible and lid only with crucible tongs as shown in Figure 1.

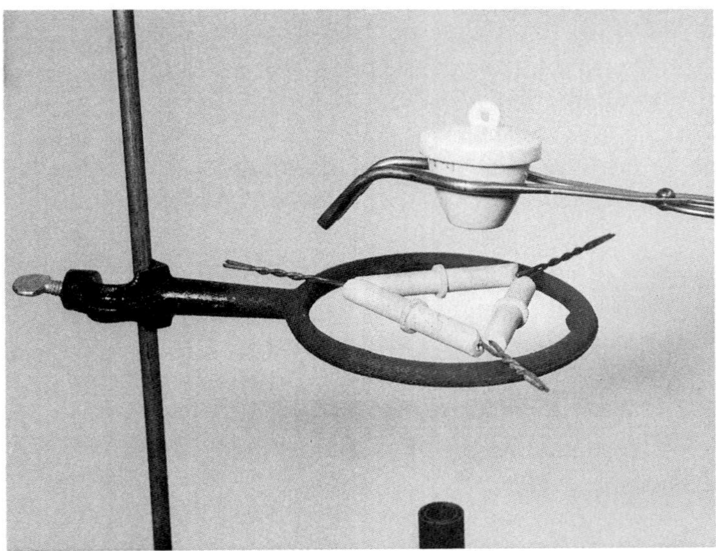

Figure 1

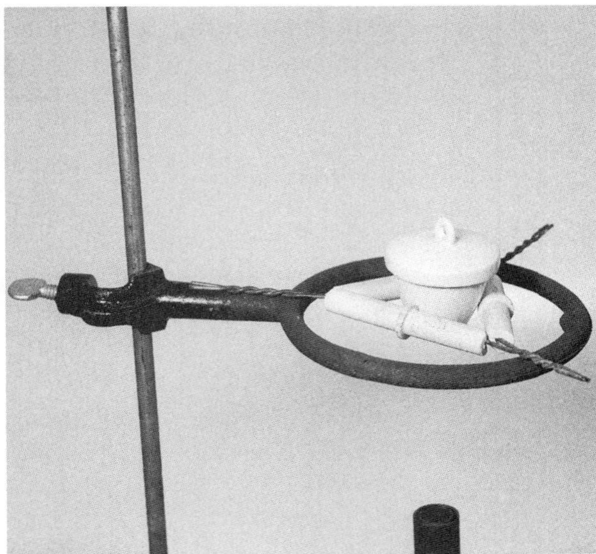

Figure 2

1. Heat the crucible and lid, supported in a clay triangle as shown in Figure 2, with the hottest flame of the burner for 5 minutes to drive moisture from the porcelain. The base of the crucible may get red-hot. Once the heating is complete, place the crucible on a clean wire gauze resting on the lab bench, and let it cool to room temperature. (Begin Section B while the crucible cools, since this may require more than 10 minutes.

2. Once the crucible and lid have cooled to room temperature, determine their mass on a zeroed balance. Record the mass on the Report Sheet to the maximum accuracy allowed by the balance. Use the same balance for all weighings.

   The crucible must be at or near room temperature when it is weighed. If it is warm, thermal updrafts will cause the measured mass to be less than the true mass.

3. Place the entire sample of the unknown hydrate in the crucible. Weigh the crucible + lid + unknown to the maximum accuracy of the balance and record this mass on the Report Sheet.

4. With the lid squarely on, return the crucible to the clay triangle and heat it for 12 to 15 minutes with a moderate flame. *If the crucible becomes red-hot, reduce the size and temperature of the flame.* If heated too strongly for several minutes, some hydrates may decompose beyond the simple loss of water. After the allotted time, return the crucible to the wire gauze and let it cool to room temperature once again.

When it is cool, weigh the crucible + lid + residue to the maximum accuracy of the balance and record this mass on the Report Sheet as the "after first heating" entry.

5. Repeat the heating-cooling-weighing cycle a second time, this time heating the crucible for about ten minutes. Record this mass on the Report Sheet as the "after second heating" entry.

6. Repeat the heating-cooling-weighing sequence until two successive weighings agree within ± 0.040 g. When this occurs you can be confident that all the water has been driven out of the hydrate and only anhydrous salt remains. Use this final mass in the calculations.

7. Calculate the percent water in the unknown hydrate. Show all calculations clearly on the back of the Report Sheet.

$$\text{percent } H_2O = \frac{\text{mass of water driven off}}{\text{mass of hydrate sample}} \times 100\%$$

8. Have your laboratory instructor check your calculations and your percent water value. If your value is satisfactory, you will be given the incomplete formula of the hydrate you analyzed. For example, if you were working with a hydrate of cobalt(II) chloride, the instructor would give you the formula $CoCl_2 \bullet X\ H_2O$. Now you must determine the value of X to complete the formula.

9. Calculate the number of moles of water lost by the hydrate as it was heated. Calculate the number of moles of anhydrous salt remaining in the crucible. Record both values on the Report Sheet. Then calculate the value of X in the formula of the hydrate.

$$X = \frac{\text{moles of water in hydrate sample}}{\text{moles of anhydrous salt in sample}}$$

Show all calculations on the back of the Report Sheet.

B. Identifying Hydrates

1. The five compounds listed below are to be tested to determine if they are true hydrates. Line up five test tubes in a test tube rack and place approximately 0.5 g (enough to just cover the bottom of the test tube) of each of substance in a test tube.

Test tube 1: nickel(II) chloride
Test tube 2: sucrose (table sugar)
Test tube 3: sodium tetraborate (borax)
Test tube 4: potassium dichromate
Test tube 5: potassium chloride

2. Heat each compound gently in the flame of the burner and carefully observe whether droplets of water condense on the cool upper regions of the test tube. If more than just a small amount of condensation appears (that from a small amount of water a*d*sorbed on the surface of the solid), it suggests that the compound may be a hydrate. These compounds are to be examined further.

3. Record on the Report Sheet the color and texture of the residue in each test tube. Add 2 to 3 mL of water to each residue and try to dissolve each one with heating. If the residue dissolves, forming a solution *identical in color* to a solution of the original substance, then the compound is a true hydrate. Prepare a solution of the original substance to check this color. If the residue forms a solution color different from that of the original substance, then extensive decomposition of the original substance has occurred beyond that of a hydrate simply losing water.

   When you are finished with the hydrate tests, properly dispose of all solutions and residues.

Name ______________________ Locker Number__________ Date__________
Please print; last name first

PRELIMINARY EXERCISES: *Experiment 6*

Hydrates and the Determination of the Formula of a Hydrate

1. How would you test a colorless crystalline compound to determine if it was a hydrate?

2. A sample of a crystalline compound, when heated in an open test tube, produced several droplets of water on the cool upper region of the tube. The residue dissolved in water forming a yellow-brown solution, while the original compound formed a colorless solution in water. Why should you conclude that the compound is *not* a true hydrate?

3. What is the meaning of the numeral represented by X in the general formula for a hydrate, salt • $X\ H_2O$?

4. Cobalt(II) chloride is commonly obtained from chemical supply houses as a hydrate with the formula $CoCl_2 \bullet 6\ H_2O$. An analysis showed that 25.0 g of this hydrate contains 11.3 g of water. What is the percent water by weight in this hydrate? Please show your calculations below.

Percent water = __________

Name ______________________ Locker Number__________ Date__________
Please print; last name first

REPORT SHEET: *Experiment 6*

Hydrates and the Determination of the Formula of a Hydrate

A. Determining the Percent Water and the Formula of a Hydrate

Unknown number:__________

Trial 2 is to be carried out if the results of trial 1 are unsatisfactory.

	Trial 1	Trial 2
Mass of crucible + lid + unknown	______	______
Mass of crucible + lid	______	______
Mass of unknown hydrate	______	______
Mass of crucible + lid + residue . . .		
. . . after first heating	______	______
. . . after second heating	______	______
. . . after third heating	______	______
Mass of anhydrous salt	______	______
Mass of water lost	______	______
Percent water in unknown hydrate	______	______

Formula of unknown hydrate (from your instructor) ______________

Moles of water in the hydrate sample ______________

Moles of anhydrous salt in the hydrate sample ______________

Complete formula of the hydrate ______________

B. Identifying Hydrates

Compound Being Tested	Color before Heating	Does Water Appear with Heating (Y/N)	Color and Texture of the Residue after Heating	Is Solution of Residue Same Color as Solution of Original Compound? (Y/N)	Is the Compound a True Hydrate? (Y/N)
nickel(II) chloride					
sucrose					
sodium tetraborate					
potassium dichromate					
potassium chloride					

Show your calculations for Section A in the space below.

Experiment 7

Simple Chemical Reactions

Purpose:

In this experiment you will observe several chemical reactions and learn to identify evidence of chemical change. For the reactions you carry out, you will complete and balance the chemical equation that describes the chemical change. In addition, you will classify each reaction as being one of four types.

Discussion:

Chemistry deals with the study of matter and its properties, structures and changes. **Properties** are the identifying characteristics of matter. The color of gold metal is one of the properties of gold, as are its softness, its electrical conductivity and its ability to be hammered into sheets so thin they are transparent. These are **physical properties** of that element since they can be determined without changing gold into another substance. **Chemical properties**, on the other hand, are determined from the chemical reactions a substance undergoes. Though gold is not very reactive chemically, it will react with a mixture of hydrochloric and nitric acids (aqua regia) to form compounds of the element. The fact that gold reacts with this mixture of acids is a chemical property of the element. Chemical properties of elements and compounds are determined by studying their chemical changes.

A. Describing Chemical Reactions

The observation of chemical changes (chemical reactions) and the proper description of these changes are important concerns of chemistry. There are many clues an observer might note that would suggest a chemical change is taking place, such as the formation of a gas or a precipitate, a color change, the disappearance of a solid, or the evolution or absorption of heat. Recording these changes demands a watchful eye and the careful use of language. Colors are very important and should be recorded accurately. A substance may be green, light green, dark green, blue-green, yellow-green, or something in between. Sometimes it is more descriptive to indicate a color in terms of its similarity to the color of some common object, such as olive green, ruby red, egg yolk yellow, robin's egg blue, silver, gold or copper. If a substance has no color, it is colorless. It may also be clear, which indicates transparency. Some colorless solids, such as sodium chloride or sucrose, are commonly described as being white. They look white because the surface of the small crystals reflect all the wavelengths of white light. You see white light reflected, but a closer examination under a magnifying glass would reveal the body of the crystals to be colorless.

Another method of describing a chemical change is with a **chemical equation**. This equation is a statement that represents the reacting species (the reactants) and those produced (the products) with formulas and symbols. A proper chemical equation is based on experimental evidence. Simply writing an equation doesn't mean that the reaction takes place. Also, to be in agreement with the Law of Conservation of Mass, the equation must be balanced. Only balanced equations can be used in chemical calculations. Symbols written as subscripts can be used to indicate the physical state of substances in equations: (aq) = in aqueous solution,

(s) = solid, (l) = liquid and (g) = gas. These symbols have already appeared in equations in previous experiments. In the following balanced equations, precipitates are indicated as solids, (s). As an alternative, the formulas of precipitates can be underlined.

$$BaCl_{2(aq)} + Na_2SO_{4(aq)} \rightarrow BaSO_{4(s)} + 2\ NaCl_{(aq)}$$

$$KOH_{(aq)} + HC_2H_3O_{2(aq)} \rightarrow KC_2H_3O_{2(aq)} + H_2O_{(l)}$$

$$NH_4Cl_{(aq)} + NaOH_{(aq)} \rightarrow NaCl_{(aq)} + NH_{3(aq)} + H_2O_{(l)}$$

B. Identifying Gaseous Products of Chemical Reactions

In the chemical reactions you will carry out, some products will be gases. To aid in identifying these gases, a few significant properties of seven common gases are given below. Only a few of these gases will be seen in this experiment.

Oxygen, O_2, is a colorless, odorless gas that supports combustion. A *glowing* splint will brighten or burst into flame when thrust into a container of oxygen.

Hydrogen, H_2, is a colorless, odorless gas that burns in air. Hydrogen is less dense than air and can be captured in an inverted vessel. A confined mixture of hydrogen and air (in an inverted test tube) will ignite with a "pop" when exposed to a flame. By contrast, pure hydrogen will burn quietly with a blue flame if it replaces natural gas in a laboratory burner.

Sulfur dioxide, SO_2, is a colorless gas that forms when sulfur burns in air and is said to have the odor of burning sulfur. This gas is toxic since it can combine with water to form an acidic solution called sulfurous acid, $H_2SO_{3(aq)}$. In small quantities, it is safely used as a preservative for dried fruit. Sulfur dioxide in the atmosphere, formed in the combustion of coal and oil, is a principal cause of acid rain.

Nitrogen dioxide, NO_2, is a red-brown gas with a heavy, burning odor. It is toxic, causing edema of lung tissue. Nitrogen dioxide and other oxides of nitrogen are significant air pollutants and can contribute to the formation of acid rain.

Dihydrogen sulfide, H_2S, is a colorless gas with the odor of rotten eggs. It is very toxic, but because its odor is so intense few people are poisoned by it. Solutions of dihydrogen sulfide are weakly acidic, and in solution it is called hydrosulfuric acid.

Ammonia, NH_3, is a colorless gas with a sharp, suffocating odor. Ammonia is toxic and will burn in air. It is very soluble in water, forming solutions that are weakly basic. In solution it is called aqueous ammonia and sometimes ammonium hydroxide.

Carbon dioxide, CO_2, is an odorless, colorless gas that does not burn nor can it support combustion. For that reason, it is used to extinguish fires. When dissolved in water, it forms a weakly acidic solution known as carbonic acid, $H_2CO_{3(aq)}$. If bubbled through a solution of barium hydroxide or calcium hydroxide, a white precipitate of the carbonate salt forms, $BaCO_3$ or $CaCO_3$. Carbonate and bicarbonate salts produce carbon dioxide when treated with acid.

C. Classifying Chemical Reactions

Another task you will be asked to do in this experiment is to classify each reaction as one of four types. The four types of reactions you will consider are diagramed below:

Combination: $A + B \rightarrow AB$ (A and B are elements or compounds; AB is a compound.)

2 REACTANTS → 1 PRODUCT

Example: $4\ Al_{(s)} + 3\ O_{2(g)} \rightarrow 2\ Al_2O_{3(s)}$

Decomposition: $AB \rightarrow A + B$ (AB is a compound; A and B are elements or simpler compounds.)

1 REACTANT → 2 OR MORE PRODUCTS

Example: $CaCO_{3(s)} \xrightarrow{\Delta} CaO_{(s)} + CO_{2(g)}$

Single replacement: $A + BC \rightarrow B + AC$ (A and B are elements; BC and AC are compounds.)

ELEMENT + COMPOUND → NEW ELEMENT + NEW COMPOUND

Example: $Zn_{(s)} + 2\ AgNO_{3(aq)} \rightarrow Zn(NO_3)_{2(aq)} + 2\ Ag_{(s)}$

Exchange: $AB + CD \rightarrow AD + CB$ (All species are compounds.)

2 COMPOUNDS → 2 NEW COMPOUNDS

Example: $BaCl_{2(aq)} + Na_2SO_{4(aq)} \rightarrow BaSO_{4(s)} + 2\ NaCl_{(aq)}$

Writing Formulas of Ionic Compounds:

There are three rules that must be followed when writing formulas for ionic compounds:

1. The positive ion (the **cation**) appears first in the formula. The negative ion (the **anion**) appears second.
2. Each positive charge from the cations must be offset by a negative charge from the anions. The formula must be electrically neutral (no charges left over).
3. The smallest set of subscripts must be used in the formula.

For example, the formula of a compound composed of Al^{3+} and Cl^- requires three chloride ions to offset the 3+ charge on one aluminum ion.

$$Al^{3+} + 3\ Cl^- = AlCl_3$$

The formula for the compound formed by combining Al^{3+} and O^{2-} requires two 3+ aluminum ions (a total of six positive charges), and three 2− oxide ions (a total of six negative charges).

$$2\ Al^{3+} \text{ plus } 3\ O^{2-} = Al_2O_3$$

The formula of aluminum oxide, Al_2O_3, is electrically neutral (6+ and 6−), and the subscripts are the smallest whole numbers that can be used to meet this requirement. Notice how the following formulas are developed from the ions that compose them:

Cation	Anion	Compound Formula	
Ag^+	NO_3^-	$AgNO_3$	(total 1+, total 1−)
Li^+	N^{3-}	Li_3N	(total 3+, total 3−)
Na^+	PO_4^{3-}	Na_3PO_4	(total 3+, total 3−)
Ba^{2+}	OH^-	$Ba(OH)_2$	(total 2+, total 2−)
Fe^{3+}	SO_4^{2-}	$Fe_2(SO_4)_3$	(total 6+, total 6−)
Ca^{2+}	HCO_3^-	$Ca(HCO_3)_2$	(total 2+, total 2−)
Al^{3+}	PO_4^{3-}	$AlPO_4$	(total 3+, total 3−)

Appendix F, *An Introduction to Chemical Nomenclature*, provides additional exercises for writing formulas of ionic compounds.

Experimental Procedure:

You are to carry out ten chemical reactions. The instructions for doing so are given on the next few pages. But before you start, review the following things that you will have to do to complete the Report Sheet:

1. Carefully observe each reaction and record your observations completely and accurately on the Report Sheet.

2. For most reactions, you must complete and balance the equation that describes the reaction. In these cases, an equation is given with one product missing and you are to fill in the missing substance. If the missing product is a gas, consult the list of gases given earlier to learn its identity. If it is not a gas, consult the displays of substances set up in the laboratory. By making a match of the color, texture and physical state of your product with one of those on display, you will learn its identity.

3. Give the reason why you selected a particular substance to complete an equation. Do this by listing the properties of the product you chose that match those you observed in the reaction. These connected properties will serve as your "proof" of product.

4. Indicate the type of reaction you carried out. This will be one of the four types described earlier.

An important aspect of this experiment is to describe what you observe accurately in proper, written English. For example, if a yellow precipitate forms in a reaction, say so; do not simply write, "it turns yellow." Choose your words carefully and express yourself in proper English.

CAUTION: Several corrosive acids and bases are used in the following tests. If you get a reagent on your skin, wash the affected area with running water for several minutes, then immediately notify the instructor.

As always, eye protection must be worn at all times.

ENVIRONMENTAL ALERT: Some of these reactions involve heavy metals or other substances that should not go down the drain. Your instructor will inform you of the proper and safe disposal of these substances.

Reaction 1: *Perform this reaction under the hood.* Place several small pieces of copper metal into a clean 15-cm test tube. Add a few drops of 6 M nitric acid, HNO_3, to the copper metal. You will quickly see a vigorous chemical reaction. The liquid in the test tube will become a blue-green color and the copper will be consumed. A red-brown gas will be evolved. What you will observe is the reaction of copper metal with nitric acid to form copper(II) nitrate, which gives the blue-green color to the solution, and nitrogen dioxide gas, NO_2, the toxic red-brown gas described in the list of gases. These observations and other data are already recorded as an example on the Report Sheet for reaction 1. You will be responsible for the Report Sheet entries for the remaining nine reactions.

Reaction 2: Place a few pieces of copper metal in another clean 15-cm test tube. Add approximately 2 mL (about 40 drops) of 0.1 M silver nitrate, $AgNO_3$, to the copper. Set the test tube aside for several minutes, then observe whether a change has occurred. Record your observations on the Report Sheet, complete and balance the equation and place the reaction in its proper class. Dispose of these reagents safely.

Reaction 3: Add a piece of mossy zinc to 5 mL of 6 M HCl in a 15-cm test tube. Hold your thumb over the mouth of the test tube until you can feel pressure developing inside the tube. *Do not shake!* Hold the mouth of the test tube near the flame of a match or burner and release your thumb. Take care not to burn your thumb! Record your observations on the Report Sheet and supply the requested information. Dispose of the reagents as instructed.

Reaction 4: Ignite a small strip of magnesium ribbon by holding it with tongs in the flame of a burner. Do not look directly at the reaction. Record your observations on the Report Sheet and complete parts b, c and d.

Reaction 5: Just cover the rounded bottom of a 20-cm test tube (the largest you have) with solid ammonium dichromate, $(NH_4)_2Cr_2O_7$. Heat the orange solid over a burner. As soon as the reaction begins, remove the test tube from the flame. Record your observations on the Report Sheet and complete parts b, c and d. Dispose of the reaction product safely.

Reaction 6: Into a 15-cm test tube place 20 drops of 0.1 M copper(II) sulfate, $CuSO_4$. To this blue solution add *1 drop* of 6 M NH_3, aqueous ammonia. Stir, and have the laboratory assistant okay the result. Record your observations on the Report Sheet.

Reaction 7: To the contents of the test tube from reaction 6, add 10 more drops of 6 M NH_3. Stir. Record your observations on the Report Sheet. Dispose of these reagents safely.

Reaction 8: Place 2 mL (40 drops) of 0.1 M lead(II) nitrate, $Pb(NO_3)_2$, in a 15-cm test tube and 2 mL (40 drops) of 0.1 M potassium iodide, KI, into a second 15 cm test tube. Place both test tubes into a 50-mL beaker and weigh the combination. Record the mass to the maximum accuracy of the balance on the Report Sheet. Now, without spilling either solution, add the contents of one test tube to the other. Return the test tubes to the 50-mL beaker and weigh both test tubes and the beaker once again. Record this second mass on the Report Sheet. Answer parts b through f on the Report Sheet. Be certain to dispose of these reagents safely.

Reaction 9: Place 5 drops of 0.1 M iron(III) chloride, $FeCl_3$, in a clean 15-cm test tube. Using a scoopula, add one small crystal of ammonium thiocyanate, NH_4SCN, to the solution. Stir. Record your observations on the Report Sheet.

Reaction 10: This is the old "soda-acid" fire extinguisher reaction.

INSTRUCTIONAL NOTE: Your instructor may wish to perform this reaction as a demonstration.

CAUTION: If *you* perform this experiment, be especially careful to direct the stream of liquid away from you and your neighbors and into a large laboratory sink. This reaction can generate a high-pressure stream of solution very quickly, so be prepared!

Attach a 30-cm length of rubber tubing to the sidearm of a 250-mL sidearm flask (also known as a suction flask). Place 4 g of solid sodium bicarbonate, $NaHCO_3$, in the flask and dissolve it in about 100 mL of tap water. Carefully lower a 10-cm test tube containing 7 mL of 6 M hydrochloric acid into the flask, taking care that the acid and soda solutions do not mix. Stopper the flask with a solid rubber stopper. Hold the flask over a sink, keeping the stopper firmly in place with your hand. While pointing the rubber tubing into the sink, turn the flask upside down and gently shake the contents. Record your observations on the Report Sheet and complete the equation.

Name ______________________________ Locker Number__________ Date___________

Please print; last name first

PRELIMINARY EXERCISES: *Experiment 7*

Simple Chemical Reactions

1. State the law of conservation of matter. In light of this law, what would you predict the results of the two weighings in reaction 8 to be?

2. Classify each of the following as a physical property (*PP*) or a chemical property (*CP*).

   ___________ a. Silver conducts an electric current.

   ___________ b. Hydrogen burns but does not support combustion.

   ___________ c. Mercury had a density of 13.6 g/mL.

   ___________ d. Sugar dissolves in a hot cup of tea.
   (Think about this one. Is the sweet taste of sugar still there?)

   ___________ e. Diesel fuel burns in a truck engine.

3. What chemical test could be used to confirm that the gas produced in a chemical reaction is carbon dioxide? Be specific, please.

4. Balance the following equations:

   a. $NaOH + (NH_4)_3PO_4 \rightarrow Na_3PO_4 + H_2O + NH_3$

   b. $C_2H_6O + O_2 \rightarrow CO_2 + H_2O$

(The Preliminary Exercises continue on the following page.)

c. $Mg_3N_2 + H_2O \rightarrow Mg(OH)_2 + NH_3$

d. $C_8H_{18} + O_2 \rightarrow CO_2 + H_2O$

e. $C_{12}H_{22}O_{11} + O_2 \rightarrow CO_2 + H_2O$

f. $H_2C_2O_4 + CaCl_2 \rightarrow CaC_2O_4 + HCl$

5. Write the correct formulas for compounds formed when the cation on the left combines with the anion listed at the top.

	NO_3^-	S^{2-}	PO_4^{3-}
K^+			
Ca^{2+}			
Fe^{3+}			

Name ______________________ Locker Number__________ Date__________

Please print; last name first

REPORT SHEET: *Experiment 7*

Simple Chemical Reactions

Reaction 1

a. Observations:

A vigorous reaction takes place. The copper metal is consumed and the solution becomes a blue-green color. A dense cloud of a red-brown gas is evolved.

b. Complete and balance:

$Cu_{(s)} + 4\ HNO_{3(aq)} \rightarrow Cu(NO_3)_{2(aq)} + 2\ H_2O_{(l)} + [\ 2\ NO_{2(g)}\ ]$

c. Proof of the product that was added to the above equation:

The red-brown color of the gas matches the description of NO_2 given in this experiment.

d. Type of reaction: *Because copper replaces hydrogen in a compound, this is a single-replacement reaction.*

Reaction 2

a. Observations:

b. Complete and balance:

$Cu_{(s)} + AgNO_{3(aq)} \rightarrow Cu(NO_3)_{2(aq)} + [\qquad\qquad]$

c. Proof of the product that was added to the equation:

d. Type of reaction:

Reaction 3

a. Observations:
b. Complete and balance: $Zn_{(s)} + HCl_{(aq)} \rightarrow ZnCl_{2(aq)} + [\qquad\qquad]$
c. Proof of the product that was added to the equation:
d. Type of reaction:

Reaction 4

a. Observations:
b. Complete and balance: $Mg_{(s)} + O_{2(g)} \rightarrow [\qquad\qquad]$
c. Proof of the product that was added to the equation:
d. Type of reaction:

Name ______________________________ Locker Number__________ Date____________
Please print; last name first

REPORT SHEET (continued): *Experiment 7*

Simple Chemical Reactions

Reaction 5

a. Observations:

b. Complete and balance:

$$(NH_4)_2Cr_2O_{7(s)} \xrightarrow{\Delta} H_2O_{(l)} + N_{2(g)} + [\qquad\qquad]$$

c. Proof of the product that was added to the equation:

d. Type of reaction:

Reaction 6

a. Observations:

b. Instructor or assistant's okay:

Reaction 7

Observations:

Reaction 8

a.	Mass of test tubes + contents + beaker before mixing:
b.	Mass of test tubes + contents + beaker after mixing:
c.	From the data recorded above, has matter been conserved? Justify your answer:
d.	Complete and balance: $KI_{(aq)}$ + $Pb(NO_3)_{2(aq)}$ → $KNO_{3(aq)}$ + [ ]
e.	Proof of the product that was added to the equation:
f.	Type of reaction:

Reaction 9

Observations:

Reaction 10

a.	Observations:
b.	Complete and balance: $NaHCO_{3(aq)}$ + $HCl_{(aq)}$ → $NaCl_{(aq)}$ + $H_2O_{(l)}$ + [ ]

Experiment 8

Analysis of a $KClO_3$-KCl Mixture

Purpose:

In this experiment you are to determine the percent potassium chlorate ($KClO_3$) in a mixture of potassium chlorate and potassium chloride (KCl). In doing so, you will perform calculations based on balanced equations using data you obtain in the experiment.

Discussion:

The area of chemistry that deals with calculations based on formulas and balanced equations is called **stoichiometry**, a term derived from Greek meaning "to measure components." This experiment will focus on calculations based on balanced equations.

A **balanced equation** has an identical number of atoms of each element on either side of the yield sign, the arrow, →. Balanced equations obey the law of conservation of mass, and only balanced equations can be used to calculate masses of products or reactants involved in the reaction. So balanced equations are important not only because they show you what species are involved, but also how much of each species is involved. The following equation, which describes the reaction of aluminum with hydrochloric acid, is balanced.

$$2\,Al_{(s)} \quad + \quad 6\,HCl_{(aq)} \quad \rightarrow \quad 2\,AlCl_{3(aq)} \quad + \quad 3\,H_{2(g)}$$

A balanced equation can be read as a mole statement. The number of moles of each substance is equal to the coefficient preceding it in the equation. The equation above can be read in this way (remember, mol is the abbreviation for mole when it is used as a unit on a number):

$$2\text{ mol }Al_{(s)} + 6\text{ mol }HCl_{(aq)} \quad \textit{yields} \quad 2\text{ mol }AlCl_{3(aq)} + 3\text{ mol }H_{2(g)}$$

The balanced equation states that 2 mol of aluminum must be consumed to produce 3 mol of hydrogen gas. The 2-to-3 mol ratio of aluminum to hydrogen will always hold. If 1 mol of aluminum is consumed, 1.5 mol of hydrogen gas will be produced, again, a 2-to-3 ratio. Several mole ratios can be obtained from the equation. Some of these are:

$$\frac{2\text{ mol Al}}{6\text{ mol HCl}} \qquad \frac{6\text{ mol HCl}}{3\text{ mol }H_2} \qquad \frac{2\text{ mol }AlCl_3}{3\text{ mol }H_2} \qquad \frac{6\text{ mol HCl}}{2\text{ mol }AlCl_3}$$

and the inverted form of each

Every substance in a balanced equation can be related *quantitatively* to every other substance. If you need quantities in grams, convert the number of moles into grams by multiplying it by the corresponding formula weight or molar mass (the mass of 1 mol of the substance).

Balanced Equations and Calculations:

The following questions demonstrate a few calculations that require a balanced equation. The reaction we will examine will be that of aluminum metal with hydrochloric acid.

Question: What mass of aluminum is required to prepare 5.00 g of hydrogen gas, H_2?

The question can be answered with a three-step calculation. You *know* the amount of hydrogen you want to make, and you *seek* the amount of aluminum needed to prepare it. Hydrogen will be labeled the *known substance* and aluminum, the *sought substance.*

$$\underset{\textit{sought}}{2\ Al_{(s)}} + 6\ HCl_{(aq)} \rightarrow 2\ AlCl_{3(aq)} + \underset{\textit{known}}{3\ H_{2(g)}}$$

Step 1: Convert the mass of H_2, the *known* substance, to moles of H_2.

$$\text{moles } H_2 = (5.00\ \cancel{g\ H_2})\left(\frac{1\ \text{mol } H_2}{2.00\ \cancel{g\ H_2}}\right) = 2.50\ \text{mol } H_2$$

Step 2: Convert the moles of the *known* substance, H_2, to moles of the *sought* substance, Al, using the coefficients of the balanced equation. A general equation to do this is:

$$\text{moles of } \textit{sought} = \frac{\text{coefficient of } \textit{sought}}{\text{coefficient of } \textit{known}} \times (\text{moles of } \textit{known})$$

Applied to the problem at hand:

coefficient of *sought* (Al) = 2 mol
coefficient of *known* (H_2) = 3 mol

$$\text{moles Al} = \frac{2\ \text{mol Al}}{3\ \text{mol } H_2}(\text{mol } H_2) = \frac{2\ \text{mol Al}}{3\ \cancel{\text{mol } H_2}}(2.50\ \cancel{\text{mol } H_2}) = 1.67\ \text{mol Al}$$

Step 3: Convert the moles of the *sought* substance, Al, to mass of Al in grams.

$$\text{g Al} = (1.67\ \cancel{\text{mol Al}})\left(\frac{26.98\ \text{g Al}}{1\ \cancel{\text{mol Al}}}\right) = 45.0\ \text{g Al}$$

This three-step sequence can be used to solve nearly all quantitative problems based on balanced equations. The steps can be viewed as three conversions: (1) grams to moles (known to known), (2) moles to moles (known to sought), and (3) moles to grams (sought to sought). The balanced equation is the "bridge" that links the known and the sought substances.

Question: If 5.00 g of hydrogen are produced in the previous reaction, what mass of aluminum chloride, $AlCl_3$, is also produced? Hydrogen is the *known* substance and $AlCl_3$ is the *sought* substance.

Step 1:

$$\text{mole of } H_2 = (5.00 \cancel{\text{ g } H_2})\left(\frac{1 \text{ mol } H_2}{2.00 \cancel{\text{ g } H_2}}\right) = 2.50 \text{ mol of } H_2$$

Step 2:

$$\text{mole of } AlCl_3 = \left(\frac{2 \text{ mol } AlCl_3}{3 \cancel{\text{ mol } H_2}}\right)(2.50 \cancel{\text{ mol } H_2}) = 1.67 \text{ mol of } AlCl_3$$

Step 3:

$$\text{mass of } AlCl_3 = (1.67 \cancel{\text{ mol } AlCl_3})\left(\frac{133.5 \text{ g } AlCl_3}{1 \cancel{\text{ mol } AlCl_3}}\right) = 223 \text{ g } AlCl_3$$

So 5.00 g of H_2 will be accompanied by the production of 223 g of $AlCl_3$.

In the following procedure, you will study the decomposition of potassium chlorate, $KClO_3$, which will be in a mixture of KCl and $KClO_3$. When this mixture is heated, $KClO_3$ decomposes, forming oxygen, O_2, while KCl is unaffected. The decomposition requires a catalyst, manganese(IV) oxide, a dark brown-black solid with the formula MnO_2.

$$2\ KClO_{3(s)} \xrightarrow{MnO_2} 2\ KCl_{(s)} + 3\ O_{2(g)}$$

$$KCl_{(s)} \xrightarrow{MnO_2} \text{no reaction}$$

The balanced equation for the $KClO_3$ decomposition shows that 2 mol of $KClO_3$ produces 3 mol of O_2. If you measure the mass of oxygen produced as the mixture is heated it will be possible to calculate the mass of $KClO_3$ that decomposed and was in the original mixture. (O_2 would be the *known* substance, $KClO_3$, the *sought* substance.) Once the mass of $KClO_3$ is calculated (following the same three-step procedure described above), the weight percent of $KClO_3$ in the mixture can be determined:

$$\%\ KClO_3 = \frac{\text{mass of } KClO_3 \text{ in mixture}}{\text{mass of the mixture}} \times 100\%$$

The mass of KCl in the analyzed mixture can be determined by difference.

$$\text{mass of KCl} = (\text{mass of mixture}) - (\text{mass of } KClO_3 \text{ in the mixture})$$

Experimental Procedure:

> **CAUTION: Eye protection must be worn at all times during the experiment.**
>
> **CAUTION: Hot $KClO_3$ can react explosively with carbon-containing compounds and several other substances. Follow the experimental procedure carefully. If you are unsure of any part of the procedure, please ask your instructor for assistance.**

1. Obtain a sample of an unknown $KClO_3$-KCl mixture from the laboratory instructor. Record the number of the unknown on the Report Sheet.

   MnO_2 is a near-black solid that resembles powdered charcoal. Make certain you are using manganese(IV) oxide and not charcoal. A serious explosion could occur if charcoal is used.

   Place 1 to 2 g of MnO_2 (enough to just cover the bottom of the test tube) into a clean 20-cm test tube, the largest test tube you have. Clamp the test tube at a 45° angle on a ring stand. Position the clamp near the mouth of the test tube as shown in Figure 1. Heat the MnO_2 with the hottest flame of the burner for about 5 minutes to drive out any moisture and to combust any organic material that may be present. Heat the upper section of the tube to completely remove any water that might condense there.

2. Allow the test tube + MnO_2 to cool to room temperature, then weigh to the maximum accuracy of the balance. Immediately record this mass on the Report Sheet. All subsequent weighings must be made on the same balance.

3. Place the $KClO_3$-KCl unknown mixture in the cool test tube with the MnO_2. Mix the two solids thoroughly by slapping the bottom of the tube against the palm of your hand. It is important that the black catalyst and the white mixture blend to form a continuous gray color.

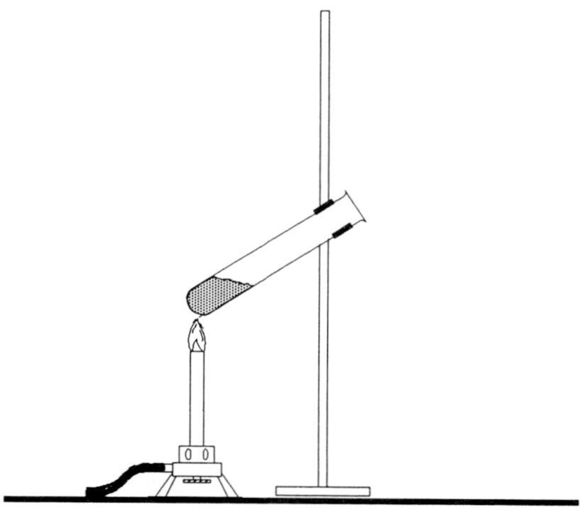

Figure 1

4. Weigh the test tube + MnO_2 + unknown mixture on the balance, and record the mass on the Report Sheet to the maximum accuracy of the balance. (Before proceeding, review the optional method below if asked to do so by your instructor.)

   Optional Method: Frequently a small amount of finely divided KCl (along with smaller amounts of $KClO_3$ and MnO_2) is carried with the oxygen out of the test tube as the sample is heated. The escaping solid appears as a white smoke. A small amount of lost solid will not appreciably affect the results of the experiment, but the loss of any solid can be greatly reduced by placing a ball of glass wool in the mouth of the test tube. If you elect this option, place the glass wool in the tube *after* the test tube + MnO_2 + unknown has been weighed in step 4. Then, with the glass wool in place, determine the mass of the test tube + MnO_2 + unknown + glass wool and record this value on the Report Sheet. Proceed to step 5.

5. Clamp the test tube at a 45° angle on the ring stand and heat the contents with a moderate flame. If at any time you see white smoke coming from the tube, remove the flame for a moment and allow the contents to cool. Then, continue heating more gently. It is important that smoking be kept to a minimum, otherwise the calculated percent $KClO_3$ will be high.

   After about eight minutes of gentle heating, heat the mixture with the hottest flame of the burner for about ten minutes. Wipe the flame along the sides of the test tube so that the entire sample is heated strongly. After heating, allow the test tube and its contents to cool to room temperature.

6. Weigh the cool test tube + MnO_2 + unknown mixture, and record the mass on the Report Sheet in the space labeled "after first heating."

7. The only way you can be certain that all the oxygen has been driven out of the mixture is to reheat it for another ten minutes and weigh the cooled test tube a second time. Use the hottest flame of the burner, and record the mass on the Report Sheet in the space labeled "after second heating." A third heating might be necessary.

   When two successive weighings agree within ± 0.05 g, you can be confident that all the potassium chlorate has decomposed and that no more oxygen will be lost. Use the mass following the last heating in the calculations.

8. From your data, calculate the mass of O_2 evolved during the heating of the unknown. Then calculate the mass of $KClO_3$ that had to decompose to produce that mass of O_2. The balanced equation is:

$$2\ KClO_{3(s)} \xrightarrow{MnO_2} 2\ KCl_{(s)} + 3\ O_{2(g)}$$

   Using the calculated mass of potassium chlorate, calculate the percent of $KClO_3$ in the mixture you were given. Complete the Report Sheet.

9. *After your results have been accepted by the instructor*, clean the residue from the test tube.

ENVIRONMENTAL ALERT: Because MnO_2 is in the solid residue, the residue must be collected for proper disposal. Your instructor will inform you where the collection container is located in the laboratory.

Name ______________________________ Locker Number__________ Date____________
Please print; last name first

PRELIMINARY EXERCISES: *Experiment 8*

Analysis of a $KClO_3$-KCl Mixture

1. Aluminum powder can burn in pure oxygen to form aluminum oxide, Al_2O_3. The balanced equation is:

$$4\ Al_{(s)} \ + \ 3\ O_{2(g)} \ \rightarrow \ 2\ Al_2O_{3(s)}$$

Write numbers in the small blanks to identify four mole ratios that exist in the above balanced equation:

$$\frac{\_\ \text{mol Al}}{\_\ \text{mol } O_2} \qquad \frac{\_\ \text{mol } Al_2O_3}{\_\ \text{mol Al}} \qquad \frac{\_\ \text{mol } O_2}{\_\ \text{mol } Al_2O_3} \qquad \frac{\_\ \text{mol Al}}{\_\ \text{mol } Al_2O_3}$$

2. Referring to the balanced equation in question 1, what mass of aluminum (1 mol Al = 27.0 g Al) is required to just completely react with 96.0 g of O_2 (1 mol O_2 = 32.0 g O_2)? Please show your calculations below.

Mass of Al = ___________

3. How many grams of oxygen, O_2, would be produced by the complete decomposition of 10.0 g of $KClO_3$ (1 mol O_2 = 32.0 g O_2; 1 mol $KClO_3$ = 122.6 g $KClO_3$)?

$$2\ KClO_{3(s)} \ \xrightarrow{MnO_2} \ 2\ KCl_{(s)} \ + \ 3\ O_{2(g)}$$

Mass of O_2 = ___________

Name ______________________ Locker Number__________ Date____________
Please print; last name first

REPORT SHEET: *Experiment 8*

Analysis of a $KClO_3$-KCl Mixture

Unknown number ___________

Trial 2 is to be performed if Trial 1 is unsatisfactory. Remember, units must be used.

	Trial I	Trial II
Mass of test tube + MnO_2 + unknown	________	________
Mass of test tube + MnO_2	________	________
Mass of test tube + MnO_2 + glass wool (optional)	________	________
Mass of unknown	________	________
Mass of test tube + MnO_2 + residue. . .		
. . .after first heating	________	________
. . .after second heating	________	________
. . .after third heating	________	________
Mass of oxygen evolved	________	________
Mass of $KClO_3$ in unknown (Please clearly show all calculations below or on the back of this page.)	________	________
Mass of KCl in unknown	________	________
Percent of $KClO_3$ in unknown	________	________
Percent of KCl in unknown	________	________

Calculations

Experiment 9

The Structure of Covalent Molecules and Polyatomic Ions

Purpose:

In this activity you will learn how to predict the three-dimensional shapes of small molecules and polyatomic ions. You will also build models of these species to see their shapes. In addition you will learn the importance of Lewis formulas in structure prediction.

Discussion:

In the early days of chemical science, investigators were mostly concerned with the composition of matter and the results of physical and chemical changes. These were the things that could be measured in the middle 1700s and early 1800s. Though the concept of the molecule was gaining acceptance from about 1803, when John Dalton proposed his atomic theory, the shape or structure of molecules was left to conjecture. At the molecular level, matter was, for all intent and purposes, invisible. The tools and techniques for determining molecular structure did not become generally available until the early 1900s. At the present time, matter at the molecular level is no longer invisible, and the complete characterization of a new substance requires not only a detailed compositional analysis, but a structural analysis as well. Today, with computer-driven instrumentation, a single chemist can determine the complete three-dimensional structure of a molecule containing 30 or more atoms in a matter of days, and sometimes in just hours. Just 25 years ago, this would have required up to two years for an experienced structural chemist.

The three-dimensional arrangement of atoms in a molecule is called its **structure**, **shape** or **spatial geometry**. Structures of small molecules can be described in terms of bond lengths and bond angles. A **bond length** is the average distance between the nuclei of two bonded atoms and is expressed in units of picometers (1 pm = 10^{-12} m) or Angstroms (1 Å = 10^{-10} m; 1 Å = 100 pm). A **bond angle** is the angle of arc (symbolized $\measuredangle$) between any two bonds that join to the same atom. If all bond angles and bond lengths are known, then the complete three-dimensional shape of that molecule can be reproduced. Frequently, large molecules require additional angles to completely describe their shapes. The bond lengths and bond angles for the formaldehyde molecule are shown below.

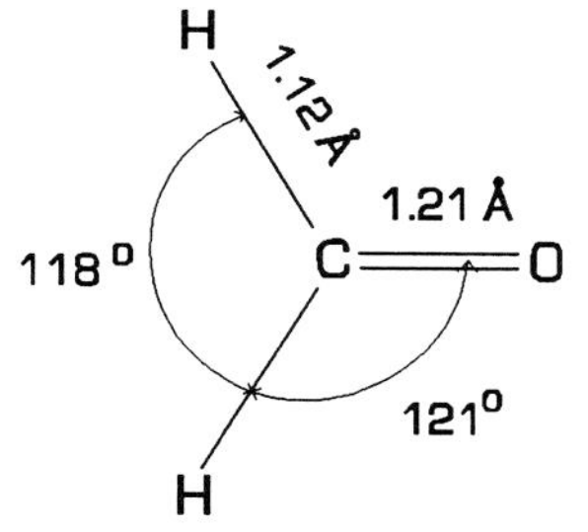

$\measuredangle$H-C-H = 118°

$\measuredangle$H-C-O = 121° (for both)

$d_{H\text{-}C}$ = 1.12 Å (for both)

$d_{C=O}$ = 1.21 Å

Carbon is the **central atom** in the formaldehyde molecule and the other atoms are bonded to it. It is the orientation of the hydrogen and oxygen atoms about the carbon atom that determines the structure of this molecule. Formaldehyde is a flat, planar molecule. The fact that the bond angles total 360° verifies this fact.

A. Understanding Three-Dimensional Models of Molecules

Many methods have been used to model the three-dimensional shape of a molecule. Three commonly used methods are shown below for carbon tetrachloride, CCl_4, which is tetrahedral.

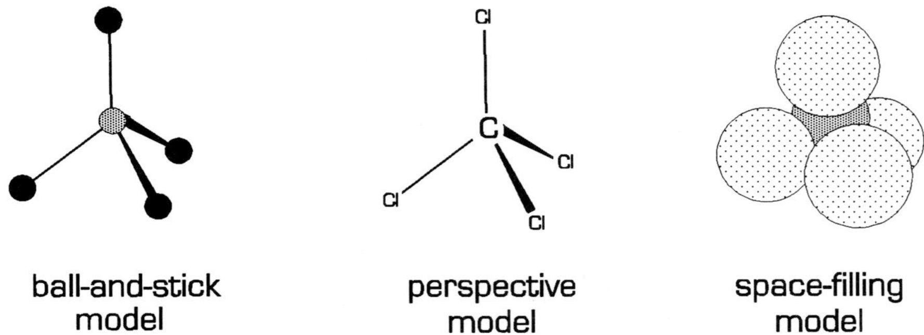

ball-and-stick model | perspective model | space-filling model

In the last several years, modeling of very large molecules has become a very sophisticated business. Proteins, nucleic acids and other biologically important substances can be modeled with great accuracy using high-speed graphic computers. The shape of many biological molecules is intimately related to their function, and pharmaceutical companies increasingly require knowledge of molecular shape to design effective drugs.

Ball-and-stick models of the eleven most common molecular structures for small molecules and polyatomic ions are shown below and on the facing page. The central atom is the light gray sphere in each structure. Your instructor may want you to learn the name of several, if not all, of these structures.

Possible shapes with three atoms:

linear | bent

Possible shapes with four atoms:

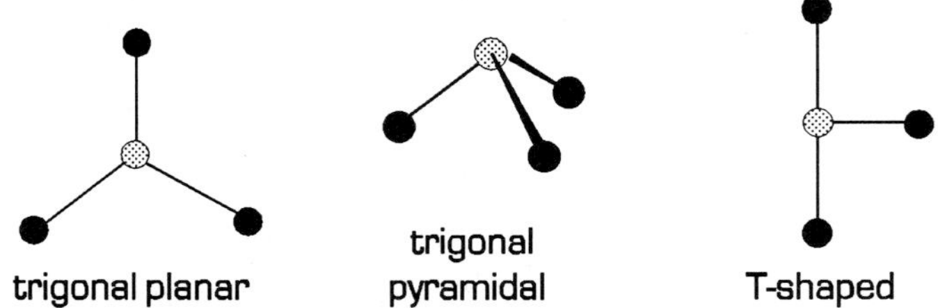

trigonal planar | trigonal pyramidal | T-shaped

Possible shapes with five atoms:

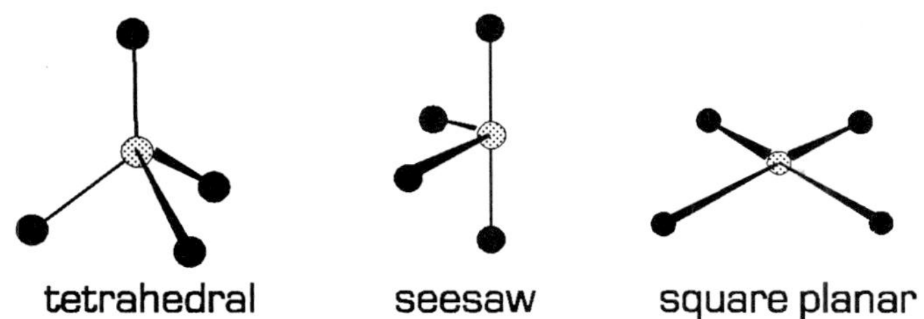

Possible shapes with six atoms:

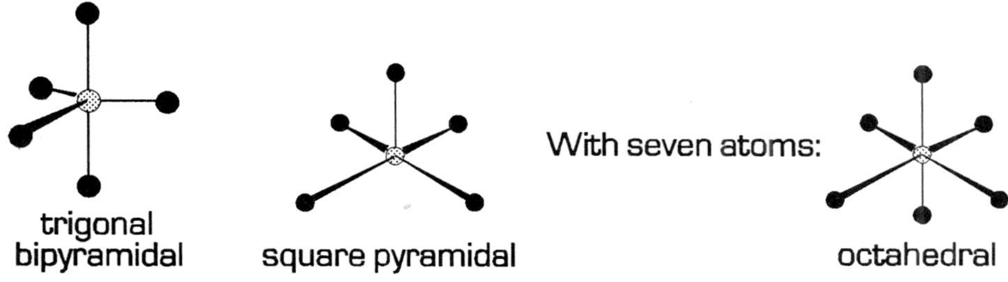

B. Writing Lewis Structures for Molecules and Polyatomic Ions

Lewis structures (or **Lewis formulas**) display how the atoms are joined together in molecules and polyatomic ions. In addition, a Lewis structure will show the location of all valence electrons for each atom in the species. The Lewis structures of water and carbon dioxide are shown below. Each dash (—) represents a shared pair of electrons in a bond. There is one bonding pair of electrons between each hydrogen and the oxygen atom in the water molecule, and two nonbonding pairs of electrons on the oxygen atom that are shown as pairs of dots. In carbon dioxide, there are two shared pairs (four electrons total) between each oxygen and the carbon atom. In addition, each oxygen atom has two nonbonding electron pairs.

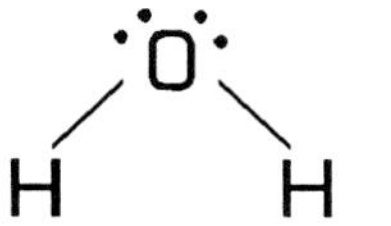

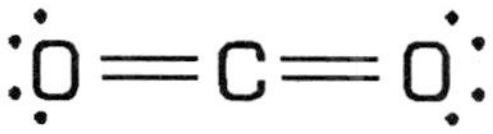

Lewis formulas can be derived for almost all small molecules and polyatomic ions by following a prescribed sequence of operations. As we go through the sequence, the Lewis formulas for the nitrate ion, NO_3^-, and formaldehyde, H_2CO, will be developed.

1. Draw the **skeleton structure** of the molecule or ion, joining the bonded atoms with single bonds (one pair of electrons).

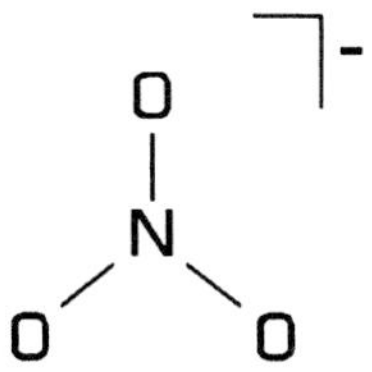

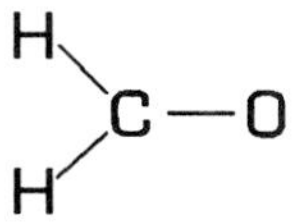

2. Total the valence electrons on all the atoms in the species and. . .
   . . .if it is an *anion*, add one electron for each negative charge.
   . . .if it is a *cation*, subtract one electron for each positive charge.

For NO_3^-: N = 5 valence e^-'s
O = 6 valence e^-'s
O = 6 valence e^-'s
O = 6 valence e^-'s
charge = 1 e^-

24 e^-'s

For H_2CO: C = 4 valence e^-'s
O = 6 valence e^-'s
H = 1 valence e^-
H = 1 valence e^-

12 e^-'s

3. From the total number of electrons obtained in step 2, subtract two electrons for each single bond drawn in the skeleton structure. Each structure has three bonds so a total of six electrons is subtracted.

For NO_3^- : (24 e^-'s) – (6 e^-'s) = 18 e^-'s For H_2CO: (12e^-'s) – (6 e^-'s) = 6 e^-'s

4. Distribute the remaining electrons as *nonbonding electron pairs* about each atom that is connected to the *central* atom until each is surrounded by eight electrons, (two electrons for H). Any extra electrons should now go on the central atom.

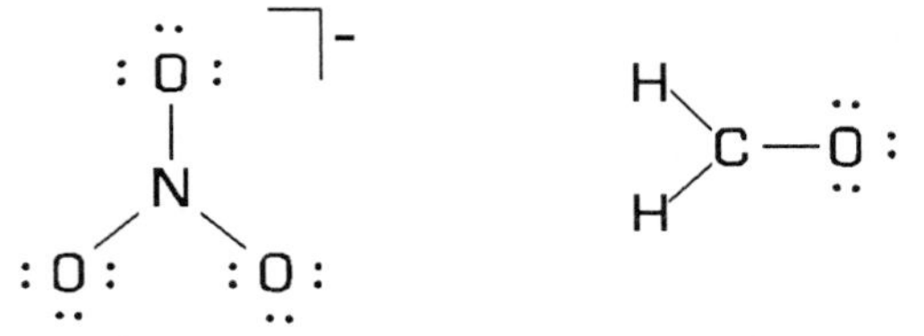

If too few electrons remain to give the central atom eight electrons, convert one or more single bonds to double or triple bonds by converting nonbonding pairs to shared pairs.

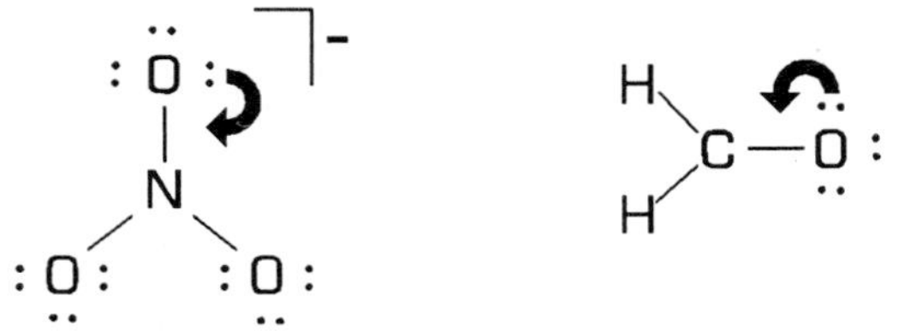

Bonding electrons are counted for *both atoms* that share them. Now each atom has eight electrons about it, except for hydrogen, which can accommodate only two.

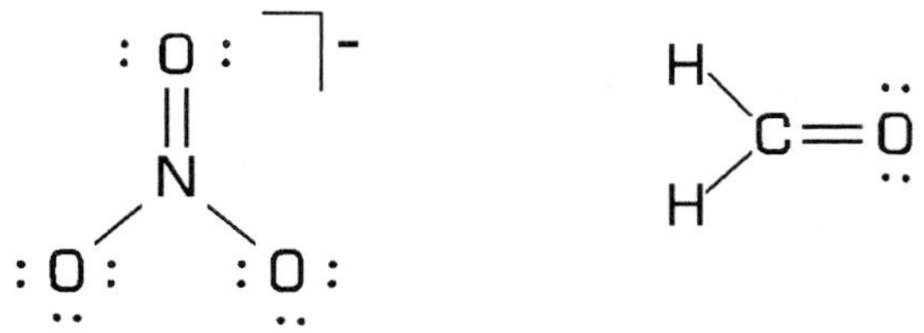

Watch for the possibility of *resonance*. In the nitrate ion, a nonbonding pair of electrons could have been shifted from any one of the three oxygen atoms to give nitrogen an octet. Whenever there are two or more identical atoms from which the nonbonding electron pair could be taken, resonance will be possible, and the Lewis formula you develop will be one of two or more resonance forms. Resonance is important because it adds stability to a molecule or ion. Since there is only a single oxygen atom from which a nonbonding pair of electrons could be converted to a shared pair in formaldehyde, resonance is not possible.

In reality, only a few elements *consistently* achieve the outer electronic configuration of the nearest noble gas, the octet of electrons (except for He), when they combine to form molecules. These elements are in the first and second periods of the periodic table, most notably H, C, N, O and F. Elements in the third and higher periods can "expand" their capacity to combine with other atoms and can accommodate more than eight electrons using empty orbitals in their valence shells. Elements like P, S, Cl and many others can accommodate as many as 10, 12 or even 14 electrons in their valence shells. For example, P in PF_5 has 10 electrons in its valence shell. Sulfur has 12 in its valence shell in SF_6. Boron only has 3 valence electrons to begin with (it is a Group IIIA element), and in BF_3, boron only has 6 electrons in its valence shell. In addition, there are some molecules with an odd number of electrons. It is impossible for these to have an even number of electrons (2 or 8) about each atom. Nitrogen dioxide, NO_2, is such a species. It has 17 valence electrons (5 e^-'s from N and 6 e^-'s from each O).

C. Predicting the Three-Dimensional Shape of a Molecule

Once an accurate Lewis structure is drawn for a molecule or polyatomic ion, it can be used to predict the shape of the species. The technique for doing this was developed in 1957 by a Canadian chemist, R. J. Gillespie.[1] He recognized the importance of the nonbonding and bonding electrons about the *central atom* of a molecule in directing the shape of the molecule. The natural, electrostatic repulsion that exists between electron pairs in the valence shell of the central atom will cause them to be as far apart as possible. It makes no difference if the electron pairs are nonbonding electrons or bonding electrons. A single, double or triple bond will have the same effect as a nonbonding pair in predicting shapes.

By orienting the bonding and nonbonding electron pairs as far apart from each other as possible, the molecule will exist in its most stable shape. Again, only the *central atom* need be considered. Here is how it is possible to predict the shape of a molecule.

1. Derive the Lewis structure of the species. The rules for deriving Lewis structures are given in Section B. If the species displays resonance, choose one resonance form.

2. Identify the central atom in the species. The central atom is the one to which all other atoms are joined. It is the N atom in NO_3^-, C in H_2CO.

3. Total the number of (1) nonbonding electron pairs and (2) bonds connected to the central atom. Remember, single, double and triple bonds all count as *one* bond here. Each total (2, 3, 4, 5 or 6) predicts a specific shape, as shown in Table 1. Each shape arranges the nonbonding pairs and bonds as far apart from one another as possible.

---

[1] Professor Gillespie called his method for predicting shapes of covalent species the valence shell electron-pair repulsion (VSEPR) theory .

Table 1. The Five Arrangements of Electron Domains About A Central Atom

Total of Nonbonding Pairs Plus Bonds	Arrangement of Nonbonding Electrons and Bonds about Central Atom	Geometry of Nonbonding Electrons and Bonds about Central Atom	Predicted Bond Angles
2	180°	linear	180°
3	120°	trigonal planar	120°
4	109.5°	tetrahedral	109.5°
5	90° 120°	trigonal bipyramidal	90° and 120°
6	90°	octahedral	90°

4. Though the *trigonal bipyramidal* and *octahedral* geometries are not as common as the first three in Table 1, they are seen in species that have a central atom from the third and later

periods, and each needs to be treated in a special way. *In the trigonal bipyramidal and octahedral geometries, nonbonding electron pairs always occupy specific locations, shown in Table 2.* In the trigonal bipyramid, nonbonding electron pairs are always located in the trigonal (or triangular) plane. In the octahedral arrangement, if two nonbonding pairs of electrons exist, they will always be placed opposite one another on the central atom. Structures based on the trigonal bipyramid and the octahedron may be used in optional exercises by your instructor. You should be aware they exist for many species.

Table 2. Location of Bonding and Nonbonding Electron Pairs in Trigonal Bipyramidal and Octahedral Geometries

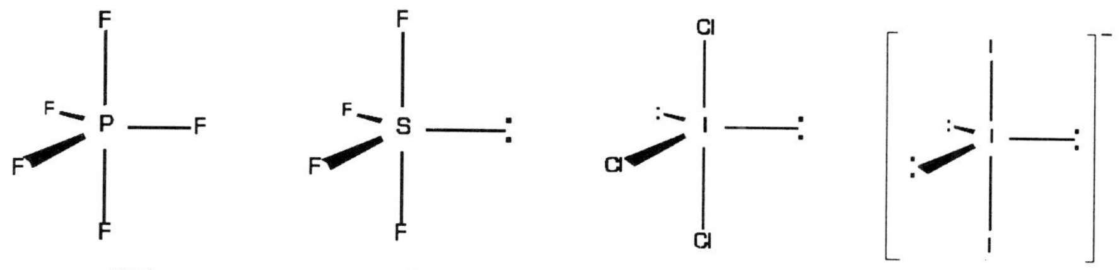

	PF_5	SF_4	ICl_3	I_3^-
Bonding pairs:	5	4	3	2
Nonbonding pairs:	0	1	2	3
Shape description:	trigonal bipyramidal	seesaw	T-shape	linear

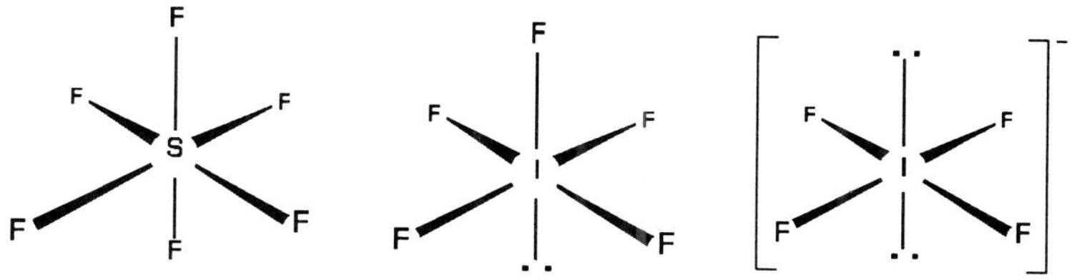

	SF_6	IF_5	IF_4^-
Bonding pairs:	6	5	4
Nonbonding pairs:	0	1	2
Shape description:	octahedral	square pyramidal	square planar

5. *The shape of the molecule or ion is determined by the arrangement of the atoms in space.* Though the nonbonding electron pairs are very important in determining the arrangement of the atoms, *only the geometric arrangement of the atoms is used to describe shapes.*

Frequently the three-dimensional shape of the molecule is *not* described by the same term that is used to describe the arrangement of the nonbonding pairs and bonds about the central atom. Table 3 compares the terms used to describe the geometry of nonbonding pairs plus bonds with the terms used to describe the arrangement of atoms in space. *Remember, the shape of the molecule is described only by the arrangement of the atoms in space.*

Table 3. Possible Arrangements of Nonbonding Pairs and Bonds and Resulting Molecular Shapes

***Total* of Nonbonding Pairs Plus Bonds**	**Number of Bonds *Plus* Nonbonding Pairs**	**Geometry of Nonbonding Pairs and Bonds about the Central Atom**	**Arrangement of the Atoms about the Central Atom, the *molecular shape***
2	2 bonds	linear	linear
3	3 bonds	trigonal planar	trigonal planar
	2 bonds + 1 nonbonding pair	trigonal planar	bent
4	4 bonds	tetrahedral	tetrahedral
	3 bonds + 1 nonbonding pair	tetrahedral	trigonal pyramidal
	2 bonds + 2 nonbonding pairs	tetrahedral	bent
	1 bond + 3 nonbonding pairs	tetrahedral	linear
5	5 bonds	trigonal bipyramidal	trigonal bipyramidal
(may be optional)	4 bonds + 1 nonbonding pair	trigonal bipyramidal	seesaw
	3 bonds + 2 nonbonding pairs	trigonal bipyramidal	T-shape
	2 bonds + 3 nonbonding pairs	trigonal bipyramidal	linear
6	6 bonds	octahedral	octahedral
(may be optional)	5 bonds + 1 nonbonding pair	octahedral	square pyramidal
	4 bonds + 2 nonbonding pairs	octahedral	square planar

Once the Lewis structure for a molecule or polyatomic ion is developed, the shape of the species can be determined using the information in Table 3. The Lewis structures of carbon dioxide, CO_2, water, H_2O, formaldehyde, H_2CO, ammonia, NH_3, the nitrate ion, NO_3^-, and fluoromethane, FCH_3, are given below with the step-by-step procedure used to determine the shape of each species. The three-dimensional shape is in bold print.

```
:O=C=O:
```

central atom = **C**
number of bonds = 2
nonbonding pairs = 0

total bonds + nb pairs = 2
geometry of bonds + nb pairs: linear
shape of molecule: **linear**

```
   :O:
  /   \
 H     H
```

central atom = **O**
number of bonds = 2
nonbonding pairs = 2

total bonds + nb pairs = 4
geometry of bonds + nb pairs: tetrahedral
shape of molecule: **bent**

```
H
 \    ..
  C = O
 /    ..
H
```

central atom = **C**
number of bonds = 3
nonbonding pairs = 0

total bonds + nb pairs = 3
geometry of bonds + nb pairs: trig. planar
shape of molecule: **trigonal planar**

```
      ..
H --- N --- H
      |
      H
```

central atom = **N**
number of bonds = 3
nonbonding pairs = 1

total bonds + nb pairs = 4
geometry of bonds + nb pairs: tetrahedral
shape of molecule: **trigonal pyramidal**

```
    : O :    ⌉-
      ||
      N
    /   \
 :O:     :O:
  ..      ..
```

central atom = **N**
number of bonds = 3
nonbonding pairs = 0

total bonds + nb pairs = 3
geometry of bonds + nb pairs: trig. planar
shape of molecule: **trigonal planar**

```
   ..
  : F :
    |
H — C — H
    |
    H
```

central atom = **C**
number of bonds = 4
nonbonding pairs = 0

total bonds + nb pairs = 4
geometry of bonds + nb pairs: tetrahedral
shape of molecule: **tetrahedral**

Experimental Procedure:

The natural arrangement of the volumes of space occupied by bonds and nonbonding electrons about an atom is simulated with balloons in the following exercises.

A. Natural Orientation of Volumes about a Central Point

1. Obtain six round balloons from the supply in the laboratory. Inflate each balloon to the *same diameter* and tie off each with a knot. Make two three-balloon sets as shown to the right by tying the three ends together. Set both three-balloon sets aside for a moment.

2. Obtain six more round balloons and inflate each to the *same diameter* as those prepared in step 1. Tie off each balloon. Make three two-balloon sets by joining two balloons together at the ends as shown.

3. Rest one of the two-balloon sets on the laboratory bench and note how the balloons naturally arrange themselves relative to one another. *Draw this arrangement on the Report Sheet and describe its shape with the appropriate term (linear, trigonal planar, etc.). Do the same for one of the three-balloon sets, recording the same information on the Report Sheet.*

4. Twist two two-balloon sets together to form a four-balloon combination. Fine tune the combination until the balloons come to their most stable arrangement, that is, the arrangement that arises naturally and that can only be changed by forcing one or more balloons to move closer to the others. *Draw the arrangement of this most stable four-balloon combination on the Report Sheet and describe its shape with the appropriate term.*

5. Twist together one two-balloon set and one three-balloon set so that all five balloons are joined by their ends to a common central point. Fine tune this five-balloon set until the most stable arrangement is found. *Draw the most stable arrangement of the five-balloon set on the Report Sheet and describe its shape with the appropriate term.*

6. Twist three two-balloon sets together, joining the six balloons at a common point. Fine tune the six-balloon set until the most stable arrangement is found. *Draw the most stable arrangement of the six-balloon set on the Report Sheet and describe its shape with the appropriate term.*

7. Match each of the five shapes formed by the balloon combinations in steps 2 through 6 to the ball-and-stick models shown in Table 1 in the discussion section.

B. Valence Shell Electron Pairs: Only Single Bonds

1. Draw the Lewis structures for $BeCl_2$, BF_3 and CH_4 in the appropriate space on the Report Sheet. PF_5 and SF_6 are optional species that your instructor may assign.

2. Record the number of bonds about the central atom in the space provided.

3. Sketch the ball-and-stick model of each compound in the space provided. This model should show *only* the arrangement of the atoms in the molecule.

4. Determine the shape of the molecule and write the name of that shape in the space provided. You may want to consult Table 3 for assistance.

C. Valence Shell Electron Pairs: Both Bonds and Nonbonding Electron Pairs

1. Draw the Lewis structures for NF_3, H_2S, H_3O^+, PH_3 and NO_2^- in the space on the Report Sheet.

2. Record the number of bonds plus nonbonding pairs about the central atom in the space provided.

3. Complete the grid on the Report Sheet with the ball-and-stick model of each species and the name of its shape, as was done in section B.

D. Valence Shell Electron Pairs: Single and Double Bonds and Nonbonding Electron Pairs

1. Draw Lewis structures for CO_3^{2-}, SO_2 (use only one resonance form), H_2CO, SO_3 (use one resonance form), SO_3^{2-}, SO_4^{2-} and PO_4^{3-} in the proper space on the Report Sheet.

2. Record the number of bonds plus nonbonding pairs about the central atom in the space provided.

3. Complete the grid on the Report Sheet with the ball-and-stick model of each species and the name of the shape of the species.

Name ______________________ Locker Number__________ Date__________
Please print; last name first

PRELIMINARY EXERCISES: *Experiment 9*

The Structure of Covalent Molecules and Polyatomic Ions

1. Define the following terms:

a. bond length
b. Lewis structure
c. nonbonding electron pair

2. Draw Lewis structures for the following species:

NH_3	H_2CCl_2

3. Fill in each blank as you analyze the Lewis structures given below. ("nb pairs" stands for nonbonding pairs of electrons.)

$[O=N=O]^+$

central atom: ____ total bonds + nb pairs: ____
number of bonds: ____ geometry of bonds + nb pairs: __________
nonbonding pairs: ____ shape of molecule: __________

Cl—S—O with Cl below S

central atom: ____ total bonds + nb pairs: ____
number of bonds: ____ geometry of bonds + nb pairs: __________
nonbonding pairs: ____ shape of molecule: __________

Cl—P—Cl with Cl below P

central atom: ____ total bonds + nb pairs: ____
number of bonds: ____ geometry of bonds + nb pairs: __________
nonbonding pairs: ____ shape of molecule: __________

Name ______________________________ Locker Number__________ Date____________
Please print; last name first

REPORT SHEET: *Experiment 9*

The Structure of Covalent Molecules and Polyatomic Ions

A. Natural Orientation of Volumes about a Central Point

Balloon Arrangement	Drawing of the Balloon Set	Description of the Shape	Matching Ball-and-Stick Model
Two-balloon set			
Three-balloon set			
Four-balloon set			
Five-balloon set			
Six-balloon set			

B. Valence Shell Electron Pairs: Only Single Bonds

Molecular Formula	Lewis Structure	Number of Bonds about Central Atom	Sketch of Ball-and-Stick Model	Shape Description
$BeCl_2$	:C̤̈l: — Be — :C̤̈l: (example)	2	●—◌—●	linear
BF_3				
CH_4				
PF_5 (optional)				
SF_6 (optional)				

Name ______________________________ Locker Number __________ Date __________
Please print; last name first

REPORT SHEET: (continued) ***Experiment 9***

The Structure of Covalent Molecules and Polyatomic Ions

C. Valence Shell Electron Pairs: Both Bonds and Nonbonding Electron Pairs

Molecular Formula	Lewis Structure	Number of Bonds and Nonbonding Pairs about Central Atom	Sketch of Ball-and-Stick Model	Shape Description
NF_3	:F̈—N̈—F̈: with :F̈: bonded below N (example)	4		*trigonal pyramidal*
H_2S				
H_3O^+				
PH_3				
NO_2^-				

D. Valence Shell Electron Pairs: Single and Double Bonds and Nonbonding Electron Pairs

Formula	Lewis Structure	Number of Bonds and Nonbonding Pairs about Central Atom	Sketch of Ball-and-Stick Model	Shape Description
CO_3^{2-}				
SO_2				
H_2CO				
SO_3				
SO_3^{2-}				
SO_4^{2-}				
PO_4^{3-}				

SECTION V:

The Chemical Behavior of Elements

Experiment 10

An Activity Series of Several Metals

Purpose:

In this experiment you will observe the chemical behavior of several elements so you can rank them in order of decreasing chemical activity. A ranking of this kind is termed an activity series of elements. Fictitious names will be given to the elements you will study so that your ranking will be based solely on your observations.

Discussion:

Each element has it own unique set of properties. One property that differs for each element is its ability to lose one or more electrons in a chemical reaction. Some elements, like sodium and potassium, do this very easily. Others like fluorine and oxygen do not; in fact, in chemical reactions, these elements tend to gain one or two electrons. In this experiment you will determine the relative ease with which several metals lose one or more electrons. This will be done by combining one metal with a compound of another metal and observing whether a reaction takes place. After observing several combinations, you will be able to arrange the metals based on their ability to lose electrons. This ranking of metals is called an **activity series**. If the metals are arranged in a column, the most active metal would be placed at the top (the metal that loses electrons most easily), and the least active metal would be placed at the bottom. The reactions you will observe are all single-replacement reactions.

In equations that describe **single-replacement reactions**, one element in its elemental form replaces a second element in a compound. An example of this type of reaction occurs between zinc metal and silver ion, Ag^+, in aqueous silver nitrate, $AgNO_{3(aq)}$.

$$Zn_{(s)} + 2\ AgNO_{3(aq)} \rightarrow Zn(NO_3)_{2(aq)} + 2\ Ag_{(s)}$$

This equation shows that zinc replaces silver in its compound, forming a compound of zinc, $Zn(NO_3)_2$. Silver, being replaced, is produced as silver metal. Zinc is then considered *more active* than silver since it is able to displace silver from its compound.

Though the reaction between zinc and silver nitrate can be regarded as a single-replacement reaction, it can also be classified as an **oxidation-reduction reaction**. **Oxidation** is the loss of electrons by a species. Zinc loses electrons as it is converted to zinc ion. Zinc is oxidized.

$$Zn_{(s)} \rightarrow Zn^{2+}{}_{(aq)} + 2\ e^- \quad \text{(oxidation)}$$

When in solution, silver nitrate is completely dissociated into silver ions and nitrate ions. The silver ion gains an electron as it is converted to silver metal. **Reduction** is the gain of electrons by a species. Silver ion is reduced to silver metal in the reaction.

$$Ag^{+}{}_{(aq)} + e^- \rightarrow Ag_{(s)} \quad \text{(reduction)}$$

Oxidation and reduction *always* occur simultaneously. If one species gives up electrons, another species receives them. As one species is oxidized, the other is reduced. Since each zinc atom loses two electrons, it requires two silver ions to accept those two electrons, with one electron going to each silver ion. This ratio of one zinc to two silver is seen clearly in the balanced equation.

$$Zn_{(s)} + 2\ AgNO_{3(aq)} \rightarrow Zn(NO_3)_{2(aq)} + 2\ Ag_{(s)}$$

Notice that the nitrate ion is not directly involved in the reaction. Its presence is necessary to maintain an equal number of positive and negative charges in solution, but it is only a spectator of the actual reaction that goes on about it in solution.

Using the language of oxidation and reduction, you can describe zinc as a metal that is able to reduce silver ion to silver metal. Zinc is able to transfer electrons from itself to the silver ion. Because zinc can do this, zinc is clearly more *active* than silver. Stated another way, *zinc atoms give up electrons more easily than silver atoms*, and as zinc atoms give electrons to silver ions, the silver atoms that form hang onto those electrons and do not give them back. Again, in terms of the ability to give up electrons, zinc is more active than silver. If you attempt to carry out the reverse reaction, adding silver metal to a solution of zinc nitrate, no reaction would be observed. Silver would not be able to force electrons from itself onto the zinc ion. Silver would be less active than zinc.

$$Ag_{(s)} + Zn(NO_3)_{2(aq)} \rightarrow \text{no reaction}$$

Even if a zinc ion was able to get an electron from a silver atom, the electron would quickly be taken back by the silver ion to form the silver atom. You would end up at the same place you started. No reaction would be observed.

So you can see, even where there is no observable reaction, a judgment can still be made concerning chemical activity. Because silver is not able to replace zinc from its compound, silver would be considered *less active* than zinc. This is the same conclusion reached earlier when a reaction was observed between zinc and silver nitrate.

In this experiment, a series of single-replacement reactions will be studied to determine the relative activities of several metals. The conclusions drawn from your observations will allow you to construct an activity series of the metals. It is important that you make accurate observations in this experiment. As you observe each combination of reactants to see if they react, look for such things as:

1. The generation of a gas. (Do not mistake small bubbles of dissolved gas on the surface of a metal for a gaseous product. If a gas is produced, you will see a continuous flow of bubbles).

2. The formation of a coating on the surface of a metal. (Note if there are changes in the texture or color of the surface.)

3. A change, usually gradual, in the color of the solution.

Do not be too hasty in your observations. If you do not see evidence of a reaction right away, set the mixture aside for a few minutes to check it later. In some cases, an oxide coating on the surface of a metal must first dissolve before a visible reaction with the metal can begin.

All elements used in this experiment, other than hydrogen, are given fictitious names in honor of six great chemists:

The Chemist		The Element
Linus Pauling	–	paulium
Gilbert N. Lewis	–	lewium
G. Willard Gibbs	–	gibbium
Svante Arrhenius	–	arrhenium
Antoine Lavoisier	–	lavoisium
Alfred Werner	–	wernium

A brief biography of each of these chemists follows the Experimental Procedure.

The name, symbol and charge on the ion formed by the element is given below, along with the name and formula of the compound of that element used in this experiment.

Element	**Symbol**	**Charge on Ion**	**Compound**	**Formula**
paulium	Pl	2+	paulium nitrate	$Pl(NO_3)_2$
lewium	Le	2+	lewium sulfate	$LeSO_4$
gibbium	Gi	2+	gibbium nitrate	$Gi(NO_3)_2$
hydrogen	H	1+	hydrochloric acid	HCl
arrhenium	Ah	2+	none used	
lavoisium	Lv	2+	lavoisium acetate	$Lv(C_2H_3O_2)_2$
wernium	We	1+	wernium nitrate	$WeNO_3$

Experimental Procedure:

CAUTION: As always, eye protection must be worn at all times during laboratory work.

ENVIRONMENTAL ALERT: Several elements used in the following tests are heavy metals and should not be discarded down the drain. Your instructor will inform you how to dispose of these substances safely.

INSTRUCTIONAL NOTE: The reaction between gibbium, Gi, and wernium nitrate, $WeNO_{3(aq)}$, in step 1 will be demonstrated in the laboratory.

1. In a convenient location in the laboratory, you will find three 15-cm test tubes labeled A, B and C. A small quantity of gibbium metal is in test tube A. A solution of wernium nitrate, $WeNO_{3(aq)}$, is in test tube B. In test tube C, gibbium has been in contact with a solution of wernium nitrate for several hours. Note the appearance of each of the samples and look for evidence of a chemical reaction. Record your observations and conclusions on the Report Sheet.

2. Place six clean 15-cm test tubes in a test tube rack. To the first, add 5 mL of a solution of lavoisium acetate, $Lv(C_2H_3O_2)_{2(aq)}$. To the second, add 5 mL of 6 M hydrochloric acid, $HCl_{(aq)}$. To the third test tube, add 5 mL of lewium sulfate solution, $LeSO_{4(aq)}$. Add 5 mL of gibbium nitrate solution, $Gi(NO_3)_2$, to the fourth test tube. To the fifth test tube, add 5 mL of wernium nitrate solution, $WeNO_{3(aq)}$. To the sixth, add 5 mL of paulium nitrate solution, $Pl(NO_3)_{2(aq)}$.

   To each solution, add a piece of arrhenium, Ah. Check for evidence of reaction. If a reaction is not seen in the first few minutes, allow the mixture to stand for about fifteen minutes and check again for evidence of reaction. Record your observations and conclusions on the Report Sheet. Dispose of all reagents safely.

3. Place five clean 15-cm test tubes in a test tube rack. Add 5 mL of lavoisium acetate solution, $Lv(C_2H_3O_2)_{2(aq)}$, to the first test tube. Place 5 mL of 6 M hydrochloric acid, $HCl_{(aq)}$, in the second. To the third, add 5 mL of gibbium nitrate solution, $Gi(NO_3)_{2(aq)}$. Add 5 mL of wernium nitrate solution, $WeNO_{3(aq)}$, to the fourth, and place 5 mL of paulium nitrate solution, $Pl(NO_3)_{2(aq)}$, in the fifth test tube.

   To each of these solutions, add a piece of lewium, Le. Check for evidence of a reaction now and again in fifteen minutes. Record your observations and conclusions on the Report Sheet. Dispose of all reagents safely.

4. Add a small piece of lavoisium to about 5 mL of 6 M hydrochloric acid, $HCl_{(aq)}$, in a small test tube. Look for evidence of a reaction immediately and again in about fifteen minutes. Record your observation and conclusion on the Report Sheet. Dispose of all reagents safely.

5. List the seven elements tested in order of decreasing chemical reactivity in the table on the Report Sheet. Place the most active element at the top of the left-hand column. Cite evidence gained from your investigation for placing each element in its particular position in the list.

The Six Great Chemists:

Linus Pauling (1901-1994)

Linus Pauling is one of the most famous chemists of the 20th century. He was born in the United States and became interested in chemistry as a boy, an interest that would win for him a Nobel Prize in chemistry and later the Nobel Peace Prize.

One of Pauling's first major contributions that changed the way chemists think about chemical bonds was the concept of electronegativity, the measure of an atom's ability to attract electrons to itself that it shares with another atom in a bond. Perhaps his most important contribution to our understanding of bonding was the development of valence bond theory, which explains how chemical bonds form and how molecules attain their three-dimensional shapes through hybridization of atomic orbitals. He developed the concept of resonance, which contributes to the stability of certain molecules. He also discovered the genetic cause of sickle-cell anemia and predicted a helical structure for DNA. For his contribution to the understanding of the chemical bond and the structure of proteins, Pauling was awarded the Nobel Prize in chemistry in 1954.

In 1958, Dr. Pauling and his wife presented an appeal to the United Nations to end nuclear testing in the atmosphere. This appeal was signed by more than 9000 scientists from 44 countries. His work for world peace was recognized when he was awarded the Nobel Peace Prize in 1962.

Gilbert Newton Lewis (1875-1946)

The American chemist, Gilbert Newton Lewis, postulated in 1902 that the atom could be envisioned as a cube. The outermost electrons would occupy the eight corners of the cube and the nucleus would be at the center. This crude model of the atom was the beginning of his octet theory. In 1916 Lewis suggested that ions were formed by either the gain or loss of electrons. He also suggested that nonionic compounds were the result of atoms *sharing* pairs of electrons in what he called a covalent bond. This concept was expanded by Irving Langmuir in 1919. Today we commonly draw Lewis structures of covalent molecules, applying many of the ideas he developed during these early days of bonding theory.

In 1923, Lewis introduced a new concept of acids and bases. According to Lewis, an acid was a substance that *accepted* an electron pair during a chemical reaction and a base was a substance that *donated* an electron pair. This greatly expanded the concept of acids and bases and allowed many nonhydrogen-containing substances, such as metal ions, SO_3 and CO_2, to be classified as acids and Cl^-, SCN^- and H_2O to be classified as bases.

Josiah Willard Gibbs (1839-1903)

J. Willard Gibbs was born in the United States and was the first student to be awarded a Ph.D. from Yale in 1863, where he remained as professor of mathematical physics until his death. He is best known for his application of thermodynamic principles to chemical equilibrium, one of the most significant intellectual achievements of his time or of any time. He defined a term called free energy, the ultimate measure of the role of energy in chemical and physical changes. Gibbs published his work in a 321-page paper in the *Transactions of the Connecticut Academy of Science* in 1876-1878. Because this was a rather obscure journal, it was not read by the more famous European chemists, and Gibbs' concepts did not receive immediate attention. In addition, his work was so profound that only a few scientists could understand it, so its acceptance was slow. History has shown, though, that Gibbs was one of the greatest minds of all time.

Svante August Arrhenius (1859-1927)

Svante Arrhenius was born in Sweden. While a student, he developed the idea that certain substances dissociate in water to produce ions. Arrhenius presented these ideas in his doctoral dissertation in 1884. His defense, in open debate, took four hours and was not well received by the faculty. The dissertation was awarded a fourth-class pass, and the defense a third-class pass. This was the minimum pass! Though Arrhenius' dissertation was not well received, time proved him correct and he received the Nobel Prize in chemistry in 1903 for his theory of electrolytic dissociation. As part of this theory, Arrhenius proposed the first useful concept of acids and bases to withstand the test of time.

It is interesting to note that a paper written by Arrhenius in 1886 entitled *On the Influence of Carbonic Acid in the Air upon the Temperature on the Ground* anticipated by decades what we today call the greenhouse effect. Arrhenius made many important contributions in several areas of chemistry, and he is regarded as one of the founders of the area of chemistry known as physical chemistry.

Antoine Laurent Lavoisier (1743-1794)

Lavoisier, a native of France, is regarded by many as the father of modern chemistry, an honor he shares with Robert Boyle (see page 166). During his life Lavoisier was involved in government, private business and science, but his first love was science. He was the first to understand the process of combustion, which ultimately led to the demise of the phlogiston theory. He named the gas required in combustion, oxygen, which was isolated by Priestly in 1774. He collaborated with others in writing *Methods of Chemical Nomenclature* in 1787, which systematized the names for various substances. In 1789 he published *Elementary*

Treatise on Chemistry, in which he clearly defined the term element and listed 33 substances that he thought were elements. The concept of conservation of mass was also stated in this book.

Lavoisier served on many government committees. One committee standardized weights and measures, the beginning of the International System of Measures. He was also a member of a committee to study the claims of a physician named Franz Anton Mesmer, who said he could cure sickness by hypnosis. Other members of this committee were Benjamin Franklin and Joseph Guillotin, the inventor of the guillotine.

Unfortunately, Lavoisier helped block the admission of the radical journalist Jean-Paul Marat to the French Academy of Science. That, coupled with his investment in a tax collecting firm in 1768, ultimately led to his execution during the French Revolution in 1794. Not to minimize the deaths of Louis XVI and Marie Antoinette, Lavoisier is considered by many to be the outstanding martyr of the French Revolution. The great mathematician Lagrange said the day after Lavoisier's death, "It took only an instant to cut off that head, and a hundred years may not produce another like it."

Alfred Werner (1866-1919)

In 1890, a young Swiss chemist named Alfred Werner became interested in compounds of transition metals that had compositions that could not be understood using the theories of the day. For example, cobalt was known to combine with three chlorine atoms, $CoCl_3$, or three bromine atoms, $CoBr_3$, or three iodine atoms, CoI_3, but how could cobalt form a compound with a formula like $[Co(NH_3)_6]Cl_3$? This composition required a fresh look at the theories of combining capacity (valence) of transition metals. Werner proposed that the valence of an element was composed of two parts, a primary valence (3 for cobalt) and an auxiliary valence (6 for cobalt). Werner proposed that six ammonia molecules "coordinated" to cobalt, and the three chlorine atoms were not joined to the metal at all but needed to be there to satisfy the primary valance. For years he studied many compounds of this type, which he called coordination compounds, and established compositions and structures for many of them. So thorough was Werner's work that at one time some chemists thought that further work in this area was unnecessary. He had explained it all. The importance and extent of his work led to Werner being named as the Nobel Laureate in chemistry in 1913. Perhaps even more remarkable is that his ideas preceded, by many years, any real understanding of the nature of the chemical bond.

Name ______________________ Locker Number__________ Date____________

Please print; last name first

PRELIMINARY EXERCISES: *Experiment 10*

An Activity Series of Several Metals

1. Describe the relative activities of metals in terms of the ease with which they lose electrons.

2. Complete and balance the following equations, which describe single-replacement reactions. Each reaction actually takes place. Draw a *circle* around the element that is oxidized and a *square* around the element (not a compound, the element) that is reduced.

   a. $2\,Na_{(s)} + 2\,HCl_{(aq)} \rightarrow 2\,NaCl_{(aq)} + H_{2(g)}$ (example)

   b. $AgNO_{3(aq)} + Cu_{(s)} \rightarrow$

   c. $Al_{(s)} + Fe_2O_{3(s)} \rightarrow$

3. Suppose a student made the observations presented below during a laboratory investigation involving the hypothetical metals G, I, M, N, O, W and Y. After each equation, note which element is more reactive; for example, if Y is more active than O, write Y > O. Then construct an activity series, placing the most active metal in the space at the top of the right-hand column.

Equation		Activity Series
$2\,Y + OCl_2 \rightarrow 2\,YCl + O$	__________	_____ (most active)
$Y + WCl \rightarrow$ no reaction	__________	_____
$N + GCl \rightarrow NCl + G$	__________	_____
$M + ICl_2 \rightarrow I + MCl_2$	__________	_____
$M + OCl_2 \rightarrow$ no reaction	__________	_____
$N + ICl_2 \rightarrow$ no reaction	__________	_____

(The Preliminary Exercises continue on the following page.)

4. In a second investigation, the hypothetical elements C, F, K, L, P, X and Y were studied. Prepare an activity table as you did in question 3, placing the most active metal at the top of the right-hand column.

Reaction		Activity Series
$P + LCl_2 \rightarrow$ no reaction	________	_____ (most active)
$2\,K + LCl_2 \rightarrow 2\,KCl + L$	________	_____
$P + FCl \rightarrow PCl + F$	________	_____
$K + CCl_2 \rightarrow$ no reaction	________	_____
$Z + FCl \rightarrow$ no reaction	________	_____
$Z + XCl_2 \rightarrow X + ZCl_2$	________	_____

Name ______________________________ Locker Number__________ Date____________
Please print; last name first

REPORT SHEET: *Experiment 10*

An Activity Series of Several Metals

1. a. Describe what you observed.

   b. Which is more reactive, gibbium or wernium? (Justify your answer with experimental evidence.)

2. a. Complete and balance the equations for reactions that took place. If no reaction occurred, write "no reaction" to the right of the arrow.

   $Ah_{(s)} + Lv(C_2H_3O_2)_{2(aq)} \rightarrow$ ______________________________

   $Ah_{(s)} + HCl_{(aq)} \rightarrow$ ______________________________

   $Ah_{(s)} + LeSO_{4(aq)} \rightarrow$ ______________________________

   $Ah_{(s)} + Gi(NO_3)_{2(aq)} \rightarrow$ ______________________________

   $Ah_{(s)} + WeNO_{3(aq)} \rightarrow$ ______________________________

   $Ah_{(s)} + Pl(NO_3)_{2(aq)} \rightarrow$ ______________________________

   b. List, by symbol, the metals that are *less reactive* than arrhenium.

3. a. Complete and balance the equations for the reactions that took place. If no reaction occurred, write "no reaction" to the right of the arrow.

   $Le_{(s)} + Lv(C_2H_3O_2)_{2(aq)} \rightarrow$ ______________________________

   $Le_{(s)} + HCl_{(aq)} \rightarrow$ ______________________________

   $Le_{(s)} + Gi(NO_3)_{2(aq)} \rightarrow$ ______________________________

   $Le_{(s)} + WeNO_{3(aq)} \rightarrow$ ______________________________

   $Le_{(s)} + Pl(NO_3)_{2(aq)} \rightarrow$ ______________________________

b. List, by symbol, the metals that are *less reactive* than lewium.

4. a. Describe your observations.

b. Which is *more reactive*, lavoisium or hydrogen? Support your answer by citing *experimental evidence.*

5. In the table below, arrange the seven elements in order of decreasing chemical activity and cite the experimental evidence to justify the placement of the element in the series.

Element	Experimental Evidence to Support Placement in the Activity Series
Pl	*In step 2, Ar replaced all elements except Pl, so Pl is the most active element of the seven elements tested.*

Experiment 11

The Preparation and Properties of Oxygen and the Properties of Oxides

Purpose:

In this experiment you will prepare oxygen by decomposing potassium chlorate, $KClO_3$, with heat. The oxygen you prepare will be collected and used to synthesize oxides of carbon, sulfur and phosphorus. Then you will determine the acidic or basic properties of these oxides, as well as those of calcium and magnesium as they are dissolved in water.

Discussion:

Oxygen is the most abundant element in the earth's crust, and it is one of the most reactive. Because of this, there are oxides known for all elements except the noble gases, though XeO_3 has been prepared. Many elements can react directly with oxygen to form oxides.

Oxygen gas will be prepared by decomposing potassium chlorate, $KClO_3$, with heat, in the presence of manganese(IV) oxide, MnO_2, a catalyst for the reaction.

$$2\ KClO_{3(s)} \xrightarrow{MnO_2} 2\ KCl_{(s)} + 3\ O_{2(g)}$$

The oxygen will be collected by water displacement and used to prepare oxides of carbon, sulfur and phosphorus. Solutions of these oxides and others will then be tested with pH or litmus paper to see if they are acidic, basic or neutral.

The soluble oxides of *nonmetals* dissolve in water to form acidic solutions and are called **acidic anhydrides**. The word "anhydride" means water free. Sulfur dioxide, SO_2, is an acidic anhydride and when dissolved in water reacts to produce sulfurous acid, $H_2SO_{3(aq)}$.

$$SO_{2(g)} + H_2O_{(l)} \rightarrow H_2SO_{3(aq)}$$

Carbon dioxide, CO_2, reacts to form carbonic acid, $H_2CO_{3(aq)}$; and phosphorous pentoxide, P_2O_5, in water produces phosphoric acid, $H_3PO_{4(aq)}$. The oxides of sulfur and nitrogen, produced in the combustion of fuels, react with moisture in the atmosphere to form acids, which eventually fall to earth as acid rain, an environmental concern in many areas of the world.

The soluble oxides of *metals* are **basic anhydrides** and form basic solutions when dissolved in water. Potassium oxide, K_2O, and barium oxide, BaO, are both basic anhydrides. In water they produce potassium hydroxide, $KOH_{(aq)}$, and barium hydroxide, $Ba(OH)_{2(aq)}$, respectively. Sodium oxide reacts with water to form a solution of sodium hydroxide:

$$Na_2O_{(s)} + H_2O_{(l)} \rightarrow 2\ NaOH_{(aq)}$$

Experimental Procedure: It is convenient to perform this experiment in pairs.

A. Preparation of Oxygen

> **CAUTION: Hot $KClO_3$ can react explosively with carbon-containing compounds and many other substances. Follow the experimental procedure carefully. If you are unsure of any step, ask your instructor for assistance.**
>
> **CAUTION: As always, eye protection must be worn at all times.**

1. Place approximately 2 g (enough to cover the bottom of the test tube) of solid manganese(IV) oxide, MnO_2, in an 20-cm test tube. *Do not mistake powdered charcoal for MnO_2. Both are black solids and look very much alike. If powdered charcoal is used by mistake, an explosion may occur when you attempt to prepare oxygen from $KClO_3$.*

2. Heat the MnO_2 for four minutes using the hottest flame of the burner to destroy any combustible impurity that may be present. All of the MnO_2 must get very hot. Wipe the flame along the sides of the test tube so you do not miss any portion of the solid. Have the instructor check to see if you are heating the solid adequately.

3. While the test tube with the MnO_2 cools, set up the oxygen-collecting apparatus as shown in Figure 1. The water in the trough should be above the bottle support bar, and rubber tubing should lead from the upper drain on the trough to the sink.

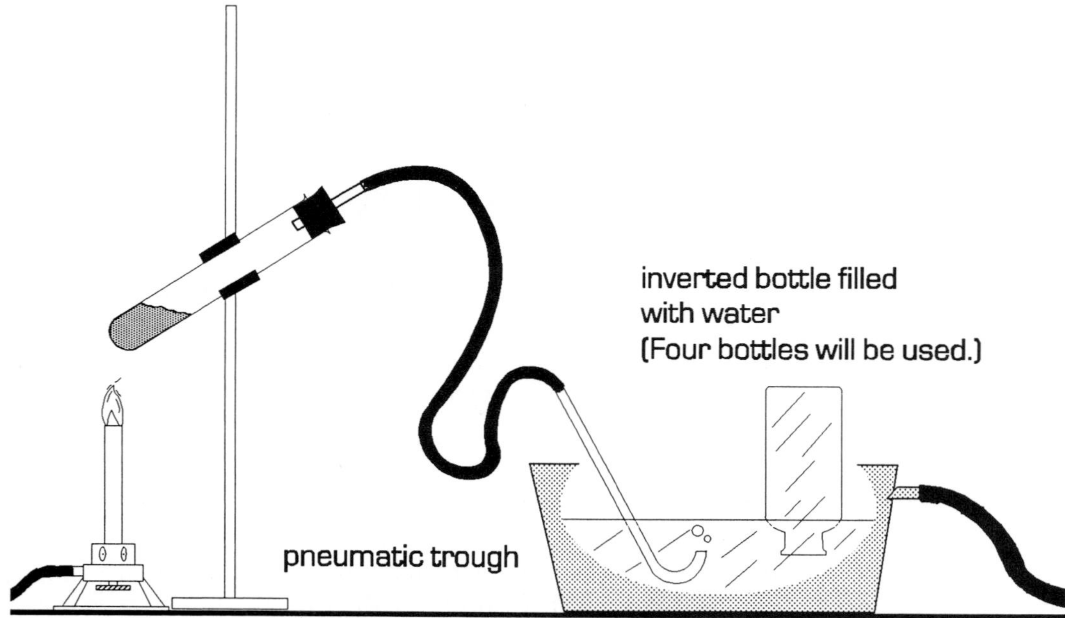

Figure 1

4. Prepare four wide-mouth bottles for collecting oxygen as directed below and shown in Figure 2. When finished you should have four inverted bottles in the trough.

   a. Fill each bottle to overflowing with water.

   b. Slide a glass plate across the mouth of the bottle. No air bubbles should be beneath the glass plate.

   c. While holding the glass plate onto the bottle, turn it upside-down and lower the mouth of the bottle beneath the water. Remove the plate and rest the bottle on the bottom of the trough.

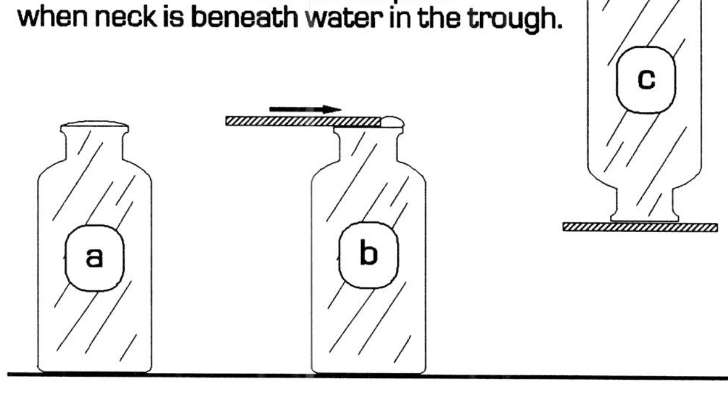

Figure 2

5. When the test tube containing the MnO_2 is cool, place 6 g of potassium chlorate, $KClO_3$, in the tube and mix the contents by slapping the side of the tube against the palm of your hand. *Good mixing is important.* Clamp the tube to the ring stand and insert the stopper with the connecting hose as shown in Figure 1. Have your instructor check your experimental setup before proceeding to step 6.

6. If you are working alone, ask another student to help you collect the oxygen gas. When all is ready, place one bottle on the bottle support bar in the trough, and begin to heat the $KClO_3$-MnO_2 mixture with a gentle flame. Control the heating carefully so that oxygen will not be evolved too rapidly. As each bottle becomes filled with oxygen, lift it from the bottle support just enough to cover the bottle mouth with a glass plate. Then remove the bottle and place it upright on the bench with the glass plate still covering the bottle mouth. Place the next bottle on the support bar and collect oxygen. Do this quickly so that you do not lose too much oxygen as the $KClO_3$ continues to decompose. Discard the first bottle of oxygen since it contains the air that was displaced from the test tube and hose.

7. Once the oxygen is collected, remove the tubing from the water bath and allow the test tube to cool. Then remove the rubber stopper and clean out the residue with the aid of a stirring rod.

ENVIRONMENTAL ALERT: Because of the presence of MnO_2 in the residue remaining in the test tube, it must be collected for safe disposal. Your instructor will inform you of the location of the disposal container in the laboratory.

B. Preparation and Properties of Oxides

In the following exercises you will test a solution with pH paper or litmus paper to determine if it is acidic or basic. To test a solution, remove a droplet of the solution with a clean glass stirring rod and touch it to a strip of pH or litmus paper. A color change will quickly be evident. If you are using litmus paper, compare your results to these test indicators:

If blue litmus paper changes to *red*, the solution is *acidic*.
If no color change is observed, the solution is *neutral*.
If red litmus paper changes to *blue*, the solution is *basic*.

If you are using pH paper you can judge the degree of acidity or basicity using these guidelines:

If the pH paper changes to *blue*, the solution is *strongly basic*.
If the pH paper changes to *green*, the solution is *weakly basic*.
If there is no color change, the solution is *neutral*.
If the pH paper changes to a *yellow-orange* color, the solution is *weakly acidic*.
If the pH paper changes to *red*, the solution is *strongly acidic*.

Before proceeding, make certain each of the three bottles of oxygen (prepared in Section A) has about 5 mL of water covering the bottom. Add some deionized water if necessary.

1. Reaction of carbon with oxygen: Place a small amount of powdered charcoal (carbon) in the cup of a deflagration spoon. The sample should be about the size of a match head. Heat the carbon in the flame of a burner until it begins to glow, then thrust it into a bottle of oxygen. Quickly replace the glass plate over the mouth of the bottle to keep the product of the reaction from escaping, as shown in Figure 3. Once the carbon has burned, quickly remove the spoon, cover the bottle with the glass plate, and shake the bottle to dissolve the gaseous product in the water. Using a clean stirring rod, touch a drop of this solution to pH or litmus paper. Record your observations and conclusion on the Report Sheet.

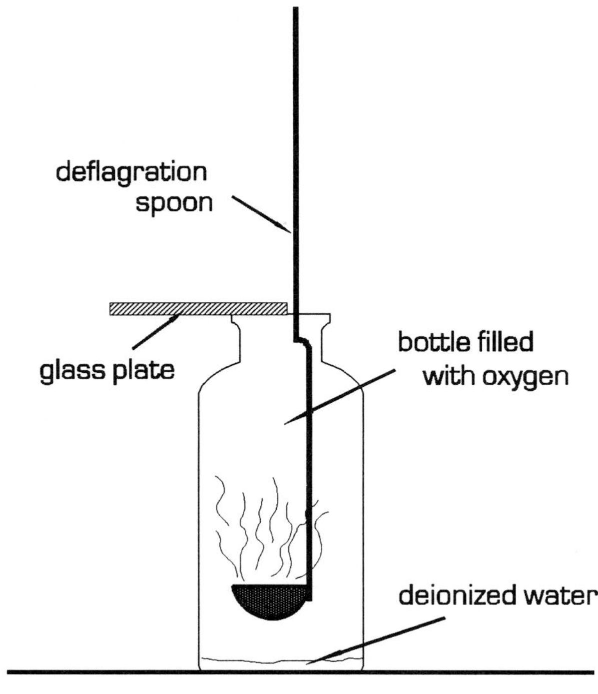

Figure 3

> **CAUTION: The reactions with sulfur and phosphorus in Parts 2 and 3 must be performed in the fume hood.**

2. Reaction of sulfur with oxygen: Clean the deflagration spoon by heating it in the flame of a burner, and when *cool*, place a small amount (an amount the size of a match head) of powdered sulfur in the cup. Then, in the hood, ignite the sulfur with the burner and burn it in the second bottle of oxygen as you did with carbon. Shake to dissolve the gaseous product in the water. Test the resulting solution with pH or litmus paper, and record your observations and conclusion on the Report Sheet.

3. Reaction of phosphorus with oxygen: Again, clean out the deflagration spoon by heating it in the flame of the burner, and when *cool*, place a small amount (an amount the size of a match head) of red phosphorus in the cup. In the hood, ignite the phosphorus and burn it in the third bottle of oxygen. Shake to dissolve the gaseous product in the water, then test the solution with pH or litmus paper. Record your observations and conclusion on the Report Sheet.

   Clean out the deflagration spoon as before in the flame of the burner, and when *cool*, return it to the supply in the laboratory.

4. Reaction of magnesium with oxygen: Using crucible tongs, hold a one-inch strip of magnesium ribbon in the flame of the burner until it ignites. *Do not look directly at the intense blue-white flame.* Collect the ash on a clean watch glass and place it in a small test tube. Add 2 mL of water, shake the mixture, and test the solution with pH or litmus paper. Record your observations and conclusion on the Report Sheet.

5. Reaction of calcium oxide with water: Place a small amount of calcium oxide, CaO, in a small test tube, and add 2 mL of water to it. Shake the mixture and test the solution with pH or litmus paper. Record your observations and conclusion on the Report Sheet.

Name ______________________________ Locker Number__________ Date____________
Please print; last name first

PRELIMINARY EXERCISES: *Experiment 11*

The Preparation and Properties of Oxygen and the Properties of Oxides

1. Classify each of the following compounds as an acidic anhydride, basic anhydride or neither. All are soluble in water.

   a. MgO ____________ d. Cl_2O_7 ____________ g. SO_3 ____________

   b. B_2O_3 ____________ e. SeO_2 ____________ h. CaO ____________

   c. I_2O_3 ____________ f. Na_2O ____________ i. NO_2 ____________

2. A 1.00-g sample of oxygen occupies about 775 mL at room temperature and pressure.

   a. What mass of oxygen is required to fill five 250-mL bottles at room temperature and pressure?

   Mass of O_2 = ____________

   b. How many grams of $KClO_3$ must be completely decomposed to prepare this mass of oxygen? (1 mol of $KClO_3$ = 122.6 g; 1 mol of O_2 = 32.0 g.) The balanced equation is:

$$2\ KClO_{3(s)} \xrightarrow{MnO_2} 2\ KCl_{(s)} + 3\ O_{2(g)}$$

   Mass of $KClO_3$ = ____________

(The Preliminary Exercises continue on the following page.)

3. Soluble metal oxides react with water to form hydroxide bases. For example, potassium oxide, K_2O, in water forms potassium hydroxide, $KOH_{(aq)}$, in solution. This is why potassium hydroxide is classified as a basic anhydride. The actual reaction is that between the oxide ion and water to form hydroxide ion:

$$2\ K^{+}{}_{(aq)} + O^{2-}{}_{(aq)} + H_2O_{(l)} \rightarrow 2\ K^{+}{}_{(aq)} + 2\ OH^{-}{}_{(aq)}$$

On the other hand, soluble oxides of nonmetals react with water to form acids, so these oxides are called acid anhydrides. Sulfur dioxide, SO_2, reacts with water to form $H_2SO_{3(aq)}$, sulfurous acid. You should notice that the formula of the acid is the sum of the formulas of sulfur dioxide and water:

$$SO_{2(aq)} + H_2O_{(l)} \rightarrow H_2SO_{3(aq)}$$

From what is given above, complete and balance the following equations in which oxides of metals or nonmetals react with water.

$Na_2O + H_2O \rightarrow$

$CaO + H_2O \rightarrow$

$CO_2 + H_2O \rightarrow$

$SO_3 + H_2O \rightarrow$

Name ______________________ Locker Number__________ Date__________
Please print; last name first

REPORT SHEET: ***Experiment 11***

The Preparation and Properties of Oxygen and the Properties of Oxides

B. Preparation and Properties of Oxides

1. Reaction of carbon with oxygen (Complete and balance all equations.)

$C + O_2 \rightarrow$	name of product:
$CO_2 + H_2O \rightarrow$	name of product:
The solution is: strongly acidic mildly acidic neutral mildly basic strongly basic	
Conclusion: (Describe the oxide of the element as an acidic or basic anhydride or neither, and justify your answer with supporting observations.)	

2. Reaction of sulfur with oxygen

$S + O_2 \rightarrow$	name of product:
$SO_2 + H_2O \rightarrow$	name of product:
The solution is: strongly acidic mildly acidic neutral mildly basic strongly basic	
Conclusion:	

3. Reaction of phosphorus with oxygen

P + O_2 →	name of product:
P_2O_5 + H_2O →	name of product:
The solution is: strongly acidic mildly acidic neutral mildly basic strongly basic	
Conclusion:	

4. Reaction of magnesium with oxygen

Mg + O_2 →	name of product:
MgO + H_2O →	name of product:
The solution is: strongly acidic mildly acidic neutral mildly basic strongly basic	
Conclusion:	

5. Reaction of calcium oxide with water

CaO + H_2O →	name of product:
The solution is: strongly acidic mildly acidic neutral mildly basic strongly basic	
Conclusion:	

SECTION VI:
The Gas Laws

Experiment 12
Boyle's Law and Charles' Law

Purpose:

In this experiment you will study the effect of pressure and temperature on a confined volume of air. The gas will be contained in a syringe that will allow its volume to be monitored easily. If careful work is done, you should verify the inverse behavior of Boyle's law and the direct relationship of Charles' law.

Discussion:

The state of matter that was first described successfully in mathematical equations was the gaseous state. As early as the mid-1600s, Robert Boyle showed how the volume of a confined quantity of air changed in a definite and predictable way as the pressure applied to it changed. The connection between volume and pressure could be stated in a mathematical equation. Nearly a century later, the connection between the volume of a gas and its temperature was revealed, again in a predictable way that could be expressed in a mathematical equation. Today, these relationships between the volume of a confined sample of gas and its temperature and pressure are stated in the classic gas laws.

A. The Effect of Temperature on Volume

In 1787, a French chemist, Jacques Charles, reported a relationship between the volume of a sample of gas and its temperature. As long as the pressure applied to a fixed mass of gas did not change, Charles found that a temperature increase of 1.00°C would cause a volume increase equal to 1/273 of the gas's volume at 0°C. A 10.000-L volume of air at 0°C would expand by 0.037 L (1/273 of 10.000 L) when its temperature was raised to 1.00°C. In a similar way, a decrease in temperature from 0°C to −1.00°C would reduce the volume by 0.037 L. Before presenting a usable volume-temperature relationship for gases here, a brief discussion of temperature scales is necessary.

Around 1845, William Thomson, later to become Lord Kelvin, concluded that the volume of a gas would reach zero, at least theoretically, if its temperature reached −273°C. Such a thing could not happen, of course, since any gas would liquefy and eventually freeze to a solid as the temperature approached −273°C, but Thomson reasoned that this very low temperature had a special significance. In time he proposed the **absolute temperature scale** that placed the zero of temperature at $-273°C$ or, to be more precise, $-273.15°C$. The absolute temperature scale begins at the lowest possible temperature, absolute zero, 0 K, and so there are no negative temperatures. The degree symbol (°) is not used with absolute temperatures. The symbol K, in honor of Lord Kelvin, indicates absolute temperature. A temperature span of 1 K is equal to the span of one degree Celsius. The absolute temperature scale, or Kelvin scale as it is commonly called, is related to the Celsius scale by a simple equation:

$$\text{Kelvin temperature} = \text{Celsius temperature} + 273.15$$

Absolute temperature allows the volume-temperature behavior of gases to be conveniently described by Charles' law.

Charles' law: The volume of a gas will vary *directly* with its absolute temperature as long as the pressure on the gas and the mass of gas remain constant.

Notice that volume varies *directly* with absolute temperature. If the absolute temperature doubles, the volume will also double (as long as both volumes are measured at the same pressure). If the absolute temperature is reduced by half, the volume will contract to half the original volume. This straight-line, direct relationship between volume and absolute temperature appears in Figure 1.

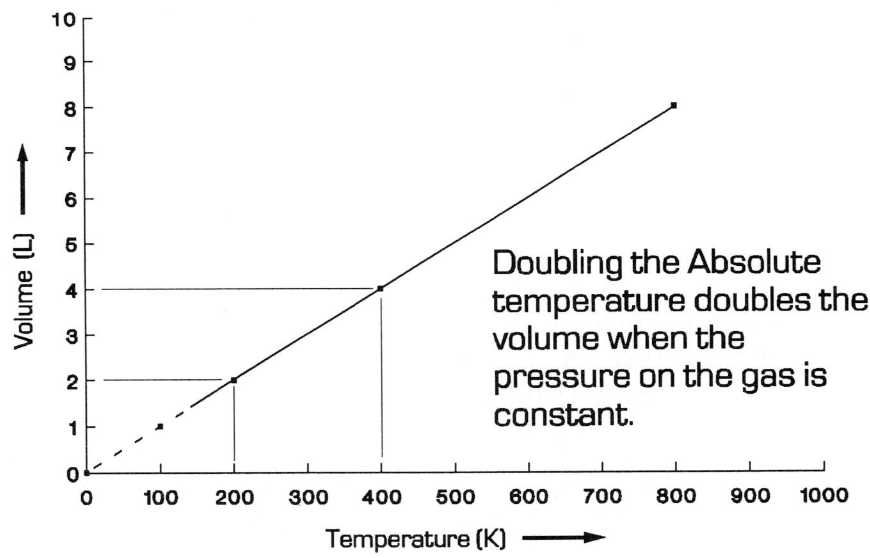

Figure 1. The volume-temperature relationship of Charles' law

Charles' law is stated below in a simple equation showing the direct relationship between volume and absolute temperature. Remember, this equation holds only under conditions of constant pressure and quantity of gas:

$$\frac{V_1}{T_1} = \frac{V_2}{T_2}$$

V_1 = initial volume
V_2 = final volume
T_1 = initial temperature (K)
T_2 = final temperature (K)

Both temperatures must be in Kelvin. Both volumes must have the same unit: liters, gallons, milliliters, etc.

B. The Effect of Pressure on Volume

The studies of Robert Boyle revealed an *inverse* relationship between the volume of a gas and its pressure. Boyle observed that the volume of a confined quantity of gas was reduced by half when the pressure applied to the gas doubled. The volume reduced to one-third of the original volume when the pressure tripled. This inverse relationship requires the mass of gas and its temperature to remain constant. The inverse nature of pressure and volume is shown in Figure 2.

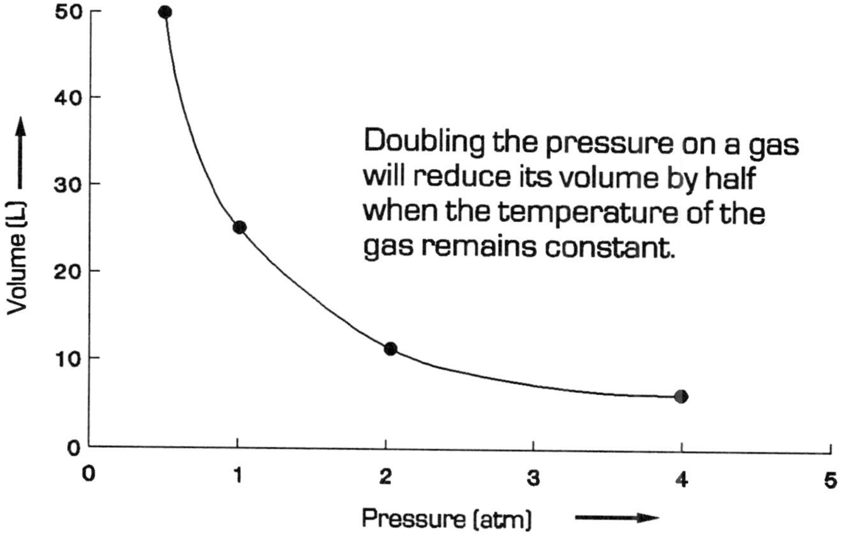

Figure 2. The volume-pressure relationship of Boyle's law

Boyle's law, which states the volume-pressure relationship of gases, is:

Boyle's law: The volume of a gas varies *inversely* with pressure as long as the temperature of the gas and the mass of gas remain constant.

Boyle's Law is restated below as an equation showing the inverse relationship between volume and pressure. Remember, this equation holds only under conditions of constant temperature and quantity of gas:

$$P_1 V_1 = P_2 V_2$$

V_1 = initial volume
V_2 = final volume
P_1 = initial pressure
P_2 = final pressure

In both Boyle's law and Charles' law, if any three terms in the equation are known, the fourth can be calculated.

An *inverse* relationship between two variables, such as exists in Boyle's law, will produce a straight-line graph when one variable is plotted as the reciprocal. In Figure 3, volume versus 1/pressure is plotted. The straight line is proof of the inverse relationship of volume and pressure for a gas.

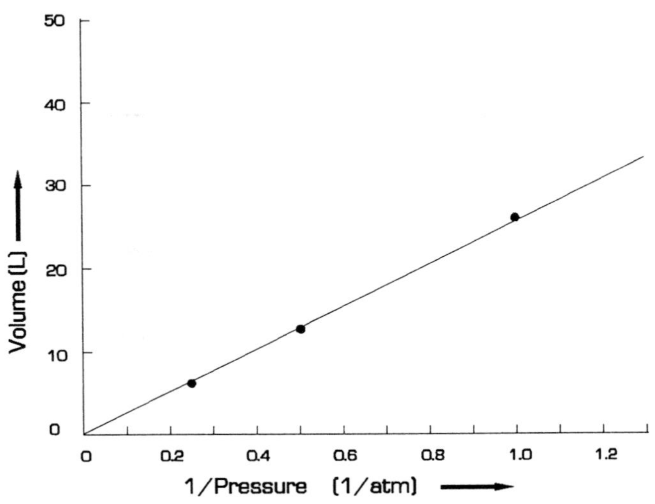

Figure 3. The inverse volume-pressure relationship of Boyle's law.

Charles' law and Boyle's law can be combined into a single equation that is frequently called the **combined gas law**. Note the *direct* relationship between volume and temperature (K) and the *inverse* relationship between volume and pressure.

Combined gas law: $$\frac{P_1 V_1}{T_1} = \frac{P_2 V_2}{T_2}$$

V_1, P_1, T_1 = initial volume, pressure and temperature (K)
V_2, P_2, T_2 = final volume, pressure and temperature (K)

The combined gas law allows for the effect of changes in temperature and pressure on the volume of a gas to be considered simultaneously. It is applied in Experiment 13.

C. The Effect of Quantity of Gas on Volume

Though the connection between the volume occupied by a gas and the mass of gas is not studied in this experiment, the relationship is important and described by Avogadro's law.

Avogadro's law: The volume a gas occupies is *directly* related to the number of moles of gas as long as all volumes are measured at the same temperature and pressure.

In this experiment, the amount of gas you will study will not change; that is, there will not (or should not) be any leaks in the syringes you will use, keeping the mass of gas constant.

Experimental Procedure:

The data you gather in the experiments are to be plotted on graph paper to see if your results verify the gas laws. Charles' law is verified if a plot of volume (V) versus temperature (T) in Kelvin is a straight line, showing the direct relationship between V and T. Boyle's law is verified if a plot of volume (V) versus the reciprocal of pressure (1/P) is a straight line, proving the inverse relationship between V and P.

Obtain a 30-mL plastic syringe with a Luer-lok® tip equipped with a plastic Luer-lok syringe cap so the tip of the syringe can be tightly closed off (no leaks). Check the syringe to make certain the plunger moves in and out smoothly. If it binds or moves only with difficulty, lubricate the inner wall of the syringe barrel with a *thin layer* of silicone oil. The plunger must move freely and not leak if good results are to be obtained.

A. Boyle's Law

1. With the Luer-lok® cap removed from the syringe, push the plunger completely in so you can locate the point on the plunger that corresponds to the 0.0-mL mark. Read all volumes using that point. This is usually where the bottom end of the plunger touches the barrel. Withdraw the plunger to the 25.0-mL mark. *Tightly* fit the Luer-lok® cap on the end to seal the air inside. Pass the end of the syringe through the hole in the center of a 6-inch square of thin plywood or fiberboard that rests on the circular top of a tripod[1], as shown in Figure 4. The hole should be just large enough to let the barrel of the syringe through while supporting the syringe by the flange on the barrel. You will measure the volume of air sealed in the syringe as pressure is applied to it. If there are no leaks, the data you gather will verify Boyle's law.

---

[1] As an alternative, the syringe can be supported by resting the wood holder on a ring supported by a sturdy ring stand.

Obtain four copies of your chemistry lecture text, borrowing some from your lab mates. Each book should weigh the same (no extra papers, pencils or paper clips). The pressure applied to the gas in the syringe will be directly related to the mass of the book or books stacked on the plunger of the syringe.[2]

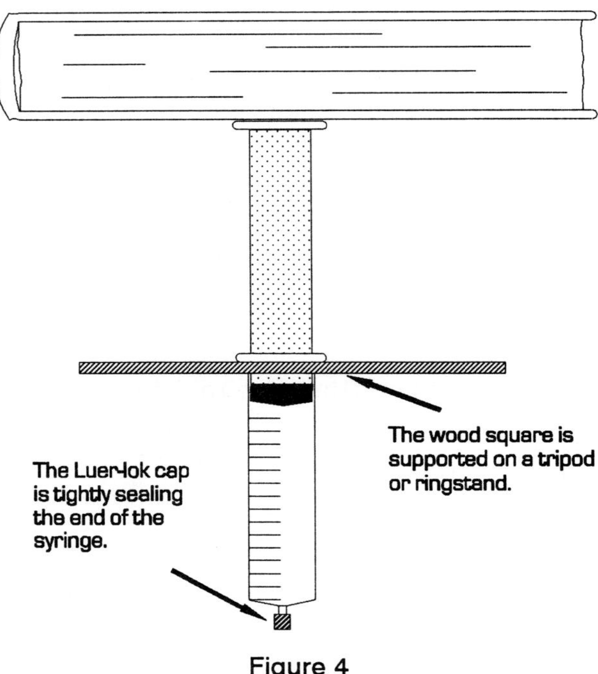

Figure 4

2. Place one book on the syringe plunger, carefully balancing the book so it does not fall. After the plunger comes to rest, record the volume of gas on the Report Sheet as trial 1. Add a second, third and fourth book to the stack, and record the new volume each time on the Report Sheet (as trial 1) in the space corresponding to the number of books. *Read each volume within a few seconds after each book is added.* This will minimize errors if there is a slow leak.

3. Use the volume-pressure data you collected to prepare a Boyle's law plot on the grid provided on the Report Sheet. First calculate the reciprocal of the number of books (corresponding to 1/pressure) for each data set. Then plot volume (in mL) on the vertical axis versus 1/pressure (in units of 1/books) on the horizontal axis. Realize that the value of "1/books" is proportional to (or parallel to) 1/pressure. Allowing for a little experimental error, if your points form a straight line like that shown in Figure 3, you have verified Boyle's law. If you do not get a straight line, consult with your instructor. You may need to repeat the experiment, recording the data as trial 2.

B. Charles' Law – This may be conveniently performed as a demonstration.

> **INSTRUCTIONAL NOTE: The Charles' law experiment may be conveniently performed as a demonstration.**

---

2 The pressure exerted on the gas equals the downward force applied to the gas divided by the cross-sectional area of the plunger: P = force/area. The downward force is directly proportional to the mass of the book(s). Therefore, as the mass of books increases, the pressure increases. Two books exert twice the pressure of one book.

Measuring the effect of temperature on the volume of gas will also involve the use of a syringe, but this time a glass syringe must be used because the plunger of a glass syringe moves much more easily than the plunger of a plastic syringe.

1. Obtain a 250-mL vacuum flask and a one-hole rubber stopper that fits tightly in the flask. Insert a thermometer through the stopper until it protrudes about 15 cm from the bottom of the stopper. Lubricate the hole in the stopper as well as the thermometer with glycerine and protect your hands with a towel. Tightly fit the stopper with the thermometer in the *clean and dry* vacuum flask.

2. Immerse the flask in a beaker of crushed ice, clamping it in place. As the flask cools, seal the sidearm with a rubber septum. (As an alternative, a short length of rubber tubing fit onto the sidearm and clamped tightly at the other end will serve as a septum.) As the air in the flask gets near 0°C, puncture the septum with a large gauge syringe needle to equate the pressure inside the flask with atmospheric pressure. Check the syringe before connecting it to the needle to ensure that the plunger moves *freely*. If it binds, clean the plunger and barrel with a soft cloth and lubricate the plunger with graphite (a No. 2 pencil will work). The Luer-lok® tip must be capable of gripping the collar of the needle tightly. Once 0°C is reached, push the plunger to the 0.0-mL volume on the syringe and *tightly* connect the Luer-lok® tip to the needle in the septum. Record the 0.0-mL volume on the Report Sheet under the trial 1 heading.

   Remove the flask and syringe from the ice bath and rest it on the base of a ring stand. Support the syringe with a utility clamp.

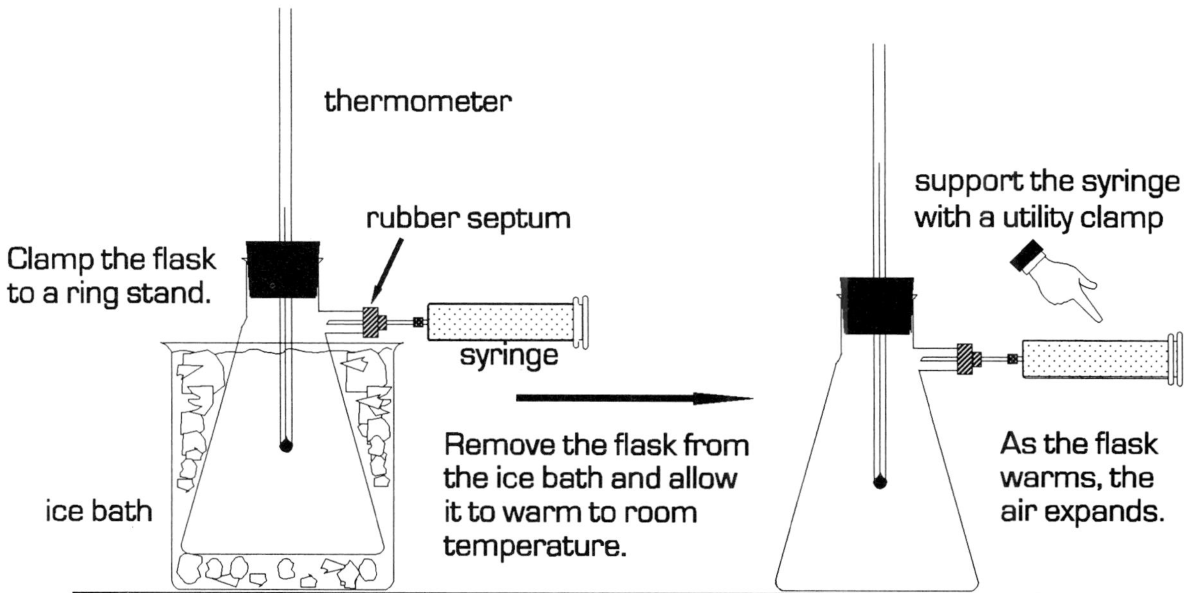

Figure 5

3. Wipe the exterior of the flask dry with a towel. As the air in the flask warms, you should notice the plunger of the syringe moving. You may have to rotate the plunger a little to overcome binding. As the air in the flask warms, the plunger will move outward in the syringe. Read the volume on the syringe to the tenths of a milliliter at 5° intervals up to 20°C, entering these volumes in the spaces on the Report Sheet as trial 1. Before each volume is read, it is important to rotate the plunger a bit to ensure it is moving freely. It will take some time to warm to 20°C, so be patient.

4. Once you have collected the volume-temperature data, you can prepare the Charles' Law plot. Convert each temperature to Kelvin and plot the syringe volume in milliliters versus Kelvin temperature on the grid provided on the Report Sheet. The data points should fall along a straight line, though there will be a little scatter due to the lack of sophistication in the experimental method. Draw the best straight line through the points. Within experimental error, if the data points lie close to or on the straight line, you have verified the nature of Charles' law. You may wish to repeat the experiment, recording volumes as trial 2 on the Report Sheet.

   Though it may not seem reasonable at first glance, you only need to plot the *increase in volume* of the gas versus temperature. Here is why. Assume you started with exactly 250.0 mL of air in the flask at 0°C. Then at 5°C, the volume may expand to 255.0 mL, and at 10°C, to 260.0 mL. You could plot 250.0 mL, 255.0 mL and 260.0 mL on the volume axis, or you could subtract 250.0 mL from each volume and plot 0.0 mL, 5.0 mL and 10.0 mL, just the changes in volume. The interval, or change, between each volume is still the same: 255.0 mL − 250.0 mL = 5.0 mL, the same as 5.0 mL − 0.0 mL = 5.0 mL.

Name ______________________ Locker Number__________ Date__________
Please print; last name first

PRELIMINARY EXERCISES: *Experiment 12*

Boyle's Law and Charles' Law

1. List the three factors that determine the volume of a gas:

   a. ______________ b. ______________ c. ______________

2. When dealing with gases, it is common to refer to volumes at standard temperature and pressure (STP). What are the values of standard temperature and pressure for gases?

   STP = ______ and ______

3.

a. State Boyle's law:
b. State Charles' law:

4. If the pressure exerted on a 100-L weather balloon changes from 1.0 atm to 2.0 atm, what would be the volume of the balloon at the higher pressure? Assume the temperature does not change and the balloon has no leaks.

   Volume = __________

5. What would happen to the volume of the 100-L weather balloon if its temperature increases from 25°C to 125°C while the external pressure remains constant? Again, there are no leaks in the balloon.

   Volume = __________

Name ______________________ Locker Number__________ Date__________
Please print; last name first

REPORT SHEET: *Experiment 12*

Boyle's Law and Charles' Law

A. Boyle's Law

Number of Books	1/(Number of books)	Volume (mL)	
		Trial 1	Trial 2
1			
2			
3			
4			

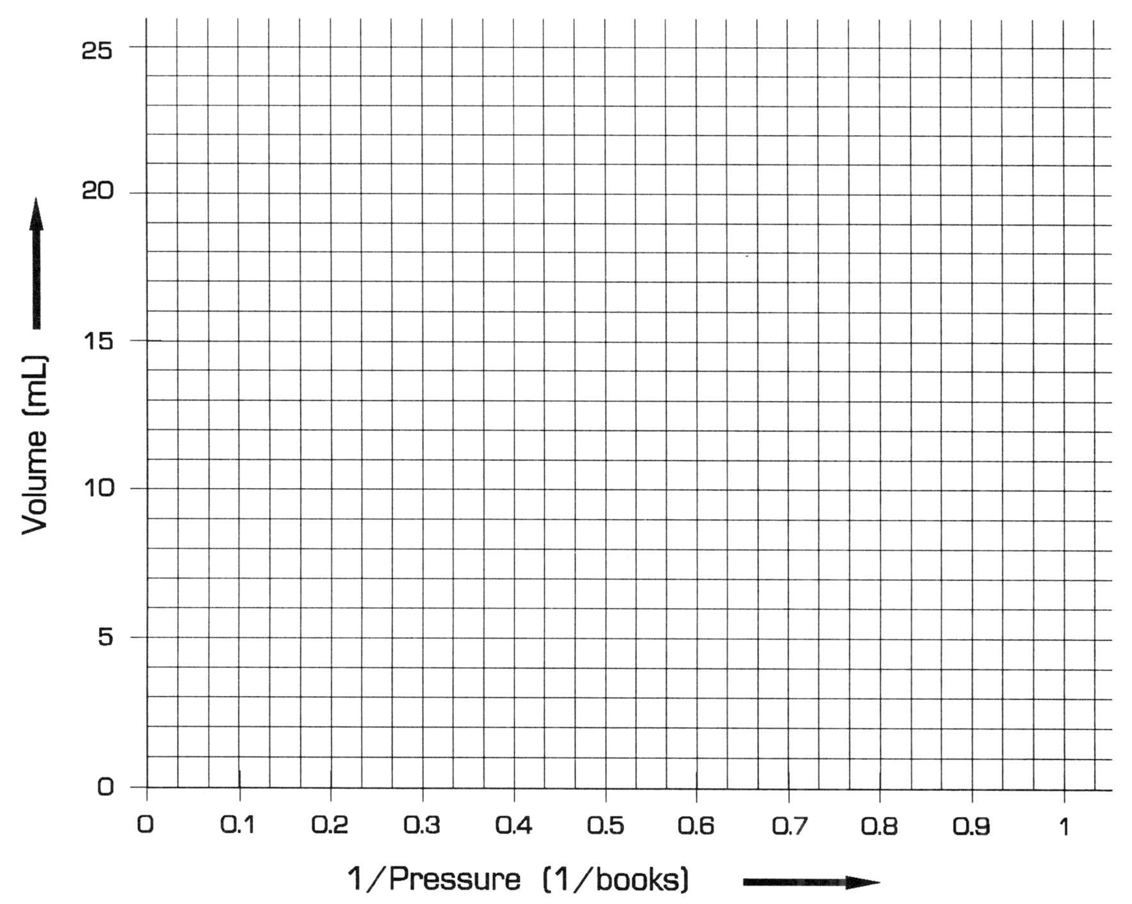

B. Charles' Law

Celsius Temperature	Kelvin Temperature	Volume from Syringe (mL)	
		Trial 1	Trial 2
0.0°C			
5.0°C			
10.0°C			
15.0°C			
20.0°C			

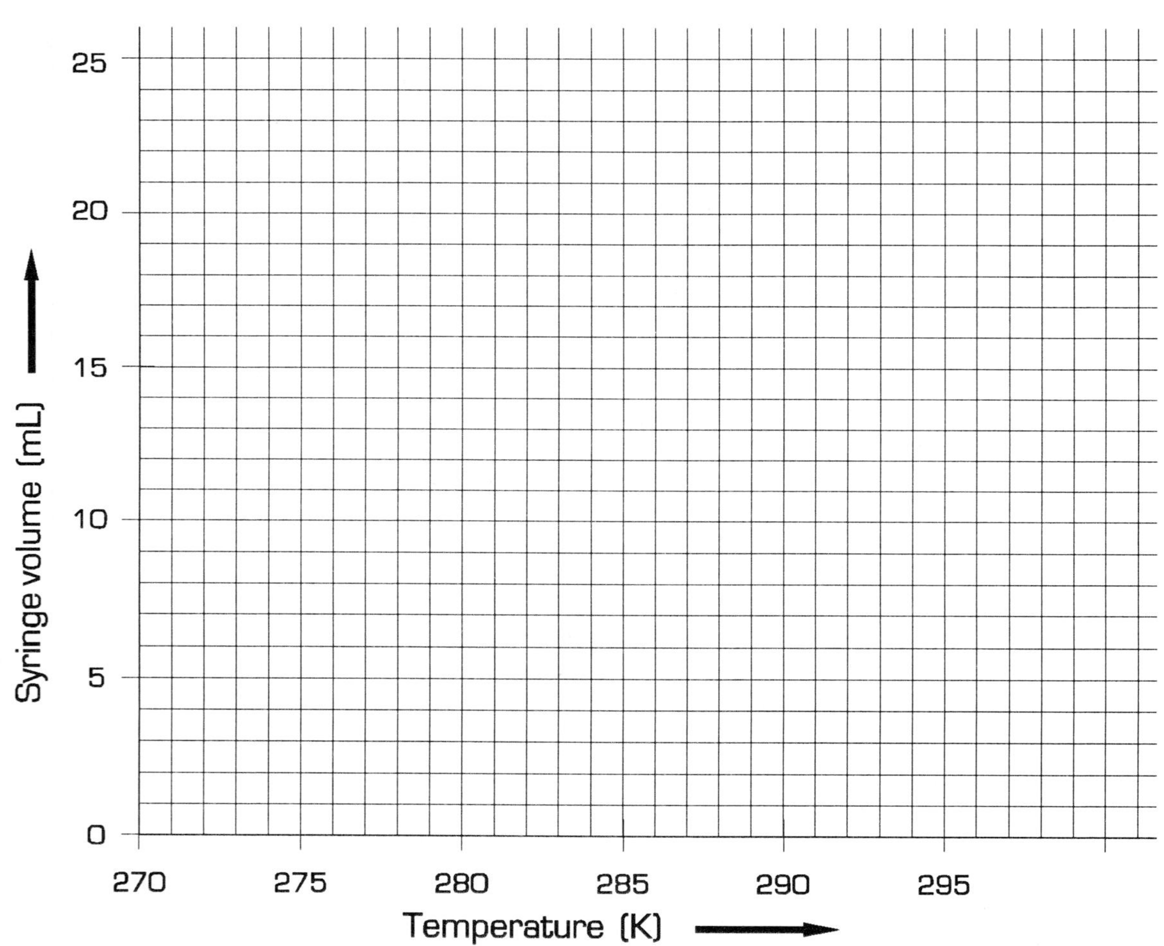

Experiment 13

The Combined Gas Law and Dalton's Law

Purpose:

In this experiment you will measure how the volume of a gas changes as its temperature changes from that of boiling water to near room temperature. Then you will calculate the theoretical change in volume using the gas laws. If your work is carefully done, you will see very good agreement between the observed and calculated volume changes, and in turn verify the accuracy of the gas laws.

Discussion:

The effect of pressure on the volume of a gas (Boyle's law), and the effect of temperature on the volume of a gas (Charles' law) were described in the discussion section of Experiment 14. Please review that discussion if you have not read it before or have not worked with these gas laws in lecture. Each is stated below:

A. Charles' Law: The Effect of Temperature on Volume

Charles' law: The volume of a gas will vary *directly* with its Absolute temperature as long as the pressure on the gas and the mass of gas do not change.

$$\frac{V_1}{T_1} = \frac{V_2}{T_2}$$

V_1 = initial volume
V_2 = final volume
T_1 = initial temperature (K)
T_2 = final temperature (K)

Both temperatures must be in Kelvin. Both volumes must be given in the same unit: liters, milliliters, gallons, etc.

Note the importance of volume varying *directly* with Absolute temperature (Kelvin temperature) in Charles' law. Doubling the Absolute temperature doubles the volume of a sample of gas; of course, both volumes must be measured at the same pressure. Cutting the Absolute temperature in half will reduce the volume of a gas to half its original volume.

B. Boyle's Law: The Effect of Pressure on Volume

Boyle's law: The volume of a gas varies *inversely* with pressure as long as the temperature of the gas and the mass of gas do not change.

$$P_1 V_1 = P_2 V_2$$

V_1 = initial volume
V_2 = final volume
P_1 = initial pressure
P_2 = final pressure

The fact that Boyle's law involves an *inverse* relationship means that if the pressure exerted on the gas doubles, its volume will be cut in half. On the other hand, if pressure is cut in half, the volume will double. Pressure and volume behave like the ends of a seesaw. If one end goes up, the other end goes down a proportional amount. Of course, the variables of temperature and amount of gas must not change if true inverse behavior is to be observed.

C. Combined Gas Law: Boyle's and Charles' Law in One Equation

Charles' law and Boyle's law can be combined into a single equation that is frequently called the **combined gas law**. Note the *direct* relationship between volume and temperature (K) and the *inverse* relationship between volume and pressure.

$$\frac{P_1 V_1}{T_1} = \frac{P_2 V_2}{T_2}$$

V_1, P_1, T_1 = initial volume, pressure and temperature (K)

V_2, P_2, T_2 = final volume, pressure and temperature (K)

The combined gas law allows for the effect of changes in temperature and pressure on the volume of a gas to be considered simultaneously.

D. Dalton's Law: The Pressure of a Mixture of Gases

Another important relationship that concerns gases is Dalton's law, which was formulated in 1803 and concerns the pressure exerted by a mixture of gases:

Dalton's law: The total pressure exerted by a mixture of gases is the *sum* of the individual pressures exerted by each gas in the mixture.

The pressure exerted by a specific gas in a mixture of gases is the **partial pressure** of that gas. In a mixture of two gases, A and B, the total pressure of the mixture, $\mathbf{P_{total}}$ or $\mathbf{P_t}$, is the sum of the pressure caused by the molecules of gas A plus the pressure caused by the molecules of gas B, referred to as their partial pressures, $\mathbf{p_A}$ and $\mathbf{p_B}$.

$$P_{total} = P_t = p_A + p_B$$

It will be necessary to apply Dalton's law in this experiment. At one point in this experiment, the total pressure of a mixture of air and water vapor (gaseous water) is measured.

$$P_t = p_{air} + p_{water\ vapor}$$

To determine only the pressure exerted by air, the pressure exerted by the water molecules must be subtracted from the total pressure. But what is the vapor pressure of water? Fortunately, the vapor pressure of water in saturated air is a known property of water that increases as the temperature of water increases. The vapor pressure of water at several temperatures is found in Appendix A. Rearranging the equation and subtracting the pressure exerted by the water molecules allows the pressure of only the air to be determined.

$$p_{air} = P_t - p_{water\ vapor}$$

Experimental Procedure:

Since Charles' law and Boyle's law apply to pure gases as well as mixtures of gases, there will be no problem using air as our gas. The volume of air we will use will be that contained in a 250-mL Erlenmeyer flask.

1. Set up the apparatus as shown in Figure 1. The inside of the flask must be *absolutely dry*. The glass tube in the rubber stopper should be as short as possible, and the rubber tubing should fit snugly onto the glass tube. The rubber tubing should be just long enough to be folded over and clamped tightly. Also, *the rubber stopper must fit tightly in the flask*. You don't want it to fall out during the experiment.

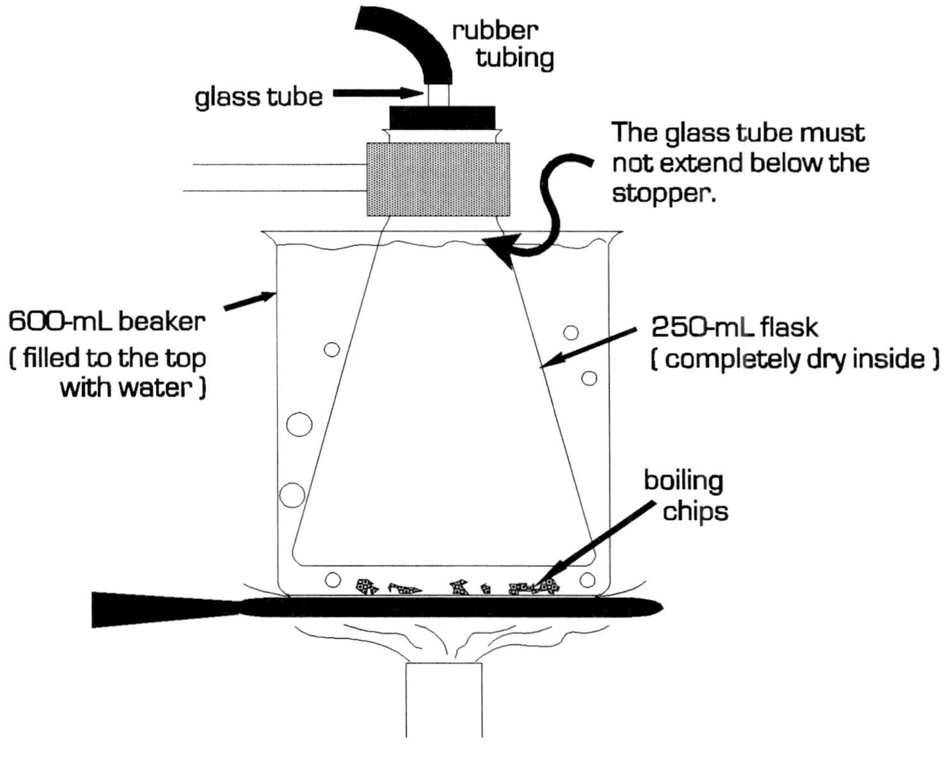

Figure 1

2. After two or three boiling chips are placed in the 600-mL beaker, clamp the flask in place and fill the beaker to nearly overflowing with water so that most of the 250-mL flask is submerged. Once all is ready, heat the water to a boil. As the air in the flask gets hotter, it will expand and escape through the open tube in the stopper.

3. After the water has boiled for 5 minutes, the temperature of the air in the flask and the temperature of the water should be the same. Measure the temperature of the boiling water to the nearest tenth of a degree and record this value on the Report Sheet as $\mathbf{T_1}$.

CAUTION: The next operation is critical to the success of this experiment. It should be done with care and as quickly as possible.

4. While the water is still boiling, *fold over the rubber tube and clamp it tightly*. Then, with the aid of a towel to protect your hands, *quickly remove the clamp holding the sealed flask from the ring stand and plunge the flask and clamp into a sink filled with cold water*. Hold the flask upside-down; you want the stopper pointing downward. *Keep the flask completely submerged in the cold water for 4 to 5 minutes*. Then, keeping the flask inverted, open the pinch clamp. Water will enter the flask. The volume of water that enters represents the decrease in volume of the entrapped air as it cooled from near 100°C to the temperature of the water in the sink.

5. Measure the temperature of the water in the sink and record this value to the nearest tenth of a degree on the Report Sheet as $\mathbf{T_2}$.

6. After 2 or 3 more minutes, carefully raise the upside-down flask until the *water level inside the flask is even with the water level in the sink* as shown in Figure 2. *Keep the neck of the flask and the tubing under water at all times*. Close the rubber tubing by clamping it near the glass tube and remove the flask from the water. Making the water levels the same makes the pressure of the air + water vapor in the flask equal to the pressure of the atmosphere, as shown in Figure 2.

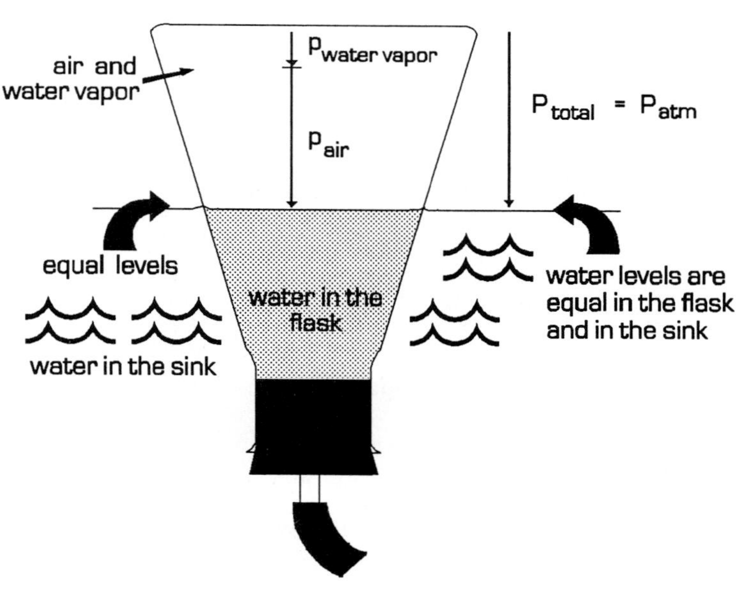

Figure 2

7. Place the flask with the stopper and clamped tube upright on the bench. Release the clamp and allow the water in the tube to drain into the flask. Measure this volume of water using a graduated cylinder and record the value to the nearest tenth of a milliliter as $\mathbf{\Delta V_{obs'd}}$ on the Report Sheet.

8. Remove the short length of rubber tubing from the glass tube in the rubber stopper. Measure the actual volume of the flask by filling it to overflowing with water, then replace the stopper in the flask. Wipe off any water that overflows, then remove the stopper and allow the small volume of water in the glass tube to enter the flask. The volume of water in the flask is measured by filling and refilling a 50-mL graduate. Record the total, actual volume of the flask on the Report Sheet as $\mathbf{V_1}$.

9. Record the barometric pressure in torr (or mmHg) on the Report Sheet as the value of $\mathbf{P_1}$. This is the pressure exerted by the gas in the flask at the boiling temperature of water when the flask was open to the atmosphere.

10. Find the vapor pressure of water at $\mathbf{T_2}$ in Appendix A. Record this pressure on the Report Sheet as $\mathbf{p_{water\ vapor}}$.

11. Calculate the pressure of just the air in the air-water vapor mixture in the flask at $\mathbf{T_2}$ by subtracting the vapor pressure of water from atmospheric pressure, $\mathbf{P_1}$. Record the pressure of dry air on the Report Sheet as $\mathbf{P_2}$.

12. Using $\mathbf{V_1}$, $\mathbf{T_1}$, $\mathbf{T_2}$, $\mathbf{P_1}$ **and** $\mathbf{P_2}$, calculate the value of $\mathbf{V_{2(calc'd)}}$ using the combined gas law. This is the volume of dry air that *should be* in the flask under the final conditions of temperature and pressure. Show this calculation on the Report Sheet.

$$V_{2\,(calc'd)} = V_1 \left(\frac{T_2\ P_1}{T_1\ P_2} \right)$$

13. Determine the observed value of $\mathbf{V_2}$, called $\mathbf{V_{2(obs'd)}}$, by subtracting the amount of water drawn into the flask, $\mathbf{\Delta V_{obs'd}}$, from $\mathbf{V_1}$.

$$V_{2\,(obs'd)} = V_1 - \Delta V_{obs'd}$$

14. The values of $\mathbf{V_{2(calc'd)}}$ and $\mathbf{V_{2(obs'd)}}$ should be nearly identical if your work was done carefully. Calculate the percent difference between the observed and calculated volumes and record this difference on the Report Sheet.

$$\textbf{percent difference} = \frac{|\,V_{2\,(calc'd)} - V_{2\,(obs'd)}\,|}{V_{2\,(calc'd)}} \times 100\,\%$$

If all went well, the percent difference should be less than 2%. If your value is much larger than this, your instructor may ask you to repeat the experiment, recording data on the Report Sheet as Trial 2.

The Men behind the Gas Laws:

Jacques Alexandre Cesar Charles (1746-1823)

In 1779 Charles set out to learn non mathematical experimental physics. In 1781 he began giving a public course of lectures and demonstrations, which soon gained him a large audience and some notoriety. Many recognize Charles' name in connection with gas laws, but he published almost nothing of scientific significance. Around 1787 he did find a relationship between the pressure exerted by a gas and its temperature, but it was Gay-Lussac who made it public in 1802. One could reasonably ask if Charles actually discovered Charles' law.

Robert Boyle (1627-1691)

Born in Ireland, Boyle has been called the founder of modern chemistry by many historians. In 1661 he wrote an outstanding book, called *The Sceptical Chymist*, that promoted chemistry as a science in its own right and insisted that theory had to be backed with experimental evidence. He championed the rigorous experimental method. Boyle gave a clear definition of an element, and explained that air is a fluid that is compressible, has weight and is able to refract light. In 1661 he showed the pressure-volume relationship in gases that we call Boyle's law.

John Dalton (1766-1844)

Dalton was born into a strict Quaker family in the small village of Eaglesfield, England. He acquired most of his education on his own. At 15 he joined his brother teaching at a small Quaker boarding school and there immersed himself in the extensive library to expand his knowledge. In 1792 he was appointed professor of mathematics and natural philosophy in the "New College" at Manchester. In a series of papers, presented between 1799 and 1801, on his studies with water vapor, he formulated his law of partial pressures. He formulated the law of multiple proportions shortly after 1803. Between 1808 and 1810 Dalton developed his ideas on atomic theory. Part of the original theory, usually not presented in textbooks, was his rule of greatest simplicity. Dalton believed that combinations of elements would always be in the simplest ratio; that is, if two elements form a compound, then in most cases the combination would be one atom to one atom. He believed water was HO.

Amedo Avogadro (1776-1856)

Avogadro was educated as a lawyer, but began studying mathematics and physics in 1800. In 1820 he became professor of natural philosophy at the College of Vercelli, and in 1820 he was appointed to the chair of mathematical physics at Turin. He worked in virtual isolation and was not well known in scientific circles. In 1811 Avogadro explained Gay-Lussac's law of combining volumes in such a way that it followed naturally that equal volumes of gases measured under the same conditions contain the same number of molecules. Also, the density of a gas had to be proportional to the weight of a molecule of the gas. His work allowed the molecular weights of gases to be determined directly. Avogadro promoted the idea that oxygen and hydrogen were molecules composed of two atoms and advanced this idea in many publications written from 1816 to 1821. His hypothesis continued to be ignored by the scientific community until 1858, when Cannizzaro explained the true significance of Avogadro's work. Many chemists immediately adopted Avogadro's hypothesis into their research and teaching.

Name ______________________ Locker Number__________ Date___________
Please print; last name first

PRELIMINARY EXERCISES: *Experiment 13*

The Combined Gas Law and Dalton's Law

1. Consider a mixture of oxygen and nitrogen. The oxygen gas exerts a pressure of 0.68 atm and the nitrogen gas, 1.20 atm. What is the pressure exerted by the mixture of the two gases?

   Total pressure = ___________

2. The term *vapor* and *gas* mean the same. Historically, vapor referred to the gas formed as a liquid evaporated. Water *vapor* is water in the gaseous state, a gas, $H_2O_{(g)}$. Air in contact with liquid water will become saturated with water molecules (water vapor). The higher the temperature of the liquid water, the higher the concentration of water molecules in the air and the higher the pressure exerted by this water vapor. Which of the gas laws can be used to subtract the pressure exerted by water molecules from the total pressure of a gaseous mixture? Besides naming the law, state it in words and as an equation.

3. What is the vapor pressure of water at 25°C? At 30°C? Where would you look to find these values?

   Vapor pressure at 25°C = ___________

   Vapor pressure at 30°C = ___________

4. Suppose you collected oxygen gas over water, allowing it to be saturated with water vapor. The total pressure of the oxygen plus water vapor equals 750.5 torr at 25°C. What is the pressure exerted by only the oxygen molecules?

   Oxygen pressure = ___________

(The Preliminary Exercises continue on the following page.)

5. Suppose a weather balloon has a volume of 3520 L at 28.5°C and 745 torr. Use the combined gas law to determine the volume, in liters, that the balloon would occupy high above the Earth where the temperature is −25.0°C and the pressure is 250 torr. Please show all your work using units on all numbers.

Volume = ____________

Name ______________________ Locker Number__________ Date__________
Please print; last name first

REPORT SHEET: *Experiment 13*

The Combined Gas Law and Dalton's Law

Trial 2 is to be carried out if the results of trial 1 are unsatisfactory.

	Trial 1	Trial 2
Temperature of boiling water: $\mathbf{T_1}$		
Temperature of water in sink: $\mathbf{T_2}$		
Volume of water drawn into flask: $\mathbf{\Delta V_{obs'd}}$		
Volume of air in flask at T_1: $\mathbf{V_1}$		
Barometric pressure: $\mathbf{P_1}$		
Vapor pressure of water at T_2: $\mathbf{p_{water\ vapor}}$		
Partial pressure of dry air: $\mathbf{P_2}$		
Calculated volume of air in flask at T_2: $\mathbf{V_{2(calc'd)}}$ (Show setup of calculation below.)		
$\mathbf{V_{2(obs'd)}}$		
percent difference between $\mathbf{V_{2(calc'd)}}$ and $\mathbf{V_{2(obs'd)}}$ (Show setup of calculation below.)		

SECTION VII:
Acid-Base Chemistry

Experiment 14

Acid-Base Titrations

Purpose:

In this experiment, you will learn the analytical procedure of acid-base titration and use this procedure to determine the molarity of acid in solutions of unknown acid concentration. In addition, you will gain experience preparing solutions, using the buret and pipet properly and carrying out titration calculations.

Discussion:

In 1884, Svante Arrhenius, a Danish chemist, became the first to describe acids and bases in terms that have stood the test of time. He defined an **acid** as a substance that increases the concentration of hydrogen ion (H^+) when dissolved in water, and a **base** as a substance that increases the concentration of hydroxide ion (OH^-) when dissolved in water. Though other definitions have been proposed since the time of Arrhenius, they all accommodate his ideas. Four commonly used acids are hydrochloric acid, $HCl_{(aq)}$, nitric acid, $HNO_{3(aq)}$, sulfuric acid, $H_2SO_{4(aq)}$, and acetic acid, $HC_2H_3O_{2(aq)}$ (also written as $CH_3COOH_{(aq)}$). Four commonly used bases are sodium hydroxide, $NaOH_{(aq)}$, potassium hydroxide, $KOH_{(aq)}$, calcium hydroxide, $Ca(OH)_{2(aq)}$, and magnesium hydroxide, $Mg(OH)_{2(aq)}$

A. Neutralization Reactions

The reaction between an acid and a base is called **neutralization**. The acid neutralizes the properties of the base and the base neutralizes the properties of the acid. The products of an acid-base neutralization reaction are water and a salt. Do not confuse the word *salt* with *table salt*, the common name we use for sodium chloride, NaCl. In acid-base chemistry, *salt* is used to categorize the ionic product of an acid-base neutralization. The other product, water, is molecular. The neutralization of potassium hydroxide with nitric acid is written below using a **formula equation** and an **ionic equation**. The ionic equation shows each species as it exists in solution.

Formula equation:	$KOH_{(aq)}$	+	$HNO_{3(aq)}$	$\rightarrow$	$KNO_{3(aq)}$	$+ H_2O_{(l)}$
	the base		the acid		the salt	water

Ionic equation: $K^+_{(aq)} + OH^-_{(aq)} + H^+_{(aq)} + NO_3^-{}_{(aq)} \rightarrow K^+_{(aq)} + NO_3^-{}_{(aq)} + H_2O_{(l)}$

In all acid-base neutralizations, the actual reaction is that between the hydrogen and hydroxide ions to form water. Crossing out duplicate ions on both sides of the preceding ionic equation gives the **net ionic equation** which shows this clearly:

Net ionic equation: $H^+_{(aq)} + OH^-_{(aq)} \rightarrow H_2O_{(l)}$

Before going further in this discussion, you must be told that the hydrogen ion, which is the nucleus of the hydrogen atom, does not exist as such in water. Rather, it is intimately associated with one or more water molecules at all times and in this form is referred to as the **hydronium ion**. The hydronium ion is commonly symbolized as $H^+_{(aq)}$ or H_3O^+.

B. Molarity

The most common method of expressing the concentration of a solute in a solution is **molarity** (M). The equation that defines molarity is:

$$\textbf{molarity} = \textbf{M} = \frac{\text{number of moles of solute}}{\text{volume of solution in liters}}$$

Molarity has units of moles per liter, mol/L, but common practice is to symbolize the two units with a capital M. If a solution is labeled 0.35 M NaOH, it means that 1.0 L of that solution contains 0.35 mol of sodium hydroxide. You will learn more about molarity in the next section in the discussion titled Working With Molarity starting on page 192.

Remember that one **mole** of any compound is an amount of that compound equal to its formula mass (or formula weight) in grams. The formula mass of sodium hydroxide is 40.0, (the sum of the atomic weights of the 3 atoms in its formula) so 40.0 g of NaOH is 1 mol of NaOH. Other terms also used to mean the same as mole are: *gram formula mass* (or gram formula weight) and *molar mass*. They all have the same units, commonly grams per mole. To convert a known mass of compound into the corresponding number of moles of the compound, divide the mass in grams by the gram formula mass of the compound.

$$\text{number of moles} = \frac{\text{mass of compound in grams}}{\text{gram formula mass of the compound in grams per mole}}$$

C. Titration

Titration is a quantitative, volumetric procedure used to determine the concentration of a solute in a solution. For example, it is possible to determine the concentration of hydrochloric acid in a solution of the acid by measuring the volume of 0.100 M NaOH needed to *just* consume the acid. The formula equation for the neutralization of hydrochloric acid with sodium hydroxide is:

$$NaOH_{(aq)} + HCl_{(aq)} \rightarrow NaCl_{(aq)} + H_2O_{(l)}$$

The point at which the acid is *just completely neutralized* by the base is called the **equivalence point** of the titration. At the equivalence point the number of moles of hydronium ion consumed (provided by the acid) and the number of moles of hydroxide ion consumed (provided by the base) are equal.

At the equivalence point: moles of $H^+_{(aq)}$ = moles of $OH^-_{(aq)}$

This equation can be rewritten in terms of the concentrations and volumes of the acid and base. Rearranging the molarity equation shows that *molarity* x *volume (in liters) = moles of solute*. For those titrations involving acids with *one* replaceable hydrogen per formula unit (such as HCl or HNO_3) and bases with *one* hydroxide ion per formula unit (such as NaOH or KOH), the condition at the equivalence point is:

$$\mathbf{V}_{\text{acid}} \times \mathbf{M}_{\text{acid}} = \mathbf{V}_{\text{base}} \times \mathbf{M}_{\text{base}}$$

where $\mathbf{V}_{\text{acid}}$ is the volume of the acid, $\mathbf{M}_{\text{acid}}$ is the molarity of the acid, $\mathbf{V}_{\text{base}}$ is the volume of the base and $\mathbf{M}_{\text{base}}$ is the molarity of the base. In a titration experiment, three of these four terms will be known, allowing the fourth, frequently the concentration of the acid or base, to be calculated. Though this equation applies to many titrations, it has its limitations. It only applies to titrations in which the acid has one replaceable hydrogen (*H*Cl, *H*NO_3 or *H*$C_2H_3O_2$) and the base has one hydroxide ion in its formula (Na*OH* or K*OH*). Acids with more than one replaceable hydrogen in their formulas (H_2SO_4 and H_3PO_4) and bases with more than one hydroxide ion ($Ca(OH)_2$ and $Ba(OH)_2$) require a modification of the $M_a \times V_a = M_b \times V_b$ equation to correctly represent the mole ratio of acid to base in neutralization reactions.

A typical titration apparatus setup is shown in Figure 1. The solution to be analyzed, the **analyte**, is in the flask. It will be the acid in this experiment. The acid is added to the flask with a *pipet* so its exact volume, $\mathbf{V}_{\text{acid}}$, is known. The buret contains the **titrant**, the base solution. The concentration of the base, $\mathbf{M}_{\text{base}}$, is accurately known. Any solution of known concentration is called a **standard solution**. The volume of base needed to reach the equivalence point, $\mathbf{V}_{\text{base}}$, can be read from the buret to the nearest 0.01 mL. With three of the four terms in the titration equation known, the fourth, $\mathbf{M}_{\text{acid}}$, can be calculated.

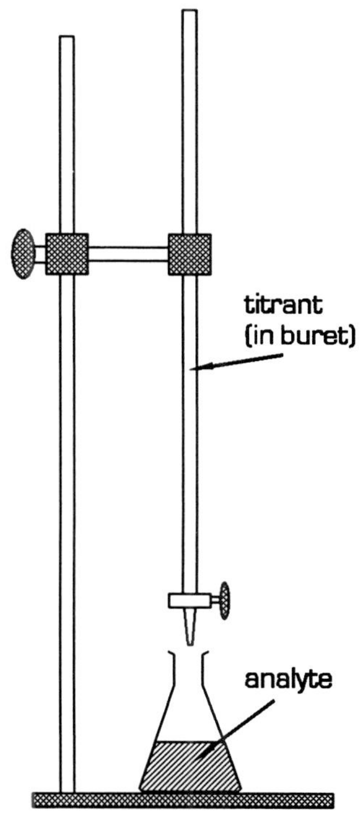

Figure 1

Three criteria must be met in any titration procedure:

1. The reaction must be fast.
2. The reactants must react completely.
3. There must be some way to know when the equivalence point has been reached, in our case, the point at which all the acid is just neutralized by the base.

These criteria are easily met in acid-base titrations. The reaction between an acid and a base is one of the fastest known, and the hydroxide and hydronium ions react completely to form water. **Indicators**, water-soluble dyes that have one color in acidic solutions and another in basic solutions, allow us to see when the equivalence point is reached. In the titration of HCl with NaOH, the solution in the flask rapidly changes from acidic to basic as the equivalence point is approached, reached and passed. This passage through the equivalence point will occur with the addition of *one* drop (or less) of base. A properly chosen indicator, added to the acid, will change color at the equivalence point in response to the rapid acid-to-base change. The indicator "indicates" to the observer when the acid is just completely neutralized. The point at which you see the indicator change color is called the **endpoint** in a titration, and care must be taken to choose the proper indicator so that the equivalence point and the endpoint occur simultaneously, or nearly so. Remember, the endpoint is a visual fact (the observed color change), while the equivalence point is a chemical fact (the point of neutralization).

There are hundreds of indicators available for use in acid-base titrations. In this experiment the indicator that works best is phenolphthalein (feen-ohl-*thay*-leen). Phenolphthalein changes color at the proper point and the color change is easy to see—colorless on the acid side to red on the basic side. The endpoint in the titration of HCl with NaOH is reached as the solution changes from colorless to red with the addition of *one drop* of base.

D. Experiment Overview

In this experiment, you are to determine the molarity of hydrochloric acid in solutions of unknown acid concentration and/or the molarity of acetic acid in vinegar. You will need a *standard* sodium hydroxide solution. If it is *not* provided to you in the laboratory, you will need to prepare it as described in Section A of the experiment. If it is provided for you, proceed directly to Section B of the experiment. The following is a brief overview of the experiment. Read this carefully before proceeding to the Experimental Procedure.

In the optional Section A of the experiment, you will be given a solution of sodium hydroxide that is approximately 0.1 M NaOH. You are to standardize the solution and determine its exact concentration to three places after the decimal by measuring the volume of base needed to just neutralize an accurately measured mass of a special acidic salt called potassium hydrogenphthalate (potassium hydrogen-*thal*-ate). For simplicity, this acid is abbreviated KHP. The formula of KHP is $KHC_8H_4O_4$ and its formula weight is 204.2. KHP is a **primary standard** for determining base concentrations. It is obtainable in a very pure state, it has a large formula weight, it is water soluble and it has *one* replaceable hydrogen, $K\mathit{H}C_8H_4O_4$. The formula and net ionic equations for the reaction of KHP with NaOH are:

Formula equation: $$KHC_8H_4O_{4(aq)} + NaOH_{(aq)} \rightarrow KNaC_8H_4O_{4(aq)} + H_2O_{(l)}$$

Net ionic equation: $$HC_8H_4O_4{}^{-}{}_{(aq)} + OH^{-}{}_{(aq)} \rightarrow C_8H_4O_4{}^{2-}{}_{(aq)} + H_2O_{(l)}$$

At the equivalence point:

$$\text{number of moles of KHP neutralized} = \text{number of moles of NaOH added}$$

This condition can be stated in an equation that holds at the equivalence point. The left side equals the number of moles of KHP and the right side equals the number of moles of NaOH:

$$\frac{\text{grams of KHP}}{204.2\ \text{g/mol}} = V_{\text{NaOH (in liters)}} \times M_{\text{NaOH}}$$

In the titration, all terms are known except $\mathbf{M_{NaOH}}$. The equation is solved for $\mathbf{M_{NaOH}}$:

$$M_{\text{NaOH}} = \frac{\text{grams of KHP}}{V_{\text{NaOH (in liters)}} \times \left(204.2\ \frac{\text{g}}{\text{mol}}\right)}$$

Once you have standard NaOH solution, it can now be used in Section B of the experiment as a standard to determine acid concentrations. The titration is done this way: A carefully measured volume of the acid to be analyzed, hydrochloric acid or acetic acid in vinegar, is placed in an Erlenmeyer flask with a few drops of phenolphthalein. Standard NaOH is *slowly* added from the buret until the endpoint is reached (phenolphthalein: colorless → red). The addition of the base from the buret must be done with great care to obtain good results. The formula equations are:

For hydrochloric acid: $NaOH_{(aq)} + HCl_{(aq)} \rightarrow NaCl_{(aq)} + H_2O_{(l)}$

For acetic acid in vinegar: $NaOH_{(aq)} + HC_2H_3O_{2(aq)} \rightarrow NaC_2H_3O_{2(aq)} + H_2O_{(l)}$

At the equivalence point:

number of moles of $HCl_{(aq)}$ neutralized = number of moles of $NaOH_{(aq)}$ added

number of moles of $HC_2H_3O_{2(aq)}$ neutralized = number of moles of $NaOH_{(aq)}$ added

So at the equivalence point for either acid titrated with NaOH:

$$V_{\text{acid}} \times M_{\text{acid}} = V_{\text{NaOH}} \times M_{\text{NaOH}}$$

The molarity of the acid, $\mathbf{M_{acid}}$, is:

$$M_{\text{acid}} = \frac{V_{\text{NaOH}} \times M_{\text{NaOH}}}{V_{\text{acid}}}$$

Working With Molarity:

Since molarity is the most common means of expressing concentration in chemistry, it is important that you become comfortable with its use. The following three problems will show how molarity is used most often in chemical calculations.

A. Determining the Molarity of a Solution

Problem: What is the molarity of sodium hydroxide in a solution prepared by dissolving 25.0 g of NaOH in sufficient deionized water to prepare exactly 1500. mL of solution?

First, recall the relationship between the solute and solution that defines molarity.

$$\text{molarity} = \text{M} = \frac{\text{number of moles of solute}}{\text{number of liters of solution}}$$

To determine the number of moles of NaOH, you will need its formula mass, which is the sum of the atomic masses of all the atoms in the formula: one sodium (23.0), one oxygen (16.0) and one hydrogen (1.0). This totals 40.0, so 1.00 mol of NaOH would be 40.0 g of that compound. This fact is required to convert 25.0 g of NaOH into moles of NaOH.

$$\text{moles of NaOH} = (25.0\ \cancel{\text{g NaOH}}) \times \left(\frac{1\ \text{mol NaOH}}{40.0\ \cancel{\text{g NaOH}}}\right) = 0.625\ \text{mol NaOH}$$

The volume of solution is exactly 1500 mL, which is 1.500 L.

$$\text{volume in liters} = (1500.\ \cancel{\text{mL}}) \times \left(\frac{1\ \text{L}}{1000\ \cancel{\text{mL}}}\right) = 1.500\ \text{L}$$

With both moles of solute and volume of solution known, the molarity of the solution is readily determined:

$$\text{M} = \left(\frac{0.625\ \text{mol}}{1.500\ \text{L}}\right) = 0.417\ \frac{\text{mol}}{\text{L}} = 0.417\ \text{M}$$

B. Determining Moles of Solute in a Known Volume of Solution

Problem: What number of moles of NaOH would be in 350. mL of 0.750 M NaOH?

The molarity of the solution tells you the number of moles of NaOH in each 1.00 L of solution.

$$0.750\ \text{M NaOH} = \begin{array}{c}0.750\ \text{mol NaOH in 1.00 L of solution} \\ \text{or} \\ 0.750\ \text{mol NaOH in 1000 mL of solution}\end{array}$$

This fact will be used to convert volume of solution to moles of NaOH.

$$\text{moles of NaOH} = 350.\,\cancel{\text{mL solution}} \times \left(\frac{0.750 \text{ mol NaOH}}{1000\,\cancel{\text{mL solution}}}\right) = 0.262 \text{ mol NaOH}$$

We now know that 350. mL of 0.750 M NaOH contains 0.262 mol of NaOH. If you needed the mass of NaOH in this volume of solution, convert moles of NaOH to mass of NaOH using its gram formula mass.

$$\text{mass of NaOH} = 0.262\,\cancel{\text{mol NaOH}} \times \left(\frac{40.0 \text{ g NaOH}}{1.00\,\cancel{\text{mol NaOH}}}\right) = 10.5 \text{ g NaOH}$$

So 350. mL of 0.750 M NaOH contains 0.262 mol of NaOH, which is equivalent to 10.5 g of NaOH.

C. Determining a Solution Volume Containing a Known Mass of Solute

Problem: How many milliliters of 0.750 M NaOH contain exactly 5.00 g of NaOH?

Again, the molarity of the solution contains the key information needed to solve this problem.

$$0.750 \text{ M NaOH} = \begin{array}{c} 0.750 \text{ mol NaOH in } 1.00 \text{ L of solution} \\ \text{or} \\ 0.750 \text{ mol NaOH in } 1000 \text{ mL of solution} \end{array}$$

We will use the fact that 1000 mL contains 0.750 mol of NaOH as a conversion factor to convert moles of NaOH to milliliters of solution. But first we will need to get our 5.00-g mass of NaOH into moles of NaOH.

$$\text{moles of NaOH} = 5.00\,\cancel{\text{g NaOH}} \times \left(\frac{1.00 \text{ mol NaOH}}{40.0\,\cancel{\text{g NaOH}}}\right) = 0.125 \text{ mol NaOH}$$

The volume of solution containing 5.00 g of NaOH is:

$$\text{volume of solution} = 0.125\,\cancel{\text{mol NaOH}} \times \left(\frac{1000 \text{ mL}}{0.750\,\cancel{\text{mol NaOH}}}\right) = 167 \text{ mL}$$

Exactly 5.00 g of NaOH is contained in 167 mL of a 0.750 M NaOH solution.

In all of these problems, fundamental definitions were used to make the necessary conversions. Notice how the units canceled as needed, leaving those appropriate for the answers we sought. Pay attention to the units and you should succeed every time.

Experimental Procedure:

> **CAUTION: Acids and bases are used in the following procedures. If you get an acid or base on your skin, wash the affected area with running water for several minutes and immediately notify the instructor.**
>
> **As always, eye protection must be worn when working in the laboratory.**
>
> **ENVIRONMENTAL ALERT: Your instructor will inform you of the proper and safe disposal of excess acid or base.**
>
> **INSTRUCTIONAL NOTE: Section A should be performed only if you must prepare standard sodium hydroxide for Section B.**

A. Preparing Standard 0.1 M NaOH (optional)

A stock solution of NaOH, which is approximately 0.1 M, will be provided to you in the laboratory. You are to determine its true molarity to *three* places after the decimal by titration of a known mass of potassium hydrogenphthalate, KHP.

1. Gather the following items from your locker and/or the supply in the laboratory: 500-mL (or 1-pint) bottle with cap, ring stand, 50-mL buret, buret clamp and buret card. Rinse the 500-mL bottle thoroughly with tap water, then deionized water. Fill the bottle with 0.1 M NaOH (the approximate molarity) from the stock in the laboratory. If you are using a 1-pint bottle, it will hold about 475 mL of solution.

2. Test the buret for cleanliness by filling it with tap water and allowing it to drain completely through the tip into a sink. If droplets of water adhere to the inner wall, the buret is dirty and it must be cleaned or exchanged for a clean one. A clean buret will drain cleanly and droplets of water will not be seen adhering to the inner wall.

3. Rinse the buret with two 10-mL portions of NaOH solution. Hold the buret nearly horizontal and rotate the buret to wet the entire inside wall, then draining each rinse through the tip. Discard the waste NaOH solution as directed by your instructor.

   Mount the buret on a ring stand using a buret clamp. Then, using a funnel, fill it with 0.1 M NaOH to just above the 0.0-mL mark. Once filled, deliver short bursts of solution from the buret to expel all the air bubbles from the tip.

4. Measure out the samples of potassium hydrogenphthalate (KHP) for the titration in this manner: Weigh a clean and dry 250-mL Erlenmeyer flask on the balance. Record its mass to the *maximum accuracy of the balance* on the Report Sheet for Section A as trial 1. Place the flask on the table and tap (or use a clean spatula) 0.700 to 0.750 g of KHP into the flask. Weigh and record the mass of the flask + KHP on the Report Sheet. Determine the mass of KHP in the flask by difference.

If your balance has a tare function, you may tare the flask (setting the balance to zero with the flask on the pan) and measure the mass of KHP directly. This requires only one weighing.

If you spilled any KHP during the weighing process, clean it up before continuing.

5. Dissolve the KHP in 50 mL of deionized water, then add 3 to 5 drops of phenolphthalein indicator to this solution.

6. Calculate the approximate volume of 0.1 M NaOH required to just neutralize the sample of KHP. You can speed up the titration by rapidly adding about 90% of the calculated volume of NaOH to the KHP solution, then proceeding slowly and cautiously to the endpoint. Use the following equation to calculate the volume of NaOH required to do this. The term $\approx M_{NaOH}$ stands for the approximate molarity of NaOH, 0.1 M.

$$\begin{array}{c}\text{90\% of calculated volume} \\ \text{of NaOH in milliliters}\end{array} = 0.9 \times \left(\frac{\text{grams of KHP}}{204.2 \text{ g/mol} \times \approx M_{NaOH}}\right)\left(\frac{1000 \text{ mL}}{1 \text{ L}}\right)$$

7. You are now ready for the titration. Make sure the buret contains at least 40 mL of NaOH solution. Record the volume of NaOH in the buret to the nearest 0.01 mL on the Report Sheet. *Remember, read down a buret.* Use a buret card (a white card with a heavy black line drawn on it) to locate the bottom of the meniscus more easily, as shown in Figure 2, which displays a volume of 7.78 mL.

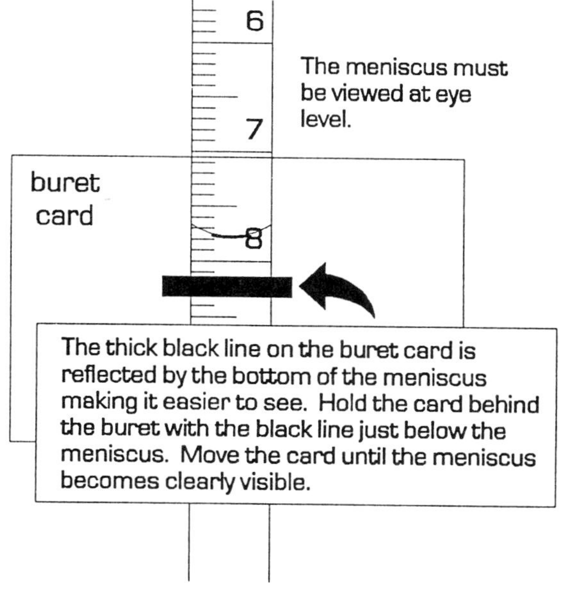

Figure 2

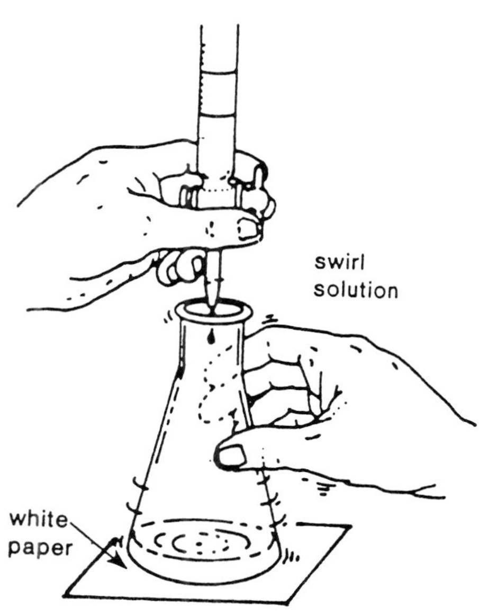

Figure 3

8. Place the flask containing the KHP solution on a white piece of paper beneath the buret and add the calculated volume of NaOH (the 90% volume) into the KHP solution. Swirl the flask to mix the solutions. From this point on, base must be added slowly until the endpoint is reached. Swirl the flask with one hand as you control the buret stopcock with the other, as shown in Figure 3. As you near the endpoint, add base one drop at a time.

   Catching the endpoint: As base is added to the acid solution, you will notice the pink color of phenolphthalein appearing at the point where the base enters the acid. As you get closer to the endpoint, the pink color spreads out further and lingers longer. Proceed with caution and add NaOH *one drop at a time* until the endpoint is reached. You've reached the endpoint when one drop of base causes the entire solution to turn pink and the color holds for at least 30 seconds. *If you pass beyond the endpoint you must repeat the analysis.*

9. Once you have reached the endpoint, record the volume of NaOH from the buret on the Report Sheet to the nearest 0.01 mL. The volume of NaOH required to reach the endpoint is the difference between the initial and final volumes. If you need to duplicate the analysis, repeat steps 4 through 9, recording your data on the Report Sheet as trial 2.

   Label the bottle containing the remaining NaOH solution, stating its molarity to the third place after the decimal.

B. Analyzing an Acid of Unknown Concentration

You will analyze either a solution of hydrochloric acid or a solution of acetic acid in a commercial brand of vinegar. Vinegar is primarily a solution of acetic acid in water. In either case, you will be determining the molarity of acid in the solution you analyze. The titration will be done at least twice (your instructor will tell you how many trials to perform), and the molarity of the acid will be taken as the average of the trials. If you have trouble with the titrations and your results are not acceptable, additional trials can be performed.

If you did not do Section A, please go back and read steps 7 and 8 from that section for important directions concerning titration, reading volumes and catching the endpoint.

1. Obtain a sample of the acid to be analyzed (HCl or vinegar) from the instructor. Record the number of the unknown where indicated in Section B of the Report Sheet.

2. Obtain a 50-mL buret, buret clamp, ring stand and buret card from the supply in the lab. Follow the procedures in steps 2 and 3 in Section A to prepare the buret for use. When clean and rinsed with the standard NaOH solution, fill it nearly to the top with *standard* 0.1 M NaOH. Mount the buret on the ring stand and drain NaOH from the buret in short bursts to expel all the air bubbles from the buret tip.

3. Obtain a pipet and pipet bulb from the supply in the lab. If you are analyzing commercial vinegar as it comes out of the bottle, a 5.00-mL pipet is recommended; a larger volume pipet can be used for the HCl samples. *Your instructor will tell you what volume pipet to use.* Check the cleanliness of the pipet by filling it with deionized water and allowing it to drain by gravity alone. If droplets of water cling to the inner surface, the pipet is dirty. Either exchange it for a clean one or clean it with warm, soapy water. It is imperative that the pipet and buret drain cleanly to ensure accuracy.

4. Rinse the pipet two times with a few milliliters of the acid solution you are to analyze. This is done by drawing a few milliliters of the acid solution into the pipet (using the pipet bulb) and tilting and rotating the pipet to wet the inner walls. Discard the rinses into a waste beaker. Then obtain the acid volume for analysis in the pipet and transfer the volume to a clean 250-mL Erlenmeyer flask. *If you have questions about the proper use of the pipet, ask your instructor for assistance.* Add about 20 mL of deionized water and 3 or 4 drops of phenolphthalein indicator to the solution and swirl to mix well.

5. Record the initial volume of NaOH in the buret to the nearest 0.01 mL on the Report Sheet under trial 1. Use a buret card as shown in Figure 2. Remember, read *down* the buret. Place the flask containing the unknown acid on a piece of white paper beneath the buret and begin to add small volumes of base. Control the buret stopcock with one hand as you swirl the solution in the flask with the other, as shown in Figure 3. Do not rush!

   Continue the addition of NaOH *slowly* until the red color of phenolphthalein begins to linger, indicating the endpoint is being approached. *From this point on, add NaOH very slowly, one drop at a time until the endpoint is reached.* Swirl the solution after each addition.

6. Once the endpoint is reached, record the final volume of NaOH to the hundredths of a milliliter on the Report Sheet. The volume of NaOH required to just reach the endpoint, $\mathbf{V_{NaOH}}$, is the difference between the initial and final volume readings.

7. Now that you know the molarity of the NaOH, $\mathbf{M_{NaOH}}$, the volume of NaOH required to reach the endpoint, $\mathbf{V_{NaOH}}$, and the volume of acid analyzed, $\mathbf{V_{acid}}$, calculate the molarity of the acid solution, $\mathbf{M_{acid}}$.

$$\mathbf{M_{acid}} = \frac{\mathbf{M_{NaOH}} \times \mathbf{V_{NaOH}}}{\mathbf{V_{acid}}}$$

   Show all calculations and record the molarity of the acid to three significant figures on the Report Sheet. Your instructor may want to check your results before attempting another trial.

8. Refill your buret with more of the standard NaOH solution and repeat steps 4 through 7 with another sample of the acid (HCl or vinegar), recording all data as trial 2 on the Report Sheet. Repeat the titration until you have completed all the trials assigned.

Name ______________________ Locker Number__________ Date__________
Please print; last name first

PRELIMINARY EXERCISES: *Experiment 14*

Acid-Base Titrations

1. Define the following terms associated with titrations:

a. standard solution
b. endpoint
c. indicator

2. Write the balanced formula equation and the net ionic equation for the reaction of sodium hydroxide with hydrochloric acid.

Formula equation: ______________________

Net Ionic equation: ______________________

3. What is the molarity of a solution prepared by dissolving 2.24 grams of NaOH in enough deionized water to obtain 500.0 mL of solution? 1 mole NaOH = 40.0 g NaOH.

M_{NaOH} = __________

4. What is the color of phenolphthalein indicator in:

a. acidic solution __________ b. basic solution __________

5. Calculate the molarity of acetic acid in a vinegar sample, knowing that 5.00 mL of vinegar requires 43.50 mL of 0.105 M NaOH to just reach the phenolphthalein endpoint in a titration.

$M_{HC_2H_3O_2}$ = __________

Name ______________________ Locker Number__________ Date__________
Please print; last name first

REPORT SHEET: *Experiment 14*

Acid-Base Titrations

A. Preparing Standard 0.1 M NaOH

	Trial 1	**Trial 2**
Mass of KHP + flask:		
Mass of Erlenmeyer flask:		
Mass of KHP:		
Number of moles of KHP:		
Final buret reading - NaOH:		
Initial buret reading - NaOH:		
Volume of NaOH needed to reach the endpoint:		
Molarity of NaOH:		
Average molarity of NaOH:		

Show all calculations for Section A below:

Trial 1	Trial 2

Name ______________________ Locker Number__________ Date__________
Please print; last name first

REPORT SHEET (continued): *Experiment 14*

Acid-Base Titrations

B. Analyzing an Acid of Unknown Concentration

Molarity of NaOH: __________ Unknown acid number: __________

	Trial 1	**Trial 2**
Volume of acid analyzed:		
Final buret reading - NaOH:		
Initial buret reading - NaOH:		
Volume of NaOH required to reach the endpoint:		
Molarity of the acid:		
Average molarity of the acid:		

Show all calculations for trials 1 and 2 below:

Trial 1	Trial 2

B. Analyzing an Acid of Unknown Concentration

Molarity of NaOH: ________ Unknown acid number: ________		
	Trial 3	**Trial 4**
Volume of acid analyzed:		
Final buret reading - NaOH:		
Initial buret reading - NaOH:		
Volume of NaOH required to reach the endpoint:		
Molarity of the acid:		
Average molarity of the acid:		

Show all calculations for trials 3 and 4 below:

Trial 3	Trial 4

Experiment 15

Acids, Bases, pH, Hydrolysis and Buffers

Purpose:

This experiment will introduce you to two methods commonly used to measure the pH of a solution. In doing so you will investigate how the hydrolysis of certain salts affects pH, measure the pH of common household chemicals, and explore the behavior of a buffer.

Discussion:

A. Acids, Bases and pH

In 1884, Svante Arrhenius described an **acid** as any substance that increases the hydrogen ion concentration when dissolved in water. Today, we realize the hydrogen ion does not exist as such in water; rather, it is intimately involved with water, existing as the hydronium ion. The hydronium ion, H_3O^+ or $H^+_{(aq)}$, forms as the proton from the acid combines with water. The equation describing the ionization of nitric acid, HNO_3, in water shows its behavior as an acid.

$$HNO_{3(aq)} + H_2O_{(l)} \rightarrow H_3O^+_{(aq)} + NO_3^-{}_{(aq)}$$

Arrhenius described a **base** as any substance that increases the hydroxide ion concentration, $OH^-_{(aq)}$, when dissolved in water. Sodium hydroxide is a base. It ionizes to form hydroxide ion when dissolved in water.

$$NaOH_{(s)} \xrightarrow[\text{water}]{\text{in}} Na^+_{(aq)} + OH^-_{(aq)}$$

Both acids and bases can be described as either *strong* or *weak*, terms that reflect the extent to which they produce either hydronium or hydroxide ions, respectively, in solution. A **strong acid** is one that is essentially 100% dissociated into ions. Hydrochloric acid, $HCl_{(aq)}$, and nitric acid, $HNO_{3(aq)}$, are both strong acids, and each will produce a hydronium ion concentration equal to the molar concentration of the acid. Acetic acid, $HC_2H_3O_{2(aq)}$, (also written as $CH_3COOH_{(aq)}$) and carbonic acid, $H_2CO_{3(aq)}$, are **weak acids**, and either will produce a hydronium ion concentration that is much lower than the concentration of the acid. Only a small fraction of the molecules of a weak acid are ionized in solution. Most of the acid is present as molecules. In a 0.1 M acetic acid solution, only 13 of every 1000 acid molecules are ionized.

Sodium hydroxide, NaOH, potassium hydroxide, KOH, and calcium hydroxide, $Ca(OH)_2$, are all **strong bases**. In solution, these compounds dissociate completely into ions. Aqueous ammonia, $NH_{3(aq)}$ (also called ammonium hydroxide, NH_4OH), and the acetate ion, $C_2H_3O_2^-$ (also written as CH_3COO^-), are **weak bases**. In solution they produce a hydroxide ion concentration that is very small compared to the concentration of the base. The way in which the acetate ion produces hydroxide ion is described in Section B.

The **pH** of a solution is a measure of its hydronium ion concentration. The defining equation of pH is:

$$pH = -\log[H^+_{(aq)}]$$ The brackets, [ ], symbolize molar concentration.

In a similar manner, a measure of the concentration of the hydroxide ion can be expressed as the **pOH** of the solution.

$$pOH = -\log[OH^-]$$

In any aqueous solution, the product of the hydronium ion concentration and the hydroxide ion concentration equals a constant, called the **ion product constant**, symbolized K_w.

$$K_w = [H^+_{(aq)}][OH^-] = 1.0 \times 10^{-14} \text{ at } 25°C$$

Because of this relationship between the hydronium and hydroxide ion concentrations in aqueous solutions, the pH and pOH of a solution are also related.

$$pH + pOH = 14$$

If the pH of a solution is 5, the pOH must equal 9. If the pH of a solution increases, the pOH decreases, and vice versa.

If $[H^+_{(aq)}]$ is greater than $[OH^-]$, the solution is considered to be acidic; thus *acidic solutions have a pH less than 7*. On the other hand, if $[OH^-]$ is greater than $[H^+_{(aq)}]$, the solution is basic; thus *basic solutions have a pH greater than 7. A pH of exactly 7 represents a neutral solution*, one in which $[H^+_{(aq)}]$ and $[OH^-]$ equal one another and equal 1.0×10^{-7} M. It is interesting to note that deionized or distilled water, if allowed to stand in contact with air for a time, will have a pH less than 7. The carbon dioxide absorbed by the water from the air will form a small concentration of carbonic acid, making the solution weakly acidic. Boiling the water will eliminate the dissolved carbon dioxide and bring the pH back to 7.

The pH scale is graphically represented below.

$[H^+_{(aq)}]$ →	10^{-1}	10^{-2}	10^{-3}	10^{-4}	10^{-5}	10^{-6}	10^{-7}	10^{-8}	10^{-9}	10^{-10}	10^{-11}	10^{-12}	10^{-13}	10^{-14}
pH →	**1**	**2**	**3**	**4**	**5**	**6**	**7**	**8**	**9**	**10**	**11**	**12**	**13**	**14**

← increasingly acidic ↑ neutral increasingly basic →

B. Hydrolysis

Any chemical reaction in which water is a reactant is termed **hydrolysis**. Some ions react with water to a small degree to produce solutions that are either acidic or basic. The ammonium ion, NH_4^+, is weakly acidic and will increase $[H^+_{(aq)}]$ in solution. In the equation below,

$H_3O^+_{(aq)}$ symbolizes the hydronium ion. In a 0.1 M NH_4^+ solution, only 75 out of every 1,000,000 ions are ionized. This produces a slightly *acidic* solution since $[H_3O^+]$ increases.[1]

$$NH_4^+{}_{(aq)} + H_2O_{(l)} \rightleftarrows H_3O^+{}_{(aq)} + NH_{3(aq)}$$

The acetate ion, $C_2H_3O_2^-$, has a great attraction for a proton. A few of these ions, at any instant, will remove protons from water, forming the molecular acid and increasing $[OH^-]$.

$$C_2H_3O_2^-{}_{(aq)} + H_2O_{(l)} \rightleftarrows HC_2H_3O_{2(aq)} + OH^-{}_{(aq)}$$

The anions of *all* weak acids undergo hydrolysis to produce solutions that are slightly basic.

When ammonium chloride is dissolved in water, only the ammonium ion undergoes hydrolysis and the solution becomes slightly acidic. The chloride ion, the anion of a strong acid, does not undergo hydrolysis.

$$NH_4^+{}_{(aq)} + H_2O_{(l)} \rightleftarrows H_3O^+{}_{(aq)} + NH_{3(aq)}$$

$$Cl^-{}_{(aq)} + H_2O_{(l)} \rightarrow \text{no reaction}$$

When sodium acetate is dissolved in water, only the acetate ion undergoes hydrolysis. The sodium ion does not react with water. The solution is slightly basic because of the hydrolysis of the anion of the weak acid.

$$C_2H_3O_2^-{}_{(aq)} + H_2O_{(l)} \rightleftarrows HC_2H_3O_{2(aq)} + OH^-{}_{(aq)}$$

$$Na^+{}_{(aq)} + H_2O_{(l)} \rightarrow \text{no reaction}$$

You may recognize $NaC_2H_3O_2$ as the *salt* of the strong base, NaOH, and the weak acid, $HC_2H_3O_2$. Similarly, NH_4Cl is formed as the weak base, $NH_{3(aq)}$, and the strong acid, $HCl_{(aq)}$, react. Ammonium acetate, $NH_4C_2H_3O_2$, is formed from ammonia, a weak base, and acetic acid, a weak acid. If we think of these ionic compounds as products of acid-base neutralization reactions, it is possible to place salts in categories based on their effect on pH when dissolved in water:

1. A solution of the salt of a strong base and strong acid will be *neutral*. Examples are NaCl, KBr, $Ca(NO_3)_2$ and $NaClO_4$.
2. A solution of the salt of a strong base and weak acid will be *basic*. Examples are $NaC_2H_3O_2$, KF and K_2CO_3.

---

[1] The double arrow indicates that a state of chemical equilibrium exists: two opposing changes occuring simultaneously at the same rate. Here, the breaking apart of the acid to form ions (→) and the recombination of the ions to form the molecular acid (←) are the two opposing changes. The hydronium ion concentration never gets very high.

3. A solution of the salt of a weak base and strong acid will be *acidic*. Examples are NH_4Cl and NH_4NO_3.
4. A solution of the salt of a weak base and a weak acid may be *neutral, slightly basic* or *slightly acidic* depending on the relative strengths of the acid and base. Examples are NH_4F and $NH_4C_2H_3O_2$.

Depending on the extent of cation and anion involvement in the hydrolysis of salts in category 4, the solution may be acidic, basic or neutral. Ammonium acetate, $NH_4C_2H_3O_2$, produces both ammonium ions and acetate ions in solution, but since both undergo hydrolysis to the same extent (which does not happen too often), the solution does not end up with an excess of either hydroxide or hydronium ion and is neutral, with a pH of 7.

C. Buffers

A **buffer** is a solution designed to resist large changes in pH when small amounts of acid or base are added to it. A buffer contains two solutes. One solute, the base, is able to consume hydronium ion, and the other, the acid, is able to consume hydroxide ion. The beauty of a buffer system is that both acid and base are linked together through an equilibrium that also involves either the hydronium ion or the hydroxide ion.

A buffer to control the pH of mildly acidic solutions can be prepared by combining a weak acid and a salt of that weak acid in solution. Acetic acid is a typical weak acid, and sodium acetate, $NaC_2H_3O_2$, is a typical salt of that acid. The salt provides an ample source of acetate ion to the solution. Acetic acid will neutralize $OH^-_{(aq)}$ added to the solution, and acetate ion will neutralize $H^+_{(aq)}$ added to it. The equilibrium in this buffer solution is that of the weak acid, but bear in mind that the two largest concentrations in solution are the molecular acid and the acetate ion from sodium acetate.

$$HC_2H_3O_{2(aq)} \rightleftarrows H^+_{(aq)} + C_2H_3O_2{}^-_{(aq)}$$

The pH of the buffer solution is determined by the hydronium ion concentration in equilibrium with the weak acid and its anion. If a small amount of acid is added to the buffer solution, it combines with the acetate ion, forming molecular acetic acid. The hydronium ion concentration returns nearly to its original value.

$$HC_2H_3O_{2(aq)} \leftarrow H^+_{(aq)} + C_2H_3O_2{}^-_{(aq)}$$

↖ added to the solution and consumed by the acetate ion

If a small amount of OH^- is added to the buffer, it is consumed by the acid to form water and acetate ion. The hydronium ion concentration, in equilibrium with the acetic acid and acetate ion, is lowered only a small amount.

$$HC_2H_3O_{2(aq)} + OH^-_{(aq)} \rightarrow H_2O_{(l)} + C_2H_3O_2{}^-_{(aq)}$$

↖ added to the solution and consumed by the acetic acid

Because the acetic acid-sodium acetate buffer is composed of a species that can consume hydronium ion and a species that can consume hydroxide ion, it is able, by virtue of the equilibrium between them, to keep the pH of the solution from changing greatly when small amounts of acid or base are added to it.

The Language of Acids and Bases:

The language of acid-base chemistry can be confusing. Some of the more important terms are described here. It is hard to follow the game if you don't know the players.

Acid: An acid is any species that produces hydronium ion, $H^+_{(aq)}$ or H_3O^+, when dissolved in water. There are two classes of acids, strong and weak.

Strong acids: Strong acids are hydrogen-containing compounds that, when dissolved in water, break apart (ionize) to form the hydronium ion and the anion of the acid. Essentially 100% of the molecules ionize in solution, and this is why they are classed as strong acids. Three commonly used strong acids are listed below.

hydrochloric acid, $HCl_{(aq)}$	$HCl_{(aq)} \rightarrow H^+_{(aq)} + Cl^-_{(aq)}$
nitric acid, $HNO_{3(aq)}$	$HNO_{3(aq)} \rightarrow H^+_{(aq)} + NO_3{}^-_{(aq)}$
sulfuric acid, $H_2SO_{4(aq)}$	$H_2SO_{4(aq)} \rightarrow H^+_{(aq)} + HSO_4{}^-_{(aq)}$

Weak acids: Weak acids produce a concentration of hydronium ion in water that is small compared to the concentration of the acid itself. They ionize to a much smaller extent than do strong acids. The ions exist in equilibrium with the molecular acid, and the ionization equations are written with a double arrow, $\rightleftarrows$, symbolizing an equilibrium. Two common weak acids are listed below.

acetic acid, $HC_2H_3O_{2(aq)}$	$HC_2H_3O_{2(aq)} \rightleftarrows H^+_{(aq)} + C_2H_3O_2{}^-_{(aq)}$
carbonic acid, $H_2CO_{3(aq)}$	$H_2CO_{3(aq)} \rightleftarrows H^+_{(aq)} + HCO_3{}^-_{(aq)}$

Base: A base is any species that produces hydroxide ion, $OH^-_{(aq)}$, when dissolved in water. As with acids, there are two classes of bases, strong and weak.

Strong bases: Strong bases are compounds that contain the hydroxide ion, OH^-, and when dissolved in water ionize completely (100%) to form the hydroxide ion and the cation of the compound. Three common strong bases are listed below.

sodium hydroxide, $NaOH_{(aq)}$	$NaOH_{(aq)} \rightarrow Na^+_{(aq)} + OH^-_{(aq)}$
potassium hydroxide, $KOH_{(aq)}$	$KOH_{(aq)} \rightarrow K^+_{(aq)} + OH^-_{(aq)}$
calcium hydroxide, $Ca(OH)_{2(aq)}$	$Ca(OH)_{2(aq)} \rightarrow Ca^{2+}_{(aq)} + 2\,OH^-_{(aq)}$

Weak bases: The most common weak bases are the nitrogen bases, such as aqueous ammonia, $NH_{3(aq)}$, or the anions of weak acids, such as the acetate ion, $C_2H_3O_2^-{}_{(aq)}$, or hydrogencarbonate ion, $HCO_3^-{}_{(aq)}$ (commonly called the bicarbonate ion). Weak bases produce a hydroxide ion concentration that is substantially less than the concentration of the base itself. Their interaction with water that produces the hydroxide ion is called **hydrolysis** (reacting with water), and the hydrolysis of weak bases is written as an equilibrium, ($\rightleftarrows$). Three weak bases are show below.

aqueous ammonia, $NH_{3(aq)}$	$NH_{3(aq)} + H_2O_{(l)} \rightleftarrows NH_4^+{}_{(aq)} + OH^-{}_{(aq)}$
acetate ion, $C_2H_3O_2^-{}_{(aq)}$	$C_2H_3O_2^-{}_{(aq)} + H_2O_{(l)} \rightleftarrows HC_2H_3O_{2(aq)} + OH^-{}_{(aq)}$
hydrogencarbonate ion, $HCO_3^-{}_{(aq)}$ (also called the bicarbonate ion)	$HCO_3^-{}_{(aq)} + H_2O_{(l)} \rightleftarrows H_2CO_{3(aq)} + OH^-{}_{(aq)}$

Salt: A salt is the ionic product of an acid-base neutralization reaction. The other product is water. The cation (the positive ion) of the salt comes from the base, and the anion (the negative ion) comes from the acid.

Salt	Parentage of the Salt
sodium acetate, $NaC_2H_3O_2$	acid: acetic acid, $HC_2H_3O_2$ base: sodium hydroxide, *Na*OH
calcium chloride, $CaCl_2$	acid: hydrochloric acid, H*Cl* base: calcium hydroxide, *Ca*(OH)$_2$
ammonium chloride, NH_4Cl	acid: hydrochloric acid, H*Cl* base: aqueous ammonia, $NH_{3(aq)}$; more easily seen if written as ammonium hydroxide, $NH_4OH_{(aq)}$

Buffer: A buffer is a solution designed to minimize changes in pH when small amounts of either acid or base are added to it. Buffers can be prepared in two ways: (1) combining a weak acid and a salt of that acid, for example, $HC_2H_3O_{2(aq)}$ and $C_2H_3O_2^-{}_{(aq)}$ (from the salt $NaC_2H_3O_2$), or (2) combining a weak base and a salt of that weak base, for example, $NH_{3(aq)}$ and $NH_4^+{}_{(aq)}$ (from the salt NH_4Cl).

In a buffer, the acidic species (which can consume added base) and the basic species (which can consume added acid) exist in equilibrium with each other and a third species that is responsible for the pH of the solution, either $H^+_{(aq)}$ or $OH^-_{(aq)}$. Equations showing the equilibria that exist in two buffers are shown below. The first is the acetic acid-sodium acetate buffer, used to control the pH of solutions in the mildly acidic range of about 4 to 6. The second is the ammonia-ammonium chloride buffer, used to control the pH of solutions in the mildly basic range of about 8 to 10. In both cases, the equations look the same as those used to describe the weak acid (acetic acid) or base (aqueous ammonia) in water, but in the acetic acid buffer there is added acetate ion (to consume acids that may enter the solution) and in the ammonia buffer, there is added ammonium ion (to consume bases that may enter the solution).

The equilibrium operating in the acetic acid-sodium acetate buffer:

$$HC_2H_3O_{2(aq)} \rightleftarrows H^+_{(aq)} + C_2H_3O_2^-{}_{(aq)}$$

The equilibrium operating in the ammonia-ammonium chloride buffer:

$$NH_{3(aq)} + H_2O_{(l)} \rightleftarrows OH^-_{(aq)} + NH_4^+{}_{(aq)}$$

Experimental Procedure:

There are two common ways to measure the pH of a solution. The easiest method is with pH paper, a paper saturated with dyes that change color in response to the pH of a solution. A more accurate (and more expensive) method is to use a pH meter, an electronic device with a special electrode that generates a small voltage proportional to the hydronium ion concentration of the solution in which it is placed. The electrical signal is amplified, converted to pH, and displayed. Though pH paper is less sensitive than the pH meter, it is satisfactory in many applications.

If you use a pH meter, follow your instructor's directions carefully. You may need to standardize the meter before use. The electrode is very fragile, so handle it with care. Before you attempt to measure the pH of a solution, the electrode should be rinsed with deionized water and gently wiped dry with a tissue. Doing this will prevent contamination of the solution you wish to measure. Never remove the electrode from one solution and place it directly into another; clean it first. Also, thoroughly rinse the electrode when you are finished with the instrument. When not in use, the electrode is commonly kept immersed in a buffer solution so the glass tip stays moist.

If you use pH paper, tear the strips of paper into pieces about 1 cm long and lay them on a *clean* glass plate. Then, to test the pH of a solution, obtain a droplet of the solution on the end of a clean stirring rod and transfer it to a piece of pH paper. Compare the color produced with the color chart on the pH paper vial. Always rinse the stirring rod with deionized water before testing the next sample.

CAUTION: Be certain to wear eye protection at all times.

A. Measuring the pH of Common Acids and Bases

1. Place about 25 mL of the following solutions in separate, clean, dry beakers. (These solutions may already be in the proper containers for your use.) Measure the pH of each solution with a pH meter. Rinse the electrode before moving it from one solution to the next. Record the pH of each solution on the Report Sheet.

   a. 0.1 M HCl b. 0.1 M H_3BO_3 c. 0.1 M $NH_{3(aq)}$ d. 0.1 M NaOH

2. Measure the pH of these same solutions using pH paper. Record the pH values on the Report Sheet.

3. Using your measured pH values, consult the pH scale on page 206 to determine the approximate hydronium ion molarity of each solution you measured. Your instructor will tell you which set of pH values (pH meter or pH paper) to use. A pH of 4.0 would indicate a hydronium ion concentration of 1×10^{-4} M, a pH of 5.0, a hydronium ion concentration of 1×10^{-5} M. A pH of 4.6 would indicate a hydronium ion concentration between 1×10^{-4} and 1×10^{-5} M.

B. Measuring the pH of Salt Solutions

1. Place about 25 mL of the following solutions in separate, clean, dry beakers. (These solutions may already be in the proper containers for your use.) Measure the pH of each using the pH meter. Record the pH of each solution on the Report Sheet.

   a. 0.1 M Na_2CO_3 b. 0.1 M $(NH_4)_2SO_4$ c. 0.1 M NaCl

C. Measuring the pH of Some Common Household Chemicals

Do not use a pH meter for these tests.

Using pH paper, determine the pH of the household chemicals assigned by your instructor. The instructor may have solutions of several household chemicals already prepared for you. If they are not, you will need to prepare your own solutions. If the chemical is a solid, dissolve about 0.1 g (an amount roughly equal to the volume of the head of a paper match) in about 2 mL of deionized water. If the chemical is a liquid, place about 10 drops in a clean, dry beaker. With a clean stirring rod, obtain a droplet of each solution and touch it to a piece of pH paper. Record the identity of the material being tested and its pH on the Report Sheet.

D. Buffer Action

Use a pH meter for these tests.

1. Pipet[2] 25.00 mL of 0.1 M sodium acetate, $NaC_2H_3O_2$, into a clean 100 mL beaker. Pipet 25.00 mL of 0.1 M acetic acid, $HC_2H_3O_2$, into the same beaker and mix the solutions. You have just prepared a buffer solution. Measure the pH of this buffer and record the value on the Report Sheet. Add 5 drops of 1.0 M HCl to the buffer *one drop at a time*. Swirl the solution after each addition. Measure the pH after each drop is added, and record the pH on the Report Sheet. You may have to wait several seconds for the displayed pH value to stabilize.

---

[2] Use a pipet bulb to draw the solution up into the pipet, never use mouth suction. Allow the pipet to drain by gravity alone. A small amount of liquid will remain in the tip. This is normal.

Before proceeding to step 2, add 5 drops of 1.0 M NaOH to the buffer solution to compensate for the 5 drops of 1.0 M HCl added before. Stir the solution well and measure its pH. It should be the same, or nearly so, as that of the original buffer solution. If the pH is not nearly the same, check with your instructor.

2. Add 5 drops of 1.0 M NaOH to the buffer solution, *one drop at a time*. Swirl the solution after each addition, and record the pH on the Report Sheet after each drop is added. Discard the buffer solution when finished. Your instructor will tell you how to dispose of the solution.

3. Place about 25 mL of *freshly boiled* deionized water into a clean 100-mL beaker. Measure the pH of the water and record the value on the Report Sheet. Add 5 drops of 1.0 M HCl *one drop at a time*. Swirl the solution after each addition, and record the pH on the Report Sheet after each drop is added. Discard the solution when finished. Your instructor will tell you how to dispose of the solution.

4. Place about 25 mL of *freshly boiled* deionized water into a clean 100-mL beaker. Measure the pH of the water and record the value on the Report Sheet. Add 5 drops of 1.0 M NaOH *one drop at a time*. Swirl the solution after each addition, and record the pH on the Report Sheet after each drop is added. Discard the solution when finished. Your instructor will tell you how to dispose of the solution.

Name ______________________________ Locker Number__________ Date____________
Please print; last name first

PRELIMINARY EXERCISES: *Experiment 15*

Acids, Bases, pH, Hydrolysis and Buffers

1. What property of a solution is described by its pH?

2. A solution has a pH of 8.7.

   a. What is the pOH of this solution? ______________

   b. Is the solution acidic, basic or neutral? ______________

3. Solution A has a hydronium ion concentration of 3.8×10^{-8} M. Solution B has a hydronium ion concentration of 2.5×10^{-4} M.

   a. Which solution has the larger hydronium ion concentration? ______________

   b. Which solution has the larger pH? ______________

4. Estimate the pH of an aqueous solution of each of the following compounds by indicating whether the pH is greater than 7 (>7), less than 7 (<7), or about 7 (~7).

   a. H_3PO_4 __________ f. $(NH_4)_2CO_3$ __________

   b. $Mg(OH)_2$ __________ g. CO_2 __________

   c. $NaNO_3$ __________ h. $KC_2H_3O_2$ __________

   d. HNO_3 __________ i. $HC_2H_3O_2$ __________

   e. $NH_{3(aq)}$ __________ j. KCl __________

5. Which solution has the *lower* pH?

   a. 0.1 M HCl or 0.1 M $HC_2H_3O_2$ ______________

   b. 0.1 M NaOH or 0.1 M $NH_{3(aq)}$ ______________

   c. 0.1 M Na_2CO_3 or 0.1 M NaCl ______________

Name ______________________ Locker Number__________ Date__________
Please print; last name first

REPORT SHEET: ***Experiment 15***

Acids, Bases, pH, Hydrolysis and Buffers

A. Measuring the pH of Common Acids and Bases

	Solution	pH (by meter)	pH (by paper)	$[H^+_{(aq)}]$
a.	0.1 M HCl	______	______	______
b.	0.1 M H_3BO_3	______	______	______
c.	0.1 M $NH_{3(aq)}$	______	______	______
d.	0.1 M NaOH	______	______	______

B. Measuring the pH of Salt Solutions

	Substance	pH (by meter)
a.	0.1 M Na_2CO_3	______
b.	0.1 M $(NH_4)_2SO_4$	______
c.	0.1 M NaCl	______

C. Measuring the pH of Some Common Household Chemicals

	Substance	pH (by paper)
a.	______________________	______
b.	______________________	______
c.	______________________	______
d.	______________________	______
e.	______________________	______

D. Buffer Action

1. Effect of HCl on the pH of a Buffer:

pH of buffer	______________
pH after 1 drop of HCl	______________
after 2 drops	______________
after 3 drops	______________
after 4 drops	______________
after 5 drops	______________

2. Effect of NaOH on the pH of a Buffer:

pH of buffer	______________
pH after 1 drop of NaOH	______________
after 2 drops	______________
after 3 drops	______________
after 4 drops	______________
after 5 drops	______________

3. Effect of HCl on the pH of Deionized Water:

pH of freshly boiled deionized H_2O	______________
pH after 1 drop HCl	______________
after 2 drops	______________
after 3 drops	______________
after 4 drops	______________
after 5 drops	______________

Name ______________________________ Locker Number__________ Date____________
Please print; last name first

REPORT SHEET (continued): *Experiment 15*

Acids, Bases, pH, Hydrolysis and Buffers

4. Effect of NaOH on the pH of Deionized Water:

pH of freshly boiled deionized H_2O	__________
pH after 1 drop of NaOH	__________
after 2 drops	__________
after 3 drops	__________
after 4 drops	__________
after 5 drops	__________

E. General Questions

1. Why is the pH of the 0.1 M acetic acid-sodium acetate buffer solution different from that of 0.1 M $HC_2H_3O_2$? Be specific please.

2. Why is the pH of deionized water that has not been boiled not exactly 7.00?

3. Show with the appropriate equation(s) how the acetic acid-sodium acetate buffer prevents each of the following:

   a. A large decrease in pH when HCl is added

   b. A large increase in pH when NaOH is added

SECTION VIII: Organic and Biochemistry

Experiment 16

The Structure of Hydrocarbons

Purpose:

In this exercise you will learn about the three-dimensional shapes of four classes of hydrocarbons: alkanes, alkenes, alkynes and an aromatic hydrocarbon, benzene. You will draw Lewis structures and construct ball-and-stick models of several compounds and, for some, you will be asked to draw their three-dimensional spatial formulas.

Discussion:

The chemical and physical properties of a covalent compound are a function of the composition of that compound (the atoms that compose it), the way the atoms are joined together (single, double or triple bonds) and the three-dimensional shape of the molecule. In the last 30 years of chemistry a great deal of work has gone into the discovery of the importance of molecular shapes. In Experiment 10, The Structure of Covalent Molecules and Polyatomic Ions, you learned the factors that determine the arrangement of atoms in molecules and the most common arrangements adopted by groups of atoms in covalent species. You might want to go back and review Experiment 10, especially the tables showing the common shapes of molecules. In this experiment you will be studying a special class of organic compounds, the **hydrocarbons**, compounds composed only of hydrogen and carbon. There are four classes of hydrocarbons: **alkanes** are hydrocarbons that contain only carbon-carbon single bonds; **alkenes** are hydrocarbons that contain at least one carbon-carbon double bond; **alkynes** are hydrocarbons that contain at least one carbon-carbon triple bond; and **aromatic hydrocarbons** are those containing at least one aromatic group, most often the benzene ring.

A. Bonding Patterns of Carbon

Though there are thousands upon thousands of hydrocarbons, their three-dimensional shapes are built from only four fundamental carbon substructures that are connected together in different sequences. One of the first things you learn about the element carbon is that it always has four bonds in its compounds. This set of four bonds can be made up of four single bonds, two single bonds and one double bond, etc., as shown in Table 1.

Table 1. The Four Bonding Patterns of Carbon

$-\overset{|}{\underset{|}{C}}-$	$>C=$	$=C=$	$-C\equiv$
4 single bonds	2 single bonds 1 double bond	2 double bonds	1 single bond 1 triple bond

Every carbon atom in a hydrocarbon, or in any organic compound, will have one of these four bonding patterns. Each can be viewed as a carbon substructure in the total structure of a compound. For example, each of the five carbon atoms in pentane, $CH_3CH_2CH_2CH_2CH_3$, a five-carbon alkane, are of the first type, having four single bonds.

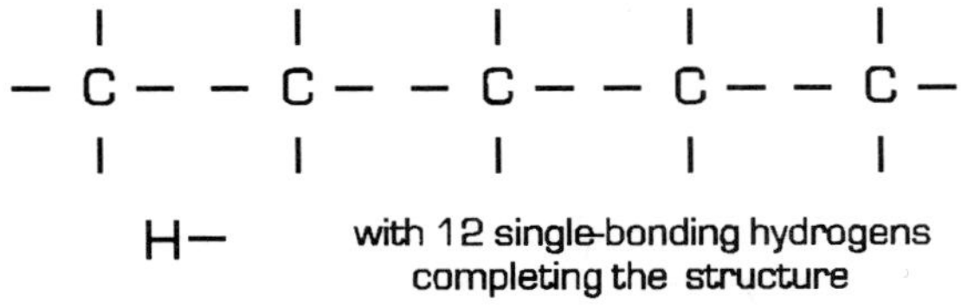

```
   |       |       |       |       |
 — C — — — C — — — C — — — C — — — C —
   |       |       |       |       |

    H—        with 12 single-bonding hydrogens
                 completing the structure
```

An alkene, such as 1-pentene, $H_2C=CHCH_2CH_2CH_2$, has two carbon atoms of the second type (each with 1 double bond and 2 single bonds) and three of the first type (each with four single bonds). Carbon atoms with the bonding patterns shown in Table 1 can be connected together like puzzle pieces and filled in with hydrogen atoms to form the Lewis structures of any hydrocarbon. To create Lewis structures in this way, certain rules must be followed:

1. Single bonds can only connect with single bonds.
2. Double bonds can only connect with double bonds.
3. Triple bonds can only connect with triple bonds.
4. There can be no "unconnected" bonds in the final Lewis structure.

The Lewis structure for ethane, a two-carbon alkane, H_3C-CH_3, is assembled below using six hydrogen atoms and two carbon atoms that each have four single bonds. This is the only bonding pattern for carbon that would allow a Lewis structure with the formula of ethane, C_2H_6, to be assembled. Remember, hydrogen can only form a single bond.

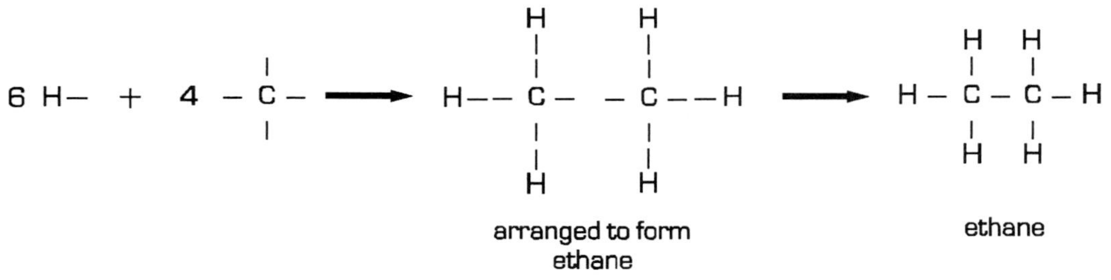

```
                         H         H
                         |         |                    H   H
               |         |         |                    |   |
6 H—  +  4  — C —  ——>  H—— C —   — C ——H    ——>    H — C — C — H
               |         |         |                    |   |
                         |         |                    H   H
                         H         H

                     arranged to form                 ethane
                          ethane
```

The Lewis structure for ethene, the alkene with two carbon atoms, $H_2C=CH_2$, is assembled using four hydrogen atoms and two carbon atoms, each with one double and two single bonds. Note how the double bonds are matched up, with hydrogen filling in the single bonds of the two carbon atoms.

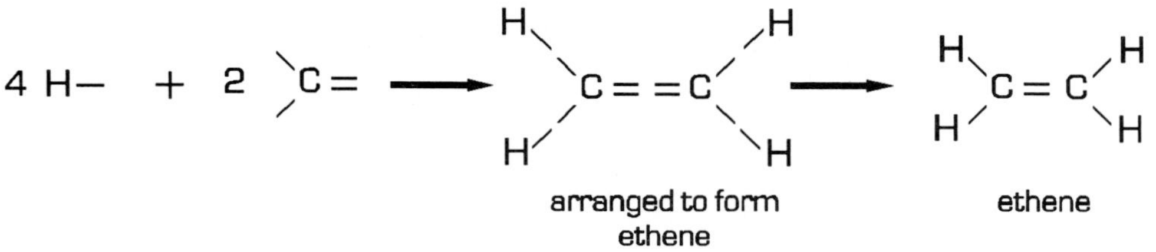

```
                       H           H
                        \         /             H       H
4 H—   +  2  >C=  ——>    C = = C       ——>       \C = C/
                        /         \             H/      \H
                       H           H

                     arranged to form            ethene
                          ethene
```

The Lewis structure for ethyne, an alkyne, HC≡CH, could be formed from two carbon atoms of the fourth type, the ones with one single bond and one triple bond, −C≡. Of course, two hydrogen atoms would be needed too: H− −C≡ ≡C− −H → H−C≡C−H.

B. Three Dimensional Structures of Hydrocarbons

Though Lewis structures show how the atoms in a molecule are joined together, they do not show the three-dimensional shape of the molecule. To develop the three-dimensional structure, you need to know how the bonds about each carbon atom are oriented in space. The bond arrangements for each bond pattern used by carbon and the resulting bond angles[1] are shown in Table 2.

Table 2. Bond Arrangements and Bond Angles for Carbon Bonding Patterns

Bonding Pattern of Carbon	Arrangement of Bonds around Carbon	Spatial Structure and Bond Angles
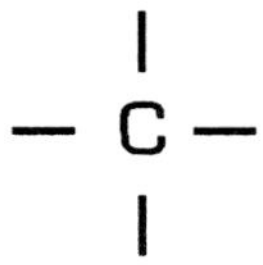 −C− (four single bonds)	tetrahedral	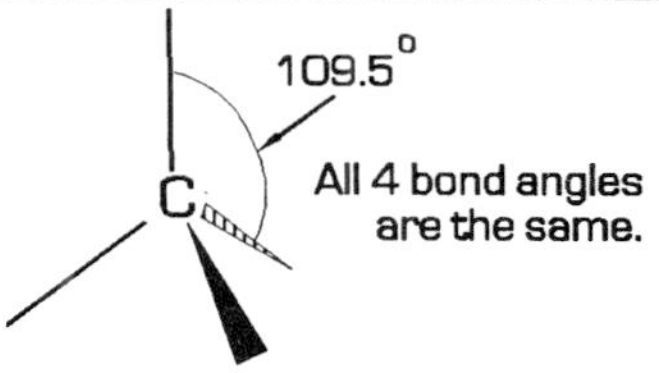 109.5° All 4 bond angles are the same.
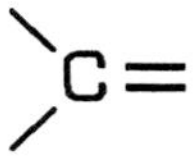 >C=	trigonal planar	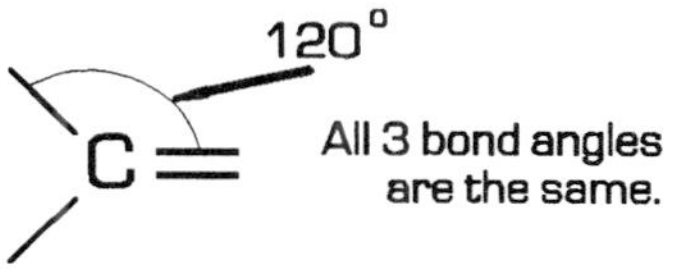 120° All 3 bond angles are the same.
=C=	linear	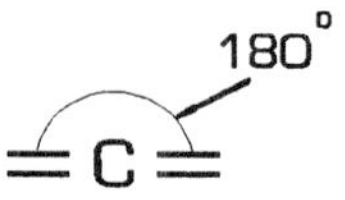 180°
−C≡	linear	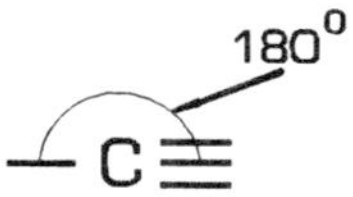 180°

---

1 The bond angle is the number of degrees between two bonds that have the same atom in common, that is, two bonds joined to the same atom. The angle between two hydrogen atoms in methane, CH_4, is 109.5°. This can be symbolized as ∠H-C-H = 109.5°. In the tetrahedral methane structure, all ∠H-C-H's are the same.

Though the Lewis structure of pentane, $CH_3CH_2CH_2CH_2CH_3$, shows how the atoms are connected, it looks flat when drawn on a sheet of paper. If the tetrahedral arrangement of bonds about each carbon is added to the picture, you create the actual three-dimensional shape of the molecule, showing the zigzag carbon atom chain. The three-dimensional structure of pentane is shown below in two versions: as one would draw it with wedges and as a ball-and-stick model. The ball-and-stick model emphasizes the tetrahedral arrangement about each carbon atom. In the drawing, bonds drawn in the plane of the paper are shown as *solid lines*, bonds coming out of the paper are *black wedges*, and bonds pointing behind the plane of the paper are *striped wedges*.

The structure of ethene, $H_2C{=}CH_2$, is planar; that is, all atoms are in the plane of the page. Both carbon atoms have the trigonal planar arrangement of bonds. The ball-and-stick model of ethene shown below clearly illustrates its planar structure. Note that the double bond is formed using connecting pieces that extend above and below the plane of the molecule. This correctly shows the location of much of the electron density in a carbon-carbon double bond.

Benzene, C_6H_6, is the most important member of the aromatic hydrocarbon family. Its carbon atoms form a planar, hexagonal ring. In its Lewis structure, the bonds joining the carbon atoms are drawn as alternating double and single bonds. Every carbon is identical, with one double bond and two single bonds, and they are arranged to form the alternating bond structure shown below. The alternating double and single bonds impart a unique stability to the benzene structure. The term *aromatic* refers to this uniquely stable bonding arrangement, not the aroma of the compound.

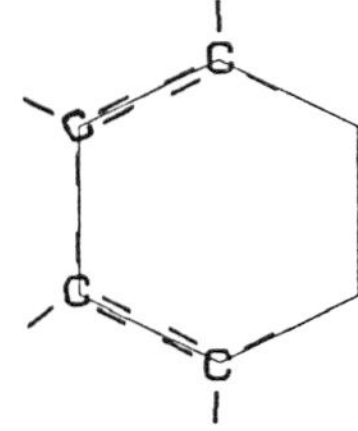

arranging the carbon atoms to form the required bonds

completed Lewis structure of benzene

The location of the alternating single and double bonds between carbon atoms is not critical, since in benzene the bonds can be regarded as exchanging positions, a property called resonance that is responsible for the greater-than-expected stability of benzene.

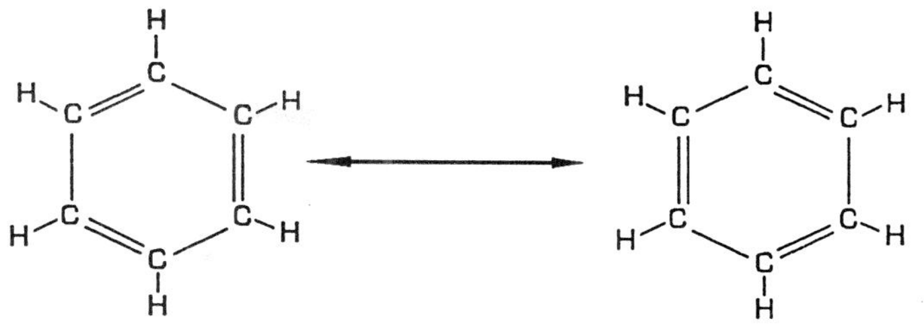

C. Structural and Geometric Isomers

An important feature of organic compounds that is displayed beautifully in hydrocarbons is **isomerism**, the existence of two or more distinctly different compounds that have the same molecular formula. Two types of isomerism are important here: structural isomerism and geometric isomerism.

Structural isomers are compounds that have the same formula but different structures, that is, with different arrangements of the same atoms. They are different compounds, each with its own set of properties; the only thing they have in common is their formulas. There are two structural isomers with the formula C_4H_{10}, and they constitute the butane family. The common name for the straight-chain isomer is butane, and for the branch-chain isomer, isobutane. Lewis structures and ball-and-stick models of these structural isomers are shown below. The systematic name (the name assigned by the International Union of Pure and Applied Chemistry (IUPAC)), for isobutane is 2-methylpropane. Each isomer has a different boiling temperature, a physical property showing they are different compounds.

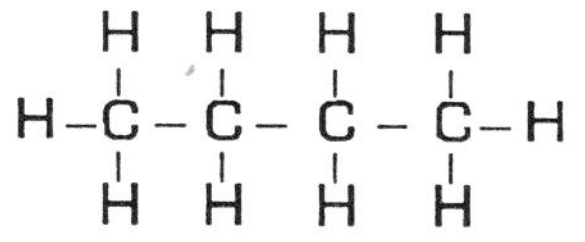

butane b.p. = 0° C

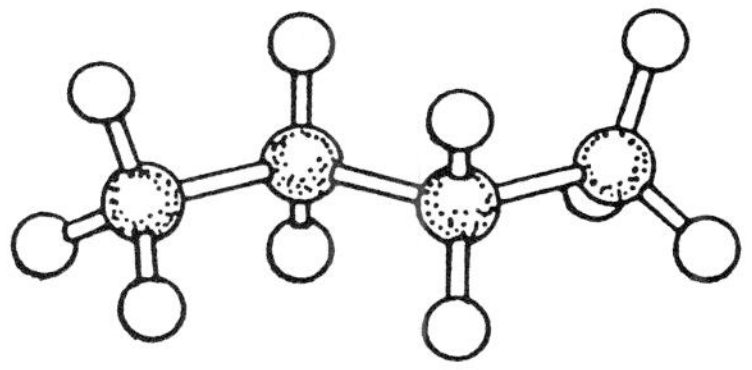

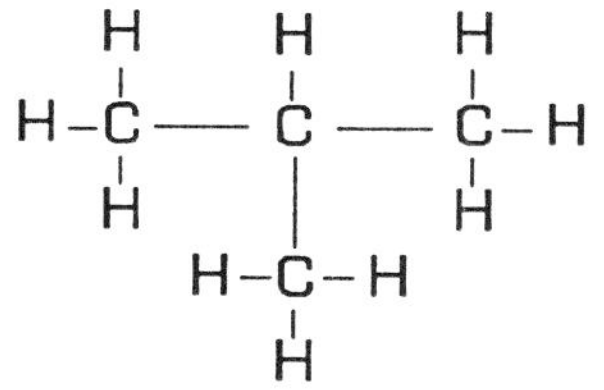

isobutane b.p. = -12° C
(2-methyl propane)

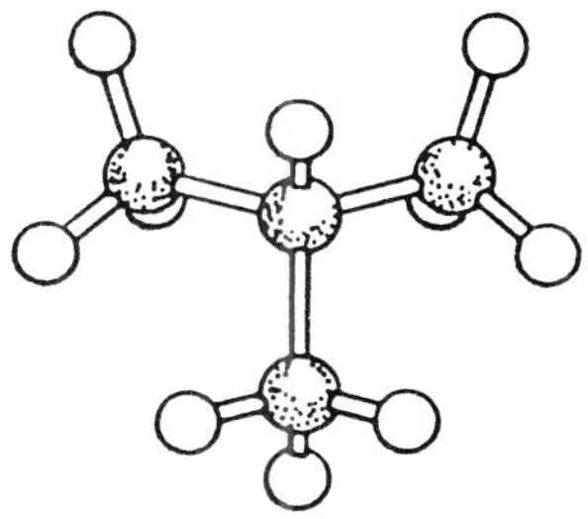

There are three structural isomers in the pentane family, the five-carbon alkanes. Each isomer has the molecular formula C_5H_{12}, though the structures of pentane, isopentane and neopentane (these are their common names) are not the same. The systematic name for isopentane is 2-methylbutane, and for neopentane, 2,2-dimethylpropane.

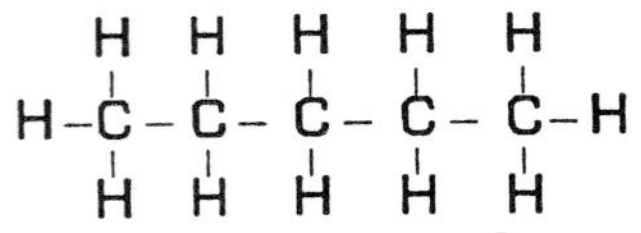

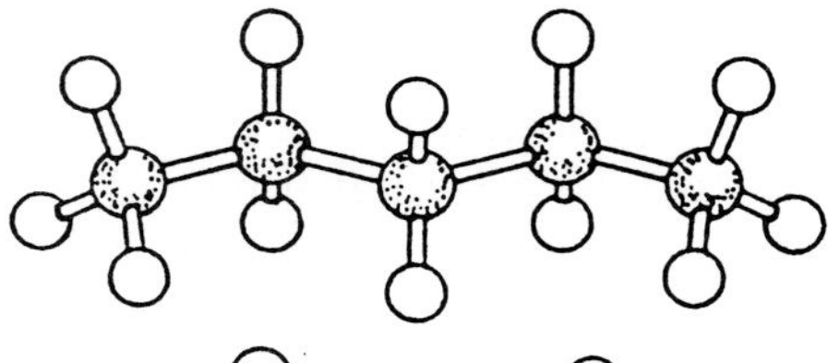

pentane b.p. = 36° C

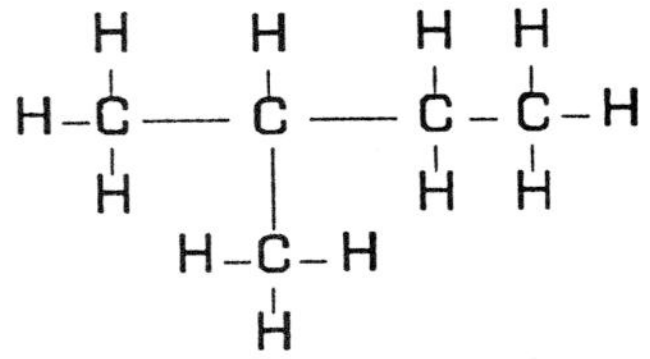

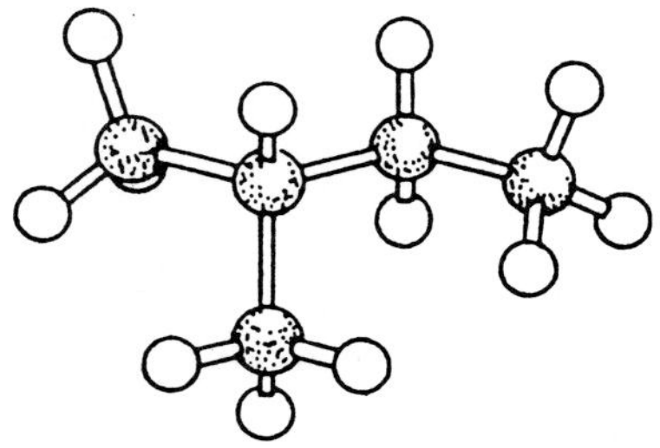

isopentane b.p. = 28° C
(2-methylbutane)

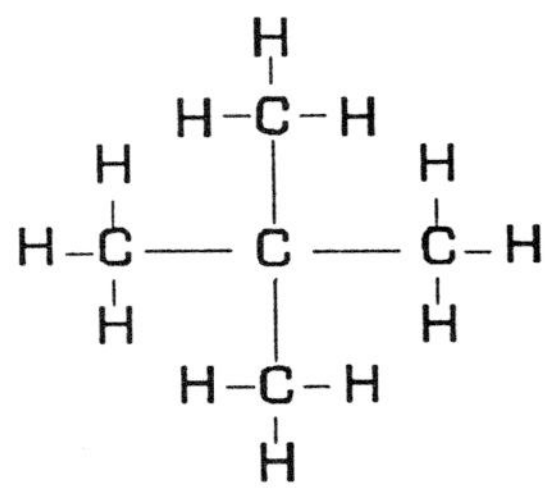

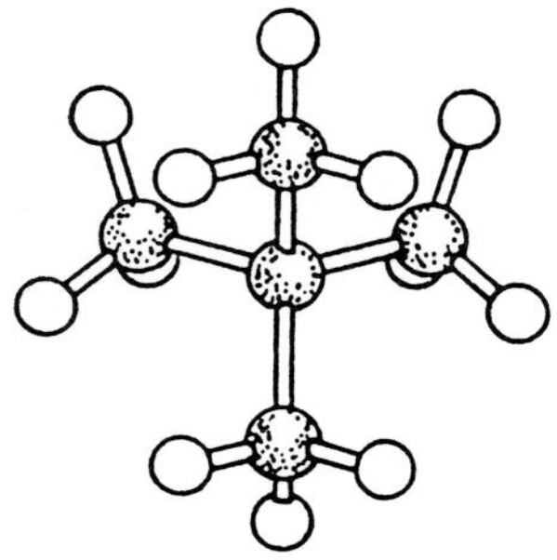

neopentane b.p. = 10° C
(2,2-dimethylpropane)

Alkenes display **geometric isomerism**, specifically ***cis-trans* isomerism**. *Cis-trans* isomerism exists because the double bond between two carbon atoms "locks" that part of the molecule around the double bond into a definite geometric arrangement. Unlike single bonds between carbon atoms, which act like swivel joints allowing free rotation, double bonds restrict free rotation. The simplest alkene that displays *cis-trans* isomerism is 2-butene, and its isomers are shown below. The structure on the left with the methyl groups on the same side of the molecule is called the *cis*-isomer. The one on the right with the methyl groups on opposite sides is the *trans*-isomer.

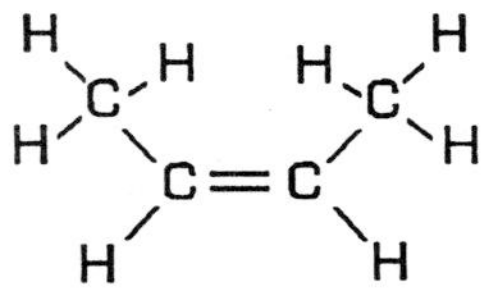

cis - 2 - butene

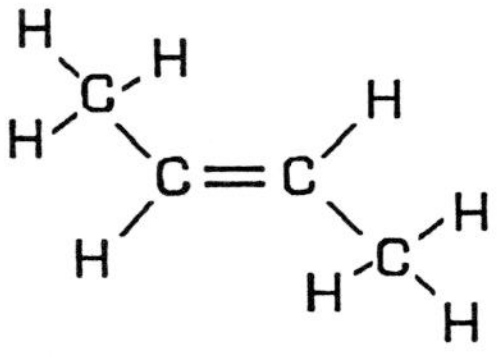

trans - 2 - butene

Not all alkenes display *cis-trans* isomerism. Only those alkenes that have two nonidentical atoms or groups on each of the double-bonded carbon atoms can exist as *cis-trans* isomers. For this reason, propene does *not* show geometric isomerism. The two atoms on the end carbon are identical, $CH_3-HC=CH_2$.

Alkynes do not display *cis-trans* isomerism. The linear arrangement of atoms or groups on either side of the triple bond eliminates the possibility.

Experimental Procedure:

> **INSTRUCTIONAL NOTE: It is convenient if students work in pairs or small groups, though each student should complete a Report Sheet.**

In the following exercises you are to draw Lewis structures and construct ball-and-stick models of several hydrocarbons. In addition, for selected compounds, you are to draw the three-dimensional structures (the spatial formulas).

A. Structures of Alkanes

1. Using the structural patterns for carbon shown in Table 1, draw the Lewis structures for each of the following alkanes in the appropriate boxes on the Report Sheet. Remember, in alkanes *all* carbon atoms have four single bonds.

methane	CH_4
ethane	CH_3CH_3
propane	$CH_3CH_2CH_3$
butane	$CH_3CH_2CH_2CH_3$, and its structural isomer
2-methypropane	CH_3CHCH_3 (also called isobutane), with CH_3 bonded to the central C

2. Obtain a ball-and-stick model kit from the stock in the lab and construct the ball-and-stick model for each of these alkanes, or examine the prepared models provided in the lab. On the Report Sheet record the expected bond angle for the angle requested.

3. To the best of your artistic ability, draw the spatial formula for propane. Remember, bonds in the plane of the page are drawn as straight lines, those going into the page as stripped wedges and those coming out of the page as solid wedges. The bond angles shown in the ball-and-stick model should be reflected in your drawing.

B. Structures of Alkenes

1. Using the structural patterns for carbon shown in Table 1, draw the Lewis structures for each of the following alkenes in the appropriate boxes on the Report Sheet. In the following alkenes, the carbon atoms involved in the double bond have one double and two single bonds all in the same plane. Carbon atoms not involved in multiple bonding have four single bonds like those in alkanes.

ethene	$H_2C{=}CH_2$
propene	$H_2C{=}CHCH_3$
1-butene	$H_2C{=}CHCH_2CH_3$
cis-2-butene	
and	$CH_3CH{=}CHCH_3$, geometric isomers
trans-2-butene	

2. Construct the ball-and-stick model for each of these alkenes, or examine the prepared models provided in the lab. On the Report Sheet record the expected bond angle for the angle requested.

3. To the best of your artistic ability, draw the three-dimensional structures for the two geometric isomers of 2-butene: *cis*-2-butene and *trans*-2-butene. As before, bonds in the plane of the page are drawn as straight lines, those going into the page as striped wedges and those coming out of the page as solid wedges. The bond angles shown in the ball-and-stick model should be reflected in your drawing.

C. Structures of Alkynes

1. Using the structural patterns for carbon shown in Table 1, draw the Lewis structures for the four alkynes listed below in the appropriate box on the Report Sheet. In the alkynes, carbon atoms involved in the triple bond each have one triple bond and one single bond in a linear arrangement. Carbon atoms not involved in multiple bonding have four single bonds like those in alkanes.

ethyne	$HC{\equiv}CH$
propyne	$HC{\equiv}CCH_3$
1-butyne	$HC{\equiv}CCH_2CH_3$
2-butyne	$H_3CC{\equiv}CCH_3$

2. Construct the ball-and-stick model for each of these alkynes, or examine the prepared models provided in the lab. On the Report Sheet record the expected bond angle for the angle requested.

3. To the best of your artistic ability, draw the three-dimensional structure of 1-butyne. As before, bonds in the plane of the page are drawn as straight lines, those going into the page as striped wedges and those coming out of the page as solid wedges. The bond angles shown in the ball-and-stick model should be reflected in your drawing.

D. The Structure of Benzene

1. Draw the Lewis structure of benzene on the Report Sheet. Each carbon in the ring will have the flat arrangement of one double and two single bonds.

2. Construct the ball-and-stick model of benzene, or examine the prepared model provided in the lab.

3. Along side the Lewis structure, draw the three-dimensional shape of benzene indicating the values of the requested bond angles.

E. Additional Structures

Your instructor may assign additional hydrocarbons to examine.

Name ______________________________ Locker Number__________ Date____________
Please print; last name first

PRELIMINARY EXERCISES: *Experiment 16*

The Structure of Hydrocarbons

1. Draw the four carbon bonding patterns commonly seen in hydrocarbons and all other organic compounds.

2. Construct the Lewis structures of the following compounds by combining the carbon atoms with the appropriate bonding patterns and the required number of hydrogen atoms.

a. hexane: $CH_3CH_2CH_2CH_2CH_2CH_3$
b. 3-hexene: $CH_3CH_2CH{=}CHCH_2CH_3$
c. 2-hexyne: $CH_3C{\equiv}CCH_2CH_2CH_3$

3. There are three distinct bond angles formed in the four bonding patterns of carbon. What are they and in which bonding pattern(s) is each seen?

(The Preliminary Exercises continue on the following page.)

4. Define the following terms:

a. isomers
b. structural isomers
c. geometric isomers

Name ______________________ Locker Number __________ Date __________
Please print; last name first

REPORT SHEET: *Experiment 16*

The Structure of Hydrocarbons

A. Structures of Alkanes

Alkane	Lewis Structure	Bond Angles
methane CH_4		∡H-C-H = ________
ethane CH_3CH_3		∡H-C-C = ________
propane $CH_3CH_2CH_3$		∡C-C-C = ________
butane $CH_3CH_2CH_2CH_3$		any ∡C-C-C = ________
2-methylpropane CH_3CHCH_3 | CH_3		any ∡H-C-C = ________

Draw the spatial (three-dimensional) formula for propane in the box to the right.	

B. Structures of Alkenes

Alkene	Lewis Structure	Bond Angles
ethene $H_2C=CH_2$		any ∡H-C=C = ________
propene $H_2C=CHCH_3$		any ∡H-C=C = ________
1-butene $H_2C=CHCH_2CH_3$		∡C=C-C = ________ ∡H-C-H = ________
cis-2-butene $CH_3CH=CHCH_3$	(Show the *cis* geometry.)	∡H-C=C = ________ ∡C=C-C = ________
trans-2-butene $CH_3CH=CHCH_3$	(Show the *trans* geometry.)	∡H-C=C = ________ ∡H-C-H = ________

Draw the spatial structures of *cis*-2-butene and *trans*-2-butene below.

cis-2-butene	*trans*-2-butene

Name ______________________ Locker Number__________ Date____________
Please print; last name first

REPORT SHEET (continued): *Experiment 16*

The Structure of Hydrocarbons

C. Structure of Alkynes

Alkyne	Lewis Structure	Bond Angles
ethyne $HC{\equiv}CH$		∡H-C≡C = ________
propyne $HC{\equiv}CCH_3$		∡C≡C-C = ________
1-butyne $HC{\equiv}CCH_2CH_3$		∡H-C≡C = ________ ∡H-C-H = ________
2-butyne $H_3CC{\equiv}CCH_3$		∡C-C-C = ________ ∡C-C≡C = ________

Draw the spatial formula of 1-butyne in the box to the right.	

D. The Structure of Benzene

Lewis Structure of Benzene	Spatial Formula of Benzene	Bond Angles
		all ∡C-C=C = ________ all ∡H-C=C = ________ all ∡H-C-C = ________

E. Additional Structures (to be assigned)

Compound	Lewis Structure	Spatial Formula
a. ____________________		
b. ____________________		
c. ____________________		

Experiment 17

Properties of Hydrocarbons

Purpose:

In this experiment you will study several physical and chemical properties of the most abundant class of organic compounds, the saturated, unsaturated and aromatic hydrocarbons.

Discussion:

Hydrocarbons are organic compounds that contain *only* carbon and hydrogen. The principal source of hydrocarbons is petroleum. All organic compounds contain carbon and hydrogen, but many also contain oxygen, nitrogen and other elements. Most organic compounds can be considered derivatives of hydrocarbons. The substitution of an organic functional group for a hydrogen atom in a hydrocarbon will produce a compound of that functional class. In this experiment, both aliphatic and aromatic hydrocarbons will be examined. **Aliphatic hydrocarbons** contain either straight chains, branched chains and/or certain cyclic arrangements of carbon atoms. **Aromatic hydrocarbons** contain at least one aromatic ring, most commonly the benzene ring.

Aliphatic hydrocarbons are divided into three classes: **alkanes** are saturated hydrocarbons that contain only carbon-carbon single bonds; **alkenes** are unsaturated hydrocarbons that contain at least one carbon-carbon double bond; and **alkynes** are unsaturated hydrocarbons that contain at least one carbon-carbon triple bond.

A. Alkanes

Members of the alkane series are termed *saturated* because all carbon-carbon bonds are single bonds. There are no double or triple bonds between carbon atoms. Each carbon atom is joined to other atoms by four single bonds. The IUPAC (International Union of Pure and Applied Chemistry) names for the members of the alkane series always end in *-ane*. The following are members of the alkane series:

```
   H         H   H          H   H   H   H          H       H       H
   |         |   |          |   |   |   |          |       |       |
 H-C-H     H-C - C-H      H-C - C - C - C-H      H-C ----- C ----- C-H
   |         |   |          |   |   |   |          |       |       |
   H         H   H          H   H   H   H          H       |       H
                                                         H-C-H
                                                           |
                                                           H
 methane     ethane             butane              2-methylpropane
                                                       (isobutane)
```

Notice that butane and 2-methylpropane (commonly called isobutane) have the same molecular formula, C_4H_{10}, but different structural formulas. Compounds with the same molecular formula but different structural formulas are **isomers** of one another.

The general molecular formula for any member of the alkane series is C_nH_{2n+2}. The small n equals the number of carbon atoms in the molecule. Butane is a four-carbon alkane, n = 4, and its molecular formula is:

$$C_4H_{2(4)+2} = C_4H_{10}$$

butane

The first ten members of the alkane family are listed in Table 1. Only the straight-chain isomers are shown for those with more than three carbon atoms.

Table 1. Some Members of the Alkane Series

Name	Molecular Formula	Condensed Structure	Number of Isomers
methane	CH_4	CH_4	1
ethane	C_2H_6	CH_3CH_3	1
propane	C_3H_8	$CH_3CH_2CH_3$	1
butane	C_4H_{10}	$CH_3CH_2CH_2CH_3$	2
pentane	C_5H_{12}	$CH_3(CH_2)_3CH_3$	3
hexane	C_6H_{14}	$CH_3(CH_2)_4CH_3$	5
heptane	C_7H_{16}	$CH_3(CH_2)_5CH_3$	9
octane	C_8H_{18}	$CH_3(CH_2)_6CH_3$	18
nonane	C_9H_{20}	$CH_3(CH_2)_7CH_3$	35
decane	$C_{10}H_{22}$	$CH_3(CH_2)_8CH_3$	75

The first four members of the alkane series are colorless, nearly odorless gases. The members of the series containing five to sixteen carbon atoms are colorless liquids that are less dense than water. Alkanes with more than sixteen carbons are waxy, colorless solids.

Alkanes are nonpolar and are not soluble in polar solvents like water. On the other hand, alkanes are very soluble in nonpolar solvents such as other alkanes, diethylether or carbon tetrachloride.

Because alkanes are saturated, they have low chemical reactivity. In fact, they are the least reactive of all organic compounds. They are not oxidized by strong oxidizing agents such as potassium permanganate ($KMnO_4$) or potassium dichromate ($K_2Cr_2O_7$). However, if sufficient oxygen is present, an alkane can undergo **combustion**, an oxidation reaction, yielding carbon dioxide, water and considerable heat energy. Alkanes can burn cleanly, producing little or no smoke or soot. Volatile alkanes can form explosive mixtures with air and should be used only in well-ventilated areas. Because alkanes are available in large quantities in crude oil and because they produce a large amount of energy per gram when they combust, they make excellent fuels. In fact, our principal use of the alkanes is for fuel.

The complete combustion of an alkane may be represented by a general equation:

General equation: $C_nH_{2n+2} + \frac{1}{2}(3n+1)\ O_{2(g)} \rightarrow n\ CO_{2(g)} + (n+1)\ H_2O_{(g)}$

For methane: $CH_{4(g)} + 2\ O_{2(g)} \rightarrow CO_{2(g)} + 2\ H_2O_{(g)}$

For octane: $2\ C_8H_{18(l)} + 25\ O_{2(g)} \rightarrow 16\ CO_{2(g)} + 18\ H_2O_{(g)}$

If the supply of air (oxygen) is limited, alkanes undergo incomplete combustion, forming carbon monoxide (instead of carbon dioxide) and water. Carbon monoxide is colorless, odorless, and tasteless, but unlike carbon dioxide, it binds strongly to the iron ion in hemoglobin, preventing oxygen transport in the blood. The result can be fatal. If oxygen is extremely limited, combustion may produce a heavy black smoke of unburned carbon particles.

Because alkanes are saturated, they do not undergo addition reactions like the alkenes and alkynes. But they can react by a **substitution reaction** in which an atom or group of atoms replaces (is substituted for) a hydrogen atom in the hydrocarbon molecule. The halogens, principally chlorine and bromine, can readily substitute for hydrogen in what are called **halogenation reactions**.

Chlorine and bromine react with alkanes at elevated temperatures or when exposed to ultraviolet (uv) light as shown below. Sunlight is a good source of ultraviolet light. The chlorination of methane produces hydrogen chloride gas as well as chloromethane.

$$CH_{4(g)} + Cl_{2(g)} \xrightarrow[\text{elevated temp}]{\text{uv light or}} CH_3Cl_{(g)} + HCl_{(g)}$$

methane chlorine chloromethane hydrogen chloride

Hydrogen chloride, $HCl_{(g)}$, dissolves readily in water to form hydrochloric acid, $HCl_{(aq)}$.

If bromine would have been used in the last equation, hydrogen bromide, $HBr_{(g)}$, would have formed. In water, hydrogen bromide produces hydrobromic acid.

$$CH_3CH_2CH_2CH_2CH_2CH_3 + Br_2 \xrightarrow{\text{uv light}} CH_3CH_2CH_2CH_2CH_2CH_2Br + HBr_{(g)}$$

hexane	bromine solution[1]	1-bromohexane	hydrogen bromide
(colorless)	(red-brown)	(colorless)	(colorless)

We can easily observe the course of a bromine substitution reaction by noting the disappearance of the red-brown color of bromine as it is consumed. The products of the reaction are usually colorless.

B. Alkenes

The alkenes are *unsaturated* hydrocarbons that contain at least one carbon-carbon double bond (C = C). Because of this double bond, the molecule contains two fewer hydrogen atoms than the corresponding alkane; it is not completely "saturated" with hydrogen, as the alkanes are. The IUPAC names for alkenes always end in *-ene*. The general molecular formula for members of the alkene series is C_nH_{2n}. The structural and molecular formulas of the first two members of the alkene series are shown below.

	$H_2C=CH_2$ or $CH_2=CH_2$	$H_3C-CH=CH_2$ or $CH_3-CH=CH_2$
IUPAC name:	ethene	propene
Common name:	ethylene	propylene

Since alkenes are nonpolar and have nearly the same atomic mass as the corresponding alkanes, it is not surprising to learn that their physical properties are very similar to the alkanes of similar molecular weight. Like all hydrocarbons, alkenes are insoluble in water, though soluble in nonpolar organic solvents. Their densities are less than that of water.

The carbon-carbon double bond is the reactive center of an alkene. Alkenes undergo **addition reactions** in which atoms or groups of atoms *add* to the double bond and, in doing so, saturate the bond. The addition of hydrogen to an alkene or alkyne is called **hydrogenation**. Frequently a catalyst such as finely divided nickel metal is required to facilitate the addition, and the hydrogen may be under high pressure.

$$CH_2=CH_2 + H_2 \xrightarrow[\text{heat, pressure}]{\text{Ni}} CH_3-CH_3$$

ethene → ethane

---

[1] Pure bromine, Br_2, is a corrosive, red-brown liquid and is difficult to use in its pure form. It is more easily handled when dissolved in cyclohexane. Cyclohexane is a good solvent for hydrocarbons as well, since both are nonpolar substances.

When a halogen, such as bromine, adds to an alkene, it saturates the bond, adding one bromine atom to each carbon.

$$CH_3-CH=CH_2 \quad + \quad Br_2 \longrightarrow CH_3-CHBr-CH_2Br$$

propene (colorless)	bromine (red-brown)	1,2-dibromopropane (colorless)

The addition of a halogen to an alkene occurs quickly at room temperature and, unlike the halogenation of saturated hydrocarbons, does not require ultraviolet light. The color change that accompanies the reaction makes a useful tool to determine if a compound is unsaturated. If an organic hydrocarbon quickly discharges the red-brown color of bromine when a few drops of Br_2 solution are added to it, then the compound must be an unsaturated hydrocarbon containing either a double bond (an alkene) or a triple bond (an alkyne) between two adjacent carbon atoms. Alkynes will be described later.

Another identifying test for alkenes is the **Baeyer permanganate test**, in which the alkene is oxidized by dilute potassium permanganate, $KMnO_4$. As the reaction proceeds, the characteristic purple color of the permanganate ion is destroyed and replaced by a brown to black suspension of finely divided manganese(IV) oxide, MnO_2. The color change of the solution from purple to brown (or black) indicates the presence of an alkene or an alkyne. A general equation for the reaction is:

$$H_2C=CH_2 \quad + \quad KMnO_4 \longrightarrow -\underset{OH}{C}-\underset{OH}{C}- \quad + \quad MnO_{2(s)}$$

alkene (colorless)	potassium permanganate (purple)	a diol (colorless)	manganese(IV) oxide (brown-black precipitate)

Alkenes, like all hydrocarbons, combust to form carbon dioxide and water. In an open container, alkenes burn with a luminous yellow flame and produce some smoke and soot. The combustion of an alkene produces slightly less heat energy than the same mass of the corresponding alkane.

$$CH_2 = CH_{2(g)} + 3\ O_{2(g)} \rightarrow 2\ CO_{2(g)} + 2\ H_2O_{(g)}$$

C. Alkynes

The alkynes are also unsaturated hydrocarbons since they contain at least one carbon-carbon triple bond. Because of the greater reactivity of the triple bond, alkynes are rarely found in nature. The general molecular formula for members of the alkyne series is C_nH_{2n-2}. The IUPAC names for alkynes always end in *-yne*. Ethyne (commonly called acetylene) is one of the most important alkynes.

Unfortunately, the common name for ethyne has an ending that suggests it is an alkene. Remember that it is not. The first two members of the alkyne series are shown below:

$$H-C\equiv C-H \qquad\qquad H-\underset{\underset{H}{|}}{\overset{\overset{H}{|}}{C}}-C\equiv C-H$$

IUPAC name:	ethyne	propyne
Common name:	acetylene	(none)

Acetylene can be prepared in small quantities by reacting calcium carbide with water:

$$CaC_{2(s)} \ + \ 2\ H_2O_{(l)} \ \rightarrow \ Ca(OH)_{2(s)} \ + \ C_2H_{2(g)}$$

This reaction was used in the early 1900s to produce the fuel in coal miners' lanterns. Because ethyne (acetylene) burns with a bright luminous flame, carbide lanterns are popular today with campers.

Ethyne (acetylene) is a colorless gas, is insoluble in water, and has a pleasant odor when pure. The combustion of ethyne in an oxyacetylene torch produces temperatures as high as 3000°C.

$$2\ C_2H_{2(g)} \ + \ 5\ O_{2(g)} \ \rightarrow \ 4\ CO_{2(g)} \ + \ 2\ H_2O_{(g)}$$

Alkynes, like alkenes, undergo addition reactions, but they do so in two steps. The first step reduces the carbon-carbon triple bond to a carbon-carbon double bond (as in an alkene); the second reduces the double bond to a carbon-carbon single bond (as in an alkane). This is shown below in the hydrogenation and bromination of ethyne:

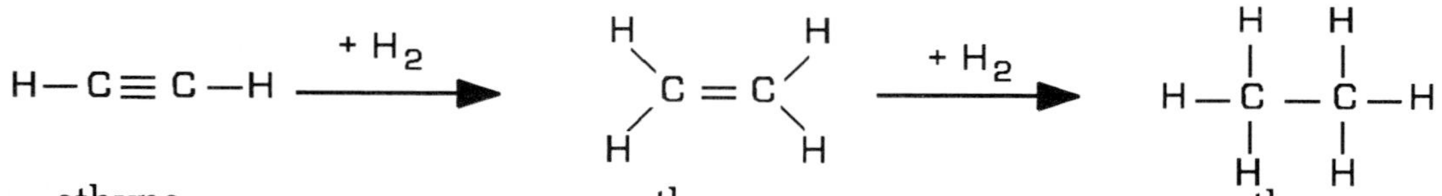

$$H-C\equiv C-H \xrightarrow{+\,H_2} HHC=CHH \xrightarrow{+\,H_2} H-\underset{\underset{H}{|}}{\overset{\overset{H}{|}}{C}}-\underset{\underset{H}{|}}{\overset{\overset{H}{|}}{C}}-H$$

ethyne   ethene   ethane

$$H-C\equiv C-H \xrightarrow{+\,Br_2} (Br)(H)C=C(H)(Br) \xrightarrow{+\,Br_2} H-\underset{\underset{Br}{|}}{\overset{\overset{Br}{|}}{C}}-\underset{\underset{Br}{|}}{\overset{\overset{Br}{|}}{C}}-H$$

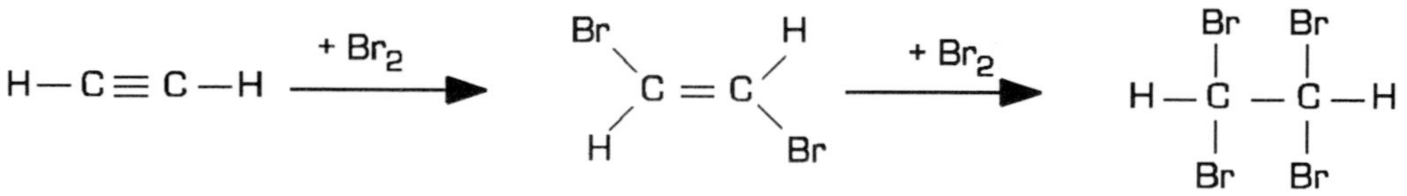

ethyne   1,2-dibromoethene   1,1,2,2-tetrabromoethane

Alkynes will also react with $KMnO_4$ in the Baeyer test, decolorizing the purple permanganate solution and producing a variety of oxidation products.

D. Aromatic Hydrocarbons

Benzene is the first and principal member of the series of aromatic hydrocarbons, and it exhibits the unique characteristics of all members of this series. Benzene was first isolated by Michael Faraday in 1825, but it was ten years before its molecular formula was determined to be C_6H_6. During the next twenty years chemists were plagued with the problem of its structure. Though the molecular formula implied a high degree of unsaturation, its chemical properties were not at all like alkenes or alkynes. It did not react with bromine and potassium permanganate, though under special conditions and with special catalysts, it could be made to react with bromine in a substitution (not addition) process.

In 1865, German chemist Frederick Kekule proposed that the benzene molecule was a cyclic, hexagonal, planar structure of six carbon atoms joined with alternating single and double bonds, with each carbon atom joined to a single hydrogen atom. He accounted for the equivalence of all six carbon atoms by suggesting that the double and single bonds were not static, but exchanged positions within the planar ring. The Kekule structures for benzene are shown below.

Kekule's explanation of the benzene structure still had a major problem: It suggested that benzene and similar aromatic compounds should still behave as highly unsaturated compounds. But they simply did not do so. By the early 1900s, it became apparent that the idea of alternating double and single bonds was not entirely correct. Rather, the six electrons (one from each carbon atom) are shared by all six carbon atoms equally in a unique and very stable bonding arrangement that is commonly symbolized by a circle drawn in the center of a hexagon of carbon atoms or simply a hexagon. This representation of the benzene molecule is shown below:

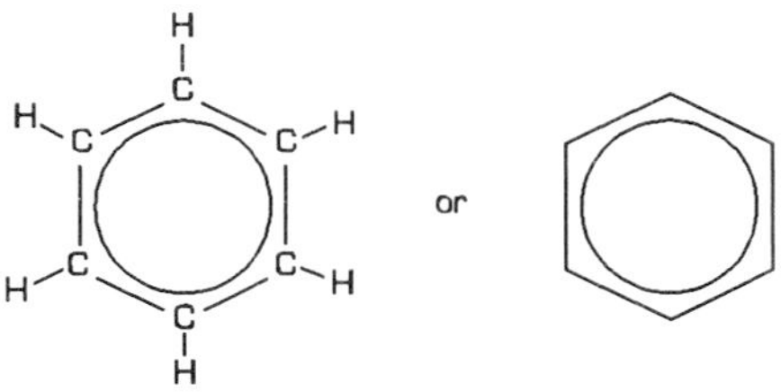

or

Benzene is a colorless liquid at room temperature. It is insoluble in water but soluble in most nonpolar organic liquids. Though benzene and its derivatives were at one time called aromatic because of their generally pleasant odors, today the term has taken on a new meaning that describes the unique benzene-like bonding in the six-member ring.

Several aromatic organic compounds are shown below. Three of them are hydrocarbons (toluene, napthalene and phenanthrene); the rest are not.

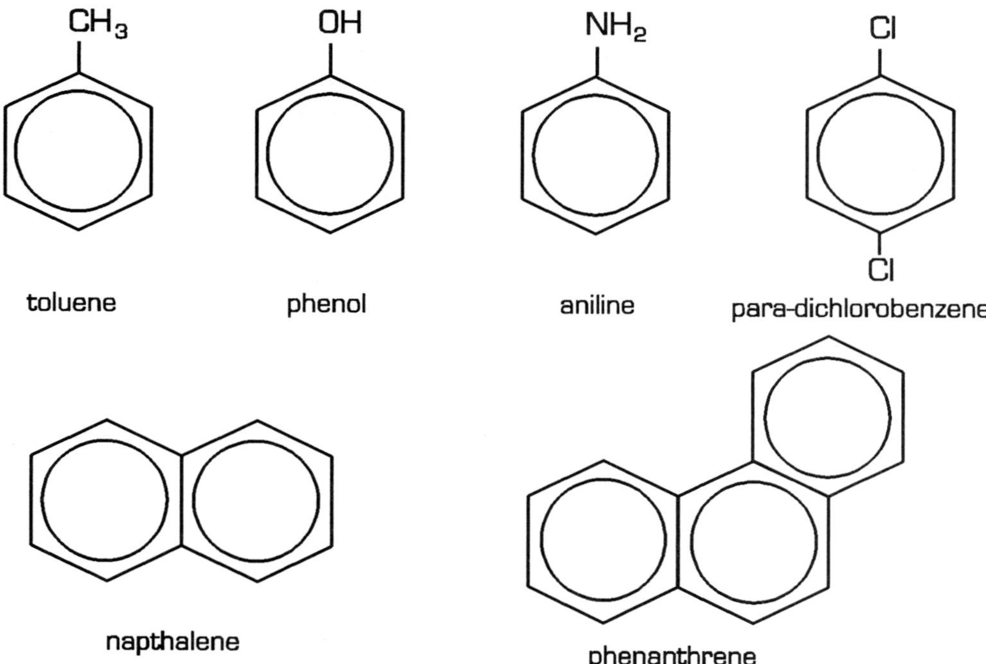

Aromatic compounds can undergo substitution reactions in which an atom or group of atoms is substituted for a hydrogen on the ring. Unlike the substitution reactions of the alkanes, those involving aromatic compounds usually require a special catalyst but do not require ultraviolet light or elevated temperatures.

If bromine is added to a solution of benzene, no reaction will occur. However, if a small amount of solid iron(III) chloride (a catalyst) is added to the mixture, the bromine substitution reaction will take place and the red-brown color of molecular bromine will disappear as it is consumed. In the equation shown below, a bromine atom is substituted for a hydrogen atom as benzene is converted to bromobenzene:

benzene + Br_2 (in cyclohexane) $\xrightarrow{FeCl_3}$ bromobenzene (Br) + HBr

Experimental Procedure:

> **CAUTION: Hydrocarbons are extremely flammable. Be especially careful when working near open flames. Do the tests *only* in the order presented. As always, wear eye protection at all times.**
>
> **ENVIRONMENTAL ALERT: Discard all waste as directed by your instructor.**

In the following tests: *hexane* will be used as a typical alkane;
1-pentene as a typical alkene;
ethyne (acetylene) as a typical alkyne, and
toluene will serve as a typical aromatic hydrocarbon.

A. Physical Properties of Hydrocarbons

1. *Odor*: Test for the odor of hexane, 1-pentene and toluene (but *not* methanol) by fanning the fumes from the corresponding reagent bottle toward your nose. Briefly describe the odor of each on the Report Sheet. If an odor is not detected, indicate "no odor detected" in the space provided.

2. *Ignition Test for Flammability*: The way in which an organic compound combusts can often provide clues as to the kind of organic compound it is. In this test you will ignite an alcohol (which is not a hydrocarbon), an alkane (a saturated hydrocarbon), an alkene (an unsaturated hydrocarbon) and an aromatic compound. The alcohol is not a hydrocarbon since it contains oxygen, CH_3OH, but its combustion is a good contrast to that of the hydrocarbons.

   For the ignition test use only 5 drops of methanol (an alcohol), hexane, 1-pentene and toluene. First add the methanol to a clean evaporating dish and ignite the liquid with a burning wood splint. Observe (a) how quickly it ignites, (b) the color of the flame and (c) whether smoke or soot forms as it burns. Watch closely since the ignition test can often be used to distinguish between an alkane and an aromatic compound. Record your observations. Clean the evaporating dish and repeat the test with hexane, then with the other two liquids, recording your observations each time.

3. *Solubility in Water*: Test for solubility in water by placing 20 drops of methanol, hexane, 1-pentene, and toluene in four different test tubes. Add about 5 mL of tap water to each. Cork or seal[2] each mixture and shake each one for a few seconds and set the test tube aside. After 30 seconds, record whether the organic compounds are

---

2 Plastic film held over the mouth of a test tube with a thumb or fingertip can be used in place of corks. Before using though, make certain the film does not react with and is not soluble in the organic compounds being studied.

soluble in water. For those that are insoluble, note the relative density of the hydrocarbon with respect to water. If the organic layer is beneath the water layer, it is more dense than water as well as insoluble in water. If corks were used, discard them to avoid contamination of other tests.

4. *The Solubility of Hydrocarbons in Each Other*: Determine the solubility of hexane, 1-pentene and toluene in each other by preparing the three mixtures described below and shaking each thoroughly. Each test tube must be dry since a few drops of water will confuse your observations. Record your observations on the Report Sheet.

   Test tube 1: 20 drops of hexane and 20 drops of 1-pentene
   Test tube 2: 20 drops of hexane and 20 drops of toluene
   Test tube 3: 20 drops of 1-pentene and 20 drops of toluene

B. Reaction of Bromine with Hydrocarbons

> **CAUTION: Handle the bromine solution with care. If bromine gets on your skin, quickly wash the area with cold water, then cover the area with glycerine. Be certain to notify your instructor.**

1. *The Reactivity of Bromine with Hydrocarbons*: Arrange three clean, dry 15-cm test tubes in a test tube rack. Place 20 drops of hexane in the first tube, 20 drops of 1-pentene in the second and 20 drops of toluene in the third. Add three drops of 5% bromine-cyclohexane solution to each sample. Cork or seal the test tubes, swirl the mixtures and observe after one minute. Record your observations on the Report Sheet. If corks were used, discard them.

2. *The Bromine Substitution Reaction with Alkanes*: Add about 20 drops of hexane to each of two dry 15-cm test tubes. Wrap a paper towel around one of the test tubes. Add 10 drops of bromine-cyclohexane solution to each tube. Cork or seal both test tubes and mix well. Expose the unwrapped test tube to sunlight (or an artificial source of ultraviolet light) for about five minutes. Record the color of each solution after this time. Place a piece of moist pH paper (or blue litmus paper) at the mouth of each of these test tubes as you first remove the seal or cork. Record your observations. If corks were used, discard them.

C. Potassium Permanganate Test for Unsaturated Hydrocarbons

Add 20 drops of 0.5% potassium permanganate solution to each of three 15-cm test tubes. Add 5 drops of hexane to the first, 5 drops of 1-pentene to the second and 5 drops of toluene to the third test tube. Cork or seal the test tubes and shake them for about one minute. Record your observations on the Report Sheet. If corks were used, discard them.

D. Preparation of Ethyne and the Test for Flammability

> **INSTRUCTIONAL NOTE: This experiment may be presented as a demonstration.**
>
> **CAUTION: Turn off all burners in the vicinity of this experiment.**

Place a small piece (the size of a green pea) of calcium carbide, CaC_2, in a 250-mL beaker. Cautiously add a few drops of water to the solid. Observe the effect of water on calcium carbide. As the gaseous acetylene rises from the bottom of the beaker, test its flammability by carefully bringing a burning wood splint to the mouth of the beaker. Record your observations on the Report Sheet.

E. Kerosene (optional)

Determine the class (or classes) of hydrocarbon (alkane, alkene or aromatic) to which kerosene belongs by treating it with (a) bromine solution and (b) potassium permanganate solution, as in tests B-2 and C. Simply substitute kerosene for hexane in each test. Also perform the ignition test on kerosene by igniting 5 drops of the liquid in a clean evaporating dish. Record all observations on the Report Sheet.

F. Evaluation of an Unknown Hydrocarbon (optional)

Obtain an unknown hydrocarbon from your instructor. Record its number on the Report Sheet. You are to determine if the hydrocarbon is an alkane, an alkene or an aromatic compound. Record all observations and your conclusion on the Report Sheet.

1. *Ignition Test*: Ignite 5 drops of the unknown as in test A-2.

2. *Bromine Test for Unsaturation*: Repeat test B-1 by substituting the unknown for hexane.

3. *Permanganate Test for Unsaturation*: Repeat test C, substituting the unknown for hexane.

Name ______________________________ Locker Number__________ Date____________
Please print; last name first

PRELIMINARY EXERCISES: *Experiment 17*

Properties of Hydrocarbons

1. Draw the structural formula of each of these hydrocarbons:

a. ethane	b. toluene	c. propene

2. Define the following terms:

a. hydrocarbon
b. alkane
c. alkene
d. alkyne
e. aromatic hydrocarbon

(The Preliminary Exercises continue on the following page.)

3. Summarize the solubility behavior of hydrocarbons in water (a polar solvent) and in nonpolar organic solvents.

4. Write the equations for the substitution reaction of methane with bromine under the following conditions.

   a. Reaction in the presence of sunlight.

   $$CH_4 + Br_2 \xrightarrow{\text{sunlight}}$$

   b. Reaction in the absence of sunlight.

   $$CH_4 + Br_2 \rightarrow$$

5. Write the equations for the addition reaction of ethene with bromine under the following conditions.

   a. Reaction in the presence of sunlight.

   $$H_2C{=}CH_2 + Br_2 \xrightarrow{\text{sunlight}}$$

   b. Reaction in the absence of sunlight.

   $$H_2C{=}CH_2 + Br_2 \rightarrow$$

Name ______________________ Locker Number__________ Date__________
Please print; last name first

REPORT SHEET: *Experiment 17*

Properties of Hydrocarbons

A. Physical Properties of Hydrocarbons

Odor, Ignition Test for Flammability, and Solubility in Water: Record observations below.

Compound Tested	Odor	Ignition Test	Solubility in Water
methanol (nonhydrocarbon)	not to be done		
hexane (saturated hydrocarbon)			
1-pentene (unsaturated hydrocarbon)			
toluene (aromatic hydrocarbon)			

Solubility of Hydrocarbons in Each Other: Record your observations and classify each combination as soluble or insoluble.

Test tube 1 (hexane and 1-pentene):
Test tube 2 (hexane and toluene):
Test tube 3 (1-pentene and toluene):
From the results of section A, what general statement can be made about the solubility of hydrocarbons in water and in other hydrocarbons?

B. Reaction of Bromine with Hydrocarbons

Reactivity of Bromine with Hydrocarbons

Why was Br_2 dissolved in cyclohexane and not in water?
Which hydrocarbon reacted immediately with the bromine solution? Write the equation for the reaction.

Bromine Substitution Reaction with Alkanes

What was observed in the test tube shielded from sunlight?
What was observed in the test tube exposed to sunlight?

Complete the equation showing the substitution of one bromine atom for hydrogen on the end carbon of hexane. The reaction is catalyzed with uv light.

$$CH_3CH_2CH_2CH_2CH_2CH_3 \ + \ Br_2 \xrightarrow{uv}$$

Is exposure to ultraviolet light required for this reaction to take place?

C. Potassium Permanganate Test for Unsaturated Hydrocarbons

Which hydrocarbons reacted with $KMnO_4$? Describe the color changes observed.
Did toluene react with $KMnO_4$?
How can the potassium permanganate test be used to distinguish between saturated and unsaturated hydrocarbons?

Name ______________________ Locker Number__________ Date__________
Please print; last name first

REPORT SHEET (continued): *Experiment 17*

Properties of Hydrocarbons

D. Preparation of Ethyne and the Test for Flammability

What gas was produced in the reaction between calcium carbide and water?
Write the balanced the equation for the reaction.

E. Kerosene (optional)

What was observed when kerosene was tested with the $KMnO_4$ solution?
What was observed when the bromine-cyclohexane solution was added to kerosene and exposed to sunlight?
What was observed when the bromine-cyclohexane solution was added to kerosene and shielded from sunlight?
What did you learn about kerosene from the ignition test?
To what class or classes (alkane, alkene or aromatic) does kerosene belong? Please justify your choice with your observations.

F. Evaluation of an Unknown Hydrocarbon (optional)

Unknown Number: ___________

Record your observations for each test below.

1. Ignition test
2. Bromine test
3. Permanganate test
To what class (alkane, alkene or aromatic) does your unknown belong? Justify your conclusion.

Experiment 18

The Synthesis of Aspirin

Purpose:

In this experiment you will synthesize and purify acetylsalicylic acid, commonly called aspirin. During the synthesis you will learn some common techniques used in the preparation of organic compounds and their purification.

Discussion:

Acetylsalicylic acid, which we commonly call aspirin, is the most widely used medicine in the world. It has the ability to reduce fever (an antipyretic), to reduce pain (an analgesic) and to act as an anti-inflammatory agent, reducing swelling, soreness and redness associated with injury or arthritis. Recent evidence suggests that, in the human body, aspirin inhibits the enzymes necessary for the formation of prostaglandins (hormones) that are associated with pain, fever and inflammation. Americans spend millions of dollars each year on headache remedies, most of which contain aspirin, and the consumption of aspirin in the United States is now approaching 50 million five-grain tablets each day.

A. The History of Aspirin

Nearly 2,500 years ago the Greeks used extracts of willow and poplar bark to relieve the pain and symptoms of illness. Before the time of Columbus, Native Americans drank teas made from the bark of the willow to reduce fever, but it was not until 1763 that the Reverend Edward Stone introduced these teas and extracts to Europeans. In the early 1800s, the active ingredient in willow bark was isolated and identified as salicylic acid, and shortly thereafter, salicylic acid was synthesized in the laboratory. By 1859, bulk quantities were available for use. It was found to be an effective preservative for meats, milk, beer and other foods. It proved useful in the treatment of gout and rheumatism, but because of its acidity, it irritated the lining of the mouth and the digestive tract and this limited its long-term use by patients. The sodium salt of salicylic acid was introduced in 1875, and though the salt was less sour to the taste, it still produced gastric discomfort. In 1866 the phenol ester of salicylic acid (known as Salol®) was introduced as a pain reliever, but its use was quickly terminated due to the toxicity of phenol, which it released when broken down in the digestive tract of those who used it heavily. Though there were many attempts to produce a suitable derivative of salicylic acid for human consumption, only one resulted in a useful product.

In 1893, Felix Hoffman, Jr., a chemist working for the Bayer Laboratories in Germany, began to study salicylic acid after hours on his own time. His father was suffering from rheumatoid arthritis and depended on salicylic acid for relief of pain. Young Hoffman learned that an ester of salicylic acid could be made that would relieve pain without causing stomach discomfort. The chemical name for that ester was acetylsalicylic acid, and in 1899, the compound was patented by Bayer and released for sale under the trade name Aspirin. In 1915, the 5-grain aspirin tablet carrying the Bayer name first appeared as a nonprescription medicine. The trade name of the drug was a combination of the "*a*" (from the *a*cetyl group) and "*sprin*" (from a common name for salicylic acid, spiric acid which refers to its presence in plants of the *Spirea* genus). Even today in Germany, Canada and other countries, the word "aspirin" can only be used on Bayer products. However, it should be noted that the Bayer Company lost its rights to the exclusive use of "aspirin" in the United States in the early 1900s. Thus, in the United States, any company can use the name "aspirin" to identify a drug that contains acetylsalicylic acid. By federal law, each 5-grain aspirin tablet must contain 0.324 g of acetylsalicylic acid. Binders, such as starch, are combined with the aspirin so that firm tablets may be produced. All 5-grain aspirin tablets contain the same active ingredient but, as you know, the price of 100 aspirin tablets may vary considerably.

B. Physiological Effects of Aspirin

Aspirin has become the wonder drug of the common person; however, it is not without its faults. Aspirin still causes stomach irritation in some people, and it is estimated that about 1 mL of blood is lost from the stomach lining for each gram of aspirin consumed. To reduce irritation, aspirin should be taken with food. Also, aspirin tablets are available that are coated with a substance that will only break down in the environment of the small intestine, where aspirin is absorbed by the body. Buffering agents, such as magnesium carbonate, which partially neutralizes the acidic nature of aspirin, can lessen stomach irritation. About one in 10,000 persons will experience allergic reactions to aspirin that may be as tolerable as a mild skin rash or as serious as asthmatic attacks that place the patient in shock. Of course, these persons should avoid all aspirin-containing medications.

Aspirin also interferes with the normal clotting of blood, which is recognized as a benefit in preventing certain cardiovascular problems. A report released in 1987 showed that patients who were given one aspirin tablet each day after their first heart attack reduced their chance of having a second heart attack by as much as 40%. More recent studies confirm that a normal middle-age man may lessen his chance of a heart attack by 20% by taking one-half an aspirin tablet every day. On the other hand, aspirin should *not* be taken immediately before or after surgery since reduced clotting ability could lead to excessive bleeding.

Aspirin, like most substances, is toxic in large doses. Aspirin is the number one cause of accidental death by poisoning in children under five years of age, and for some individuals, the lethal dose of aspirin is as low as 50 mg per kilogram of body weight. Reports since 1989 have confirmed that some children under the age of 15 who suffer from viral infections have developed the rare but often fatal brain disorder called Reye's Syndrome. Thus, any child suspected of having a viral infection should not be given aspirin-containing products.

Acetaminophen and Ibuprofen® are frequently used as substitutes for aspirin though neither, unlike aspirin, inhibits blood clotting. Though acetaminophen is an analgesic and an antipyretic it does not seem to equal the anti-inflammatory property of aspirin.

C. Salicylic Acid and Its Derivatives

Aspirin is a derivative of salicylic acid. The structural formula of salicylic acid is:

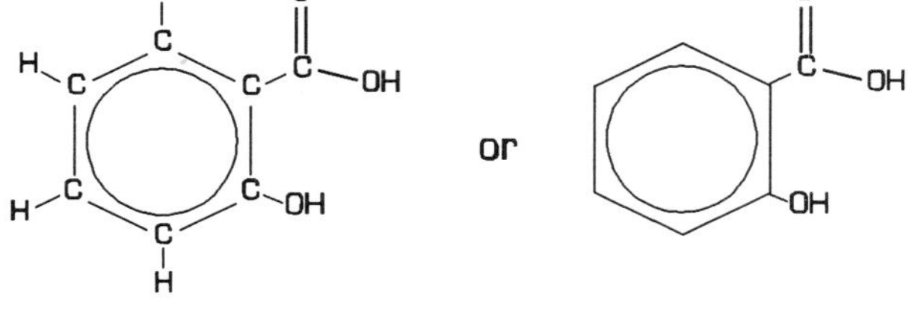

or

The salicylic acid molecule can be viewed as being composed of three units:

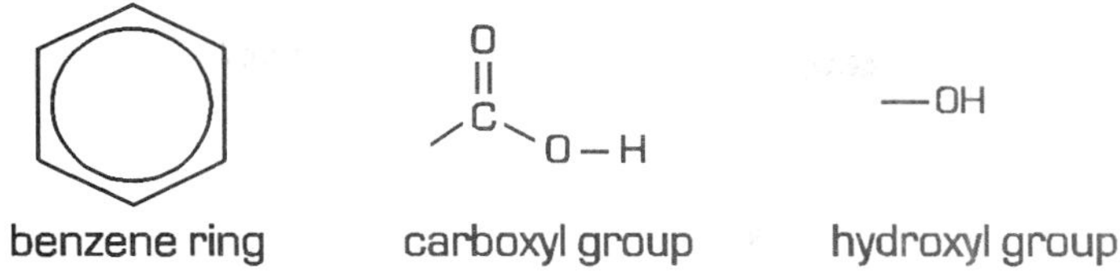

1. A benzene ring, symbolized with a hexagon enclosing a circle
2. A carboxyl group, the functional group of carboxylic acids, also written as $-COOH$
3. A hydroxyl group, which when joined to a benzene ring, classifies the compound as a phenol.

Salicylic acid is both a carboxylic acid and a phenol because it contains both functional groups. It is a colorless, crystalline solid from which many important compounds having medical applications have been derived.

When salicylic acid reacts with sodium hydroxide, the salt, sodium salicylate, is formed. Sodium salicylate is used both as a food preservative and in the treatment of arthritis.

$$C_6H_4(OH)COOH + NaOH \longrightarrow C_6H_4(OH)COO^- Na^+ + H_2O$$

salicylic acid + sodium hydroxide → sodium salicylate + water

When a carboxylic acid, (such as salicylic acid) reacts with an alcohol (or phenol) in the presence of a strong acid, an ester is formed. In the equation below, sulfuric acid is the catalyst and R symbolizes an organic group.

$$R{-}C({=}O){-}O{-}H + HOR' \xrightarrow{H_2SO_4} R{-}C({=}O){-}O{-}R' + H_2O$$

carboxylic acid + alcohol → ester + water

The methyl ester of salicylic acid, methyl salicylate, commonly called "oil of wintergreen," is prepared by the reaction of methyl alcohol, CH_3OH, with salicylic acid. It is used in perfumes and as a flavoring agent in candy, gum and foods. Because of its mild antiseptic property, it is also used in throat lozenges and in mouth washes. Methyl salicylate is also used in "pain-relieving" creams and ointments, such as Ben-Gay®. As it is rubbed on the skin it is rapidly absorbed and breaks down (it reacts with water, a process known as hydrolysis) to form salicylic acid which relieves local pain. In addition, methyl salicylate causes a mild inflammation of the tissue, which increases blood circulation to the area. The equation for the preparation of methyl salicylate is:

salicylic acid + CH_3OH (methyl alcohol) $\xrightarrow{H_2SO_4}$ methyl salicylate + H_2O (water)

Notice that the hydrogen of the carboxylic acid is replaced by the methyl group, $-CH_3$. Methyl salicylate will be prepared but not isolated in this experiment. Its presence will be indicated by its characteristic wintergreen aroma.

The phenol ester of salicylic acid, also called Salol®, is used as an intestinal antiseptic. It passes through the stomach unchanged and is slowly hydrolyzed to phenol and salicylic acid in the small intestine. Salol® is also used to coat enteric pills that are to pass through the stomach unchanged.

Salol

When aspirin is prepared from salicylic acid, the hydroxyl group, $-OH$, on the salicylic acid molecule is the point of reaction. Salicylic acid then, is taking the role of the alcohol in the formation of the acetate ester. In the equation below, note that it is the hydroxyl group of salicylic acid that reacts with acetic acid.

salicylic acid + acetic acid $\xrightarrow{H_2SO_4}$ acetylsalicylic acid (aspirin) + H_2O

The reaction between acetic acid and salicylic acid to form aspirin, though suitable, does not proceed at a convenient rate for a brief laboratory period, so in this experiment, aspirin will be prepared using acetic anhydride in place of acetic acid, as shown below. This is the same reaction used to produce aspirin commercially. The molecular formulas and molecular weights for each species in the reaction are given in Table 1. These will be useful later.

Table 1. Molecular Formulas and Weights

	Salicylic Acid	Acetic Anhydride	Acetylsalicylic Acid	Acetic Acid
Molecular formula	$C_7H_6O_3$	$C_4H_6O_3$	$C_9H_8O_4$	$C_2H_4O_2$
Molecular weight	138 g/mol	102 g/mol	180 g/mol	60 g/mol

Even though this reaction proceeds at a convenient rate at 85°C, not all of the salicylic acid will be converted into aspirin in 20 minutes. Because of this, it will be necessary to purify the crude product to obtain pure aspirin. Fortunately, this can be done by simply washing the impure aspirin with ice-cold deionized water. Aspirin is not very soluble in ice-cold water, but the impurities are. By washing the product with several small portions of ice-cold water, very pure aspirin can be obtained.

As a check for purity, a test will be carried out to determine if any unreacted salicylic acid contaminates the aspirin. Salicylic acid has a phenol group, –OH, while aspirin has none. A specific chemical test for the phenol group would allow us to see if the aspirin contains any salicylic acid. The iron(III) chloride test will do this. If a drop of iron(III) chloride solution, $FeCl_{3(aq)}$, is added to an aqueous solution containing traces of a phenol group, a color change occurs. The color varies from red to green to blue, depending on the particular phenol. If your aspirin produces the same color as salicylic acid, then you know salicylic acid is present in your aspirin as an impurity.

A test for starch will also be performed on your synthesized aspirin as well as on commercial aspirin. Starch is often used as a binder in aspirin tablets. The positive test for the presence of starch will be the blue color that forms when starch is treated with iodine-potassium iodide solution.

Calculating the Percent Yield:

One way of measuring the success of a chemical synthesis is to calculate the percent yield of the process. For example, suppose you calculated that a synthesis could produce a maximum of 10.0 g of product, but when you carried out the reaction you obtained only 5.0 g. The percent yield would then be 50%. The actual yield was only half of the calculated yield, that is, the theoretical yield for the synthesis.

The important terms for understanding percent yield are defined below:

The actual yield is the mass of product actually obtained in a synthesis. This can never be greater than the theoretical yield.

The theoretical yield is the mass of product calculated from the balanced equation for the synthesis. This is the maximum possible mass of product that could be made from a given amount of reactant.

The percent yield is the fraction of the theoretical yield actually obtained, expressed as a percent. It is calculated as shown below.

$$\text{percent yield} = \frac{\text{actual yield}}{\text{theoretical yield}} \times 100\%$$

In this experiment, aspirin is produced by reacting salicylic acid with an excess of acetic anhydride, that is, with a greater amount of acetic anhydride than is necessary to completely react with *all* the salicylic acid. The balanced equation at the top of the previous page shows that 1.0 mol (138 g) of salicylic acid should theoretically produce 1.0 mol (180 g) of aspirin.

1.0 mol salicylic acid	should theoretically produce	1.0 mol aspirin
138 g salicylic acid	should theoretically produce	180 g aspirin

In this experiment you will start with 2.00 g of salicylic acid (0.0145 mol of salicylic acid). Theoretically, you should then obtain 0.0145 mol, or 2.61 g of aspirin.

0.0145 mol salicylic acid	should theoretically produce	0.0145 mol aspirin
2.00 g salicylic acid	should theoretically produce	2.61 g aspirin

Suppose you started with 2.00 g of salicylic acid and obtained 1.50 g of aspirin instead of the theoretical yield of 2.61 g. The percent yield of your synthesis would be:

$$\text{percent yield} = \frac{1.50\ \text{g}}{2.61\ \text{g}} \times 100\% = 57.5\%$$

It is not common for the percent yield of a synthesis to exceed 90 or 95%, yet there are a variety of factors that can keep the actual yield less than 90%. You may lose some product during the filtering procedure, or there may be some decomposition if reaction conditions are not carefully controlled, and some reactions simply do not go to completion.

Experimental Procedure:

> **CAUTION: Acetic anhydride and concentrated sulfuric acid are used in this experiment. If either reagent is spilled on your skin, immediately wash the area with water thoroughly and inform your instructor.**
>
> **As always, eye protection must be worn at all times.**

A. Preparation of Aspirin

1. Weigh out *exactly* 2.00 g (0.0145 mol) of salicylic acid (a white solid) and transfer it to a clean, dry 20-cm test tube.

   Have your instructor or laboratory assistant dispense exactly 5.00 mL (0.0529 mol) of *fresh* acetic anhydride into the test tube.

   Add 5 drops of concentrated sulfuric acid (18 M H_2SO_4) to the test tube. *(Caution, handle sulfuric acid with care.)*

   Use a long stirring rod to stir the mixture until the salicylic acid dissolves. Keeping the stirring rod in the test tube, place the test tube in a test tube rack or suitable holder while you proceed with the next activity.

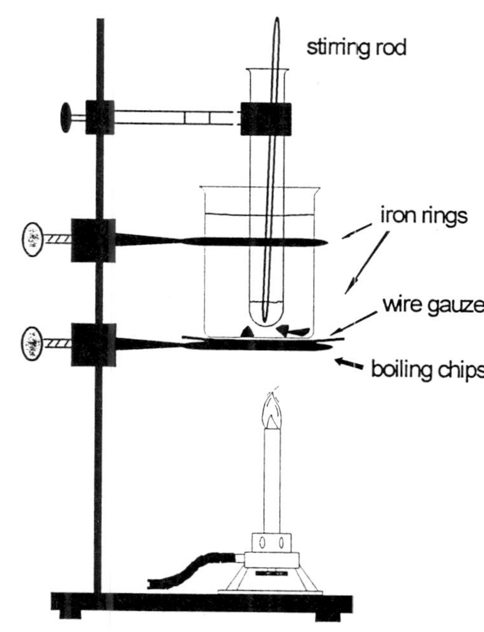

Figure 1

2. Bring a 250-mL beaker three-fourths full of tap water to boiling. Support the beaker on a ring stand as shown in Figure 1.

   Fill a 250-mL plastic wash bottle with deionized water and pack it in ice contained in a large beaker or pneumatic trough.

3. When the water in the beaker begins to vigorously boil, *turn off the burner* and clamp the test tube in the hot water bath as shown in Figure 1. Stir the contents of the test tube well for 1 minute. The reaction mixture should remain in the hot water bath for at least 20 minutes. To ensure this, record the time of day now in the blank below.

   Time of day: ______________

To use your time efficiently, you should proceed with the iron(III) chloride test (Section B) on samples of commercial aspirin and salicylic acid while the reaction mixture is heated in the hot water bath. Because you will test your synthesized aspirin later, record your observations accurately on the Report Sheet so a valid comparison with your aspirin can be made.

Also, do the starch test on commercial aspirin and on the control as instructed in Section C. You will test your synthesized aspirin later, so record your observations accurately on the Report Sheet.

If you have time before the 20-minute heating period is over, you can do any of the optional tests assigned by your instructor.

4. After making certain the reaction mixture has been heated at least 20 minutes, remove the test tube from the hot water bath. Then, *slowly and cautiously*, pour the contents of the test tube into 5 mL of deionized water in a 150-mL beaker. As the hot, unreacted acetic anhydride combines with the water it reacts to form acetic acid. Because this can be a vigorous reaction, the solution in the test tube must be added *slowly* to the water in the beaker to avoid spattering.

5. When the reaction in the beaker subsides, stir the mixture for 1 minute then slowly add room-temperature deionized water, a few drops at a time, until a *maximum* of 40 drops has been added *or* until the solution becomes cloudy. If the solution becomes clear (transparent) again, continue to add water dropwise until the cloud of small crystals of aspirin reappears. If you do not have crystals of aspirin at this point, contact your instructor for advice.

   Once crystals of aspirin appear in the beaker, rinse the test tube with *small* volumes of ice-cold deionized water from your wash bottle, adding the rinse to the solution in the 150 mL beaker.

   If, at this point, you do not have crystals of aspirin forming in the beaker, scratch the inner wall and bottom of the beaker to induce crystal formation (the vibrations produced when glass scratches glass can induce the formation of crystals). If you still do not have crystals, contact your instructor.

6. Measure 20 mL of *ice-cold* deionized water in a graduated cylinder and add it to the 150-mL beaker. Stir and then pack the beaker in ice to allow the aspirin to crystalize. Cold temperatures encourage the formation of crystals. Remember to keep your plastic wash bottle packed in ice when not in use.

   The beaker should remain packed in ice for at least 10 minutes. During this time, the cloud of white aspirin crystals should grow to fill nearly half the volume of the beaker.

7. The aspirin will by isolated by filtration. Either gravity filtration or suction filtration can be used, though suction filtration is faster. If you use suction filtration, assemble the apparatus shown in Figure 2. Either a sintered glass crucible (a funnel using a porous glass plate through which liquid is drawn) or a Buchner funnel (a funnel with a flat inner bottom punctured with holes, which are covered by filter paper) can be used. The figure shows a Buchner funnel.

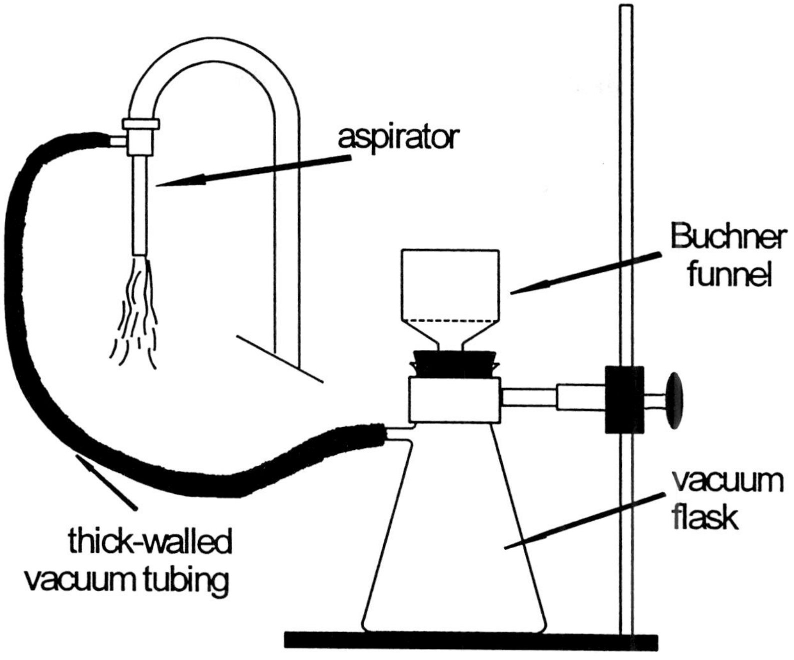

Figure 2

8. Attach one end of a length of thick-walled vacuum tubing to the sidearm of a vacuum flask and the other end to an aspirator on a water tap. Support the flask on a ring stand so it does not tip over. The Buchner funnel should fit tightly into a rubber stopper, which in turn fits tightly into the vacuum flask. Turn on the water to ensure that suction is created in the flask. Press the palm of your hand over the funnel and you should feel the suction building. If you do not, check your apparatus.

9. Place a disk of filter paper in the Buchner filter so that it covers all the holes in the flat bottom, and wet it with deionized water from the wash bottle to seal it to the bottom. Then, using suction, slowly pour the contents of the chilled 150-mL beaker onto the center of the filter paper, transferring as much solid as you can using the stirring rod and bursts of ice-cold water from the wash bottle. Allow the suction to remove as much liquid as possible from the solid.

10. Measure out 10 mL of ice-cold water from your wash bottle and pour it slowly over the entire surface of the solid on the filter paper to wash out impurities. In 2 to 3 minutes, repeat the washing with a second 10-mL volume of ice-cold water. Then, 2 or 3 minutes later, use your wash bottle to gently wash the aspirin with sprays of ice-cold deionized water for 30 seconds.

    Allow as much liquid as possible to be drawn away from the solid by allowing it to remain under suction for 3 to 5 more minutes.

    While the water is still flowing through the aspirator, remove the rubber tubing from the sidearm of the vacuum flask, then turn off the water. Do not shut off the water before you remove the tubing. If you do, water may be drawn back into the flask with such force that it may rewet the aspirin sample.

11. Remove a small sample of your synthesized aspirin from the filter and perform the iron(III) chloride test (Section B). Show your results to your instructor and record your observations on the Report Sheet. Also perform the starch test (Section C) and any optional tests (Section D), if assigned.

12. Weigh a disk of clean, dry filter paper on the balance, recording its mass on the Report Sheet. Transfer all the aspirin from the Buchner funnel onto this disk of filter paper, then remove the paper with the aspirin on it to a clean watch glass to dry. Your instructor will tell you how to dry your sample of aspirin. This may require setting out the sample to air-dry until the next laboratory period.

13. Discard the liquid in the vacuum flask and the wet filter paper as directed by your instructor.

14. When your aspirin sample is dry, weigh the paper + aspirin on the balance. Subtracting the mass of the paper from the mass of the paper + aspirin will give the mass of aspirin you prepared. Complete the calculations on the Report Sheet to determine the percent yield of your synthesis.

B. Iron(III) Chloride Test

The presence of unreacted salicylic acid in the synthesized aspirin can be detected with the iron(III) chloride test.

Add about 20 drops (1 mL) of deionized water to each of four clean 10-cm test tubes. Using a clean stirring rod, place a crystal of salicylic acid into the first test tube. In the second, place a crystal of powdered commercial aspirin, and in the third, place a crystal of your synthesized aspirin. The fourth test tube is the control. To each test tube add 1 drop of iron(III) chloride solution. Shake each test tube a little, and observe the colors produced. Record your observations and conclusions on the Report Sheet.

C. Starch Test

Add about 2 mL (40 drops) of deionized water to each of three 10-cm test tubes. To the first, add enough powdered commercial aspirin to just cover the bottom of the test tube. To the second, add a small amount of your synthesized aspirin. The third test tube is the control. To each test tube, add 1 drop of iodine solution. A blue or blue-black color will indicate the presence of starch. Record the colors produced and your conclusions on the Report Sheet.

D. Optional Tests

1. Place a few crystals of your aspirin on a moist piece of pH paper. Do the same with a few crystals each of salicylic acid and commercial aspirin on separate strips of moist pH paper. Record your observations on the Report Sheet.

2. Cover the bottom of a clean 15-cm test tube with powdered commercial aspirin. Add about 20 drops of 10% sodium bicarbonate ($NaHCO_3$) solution to the aspirin. What do you observe when the sodium bicarbonate solution is first added? Complete the equation for this reaction on the Report Sheet.

3. Cover the bottom of a clean 15-cm test tube with powdered commercial aspirin. Add 3 drops of deionized water to the aspirin. Slowly heat the mixture over a burner until the sample melts, then continue heating for 1 more minute. Turn off the burner and carefully fan the vapors from the test tube toward your nose.[1] What do you smell? Complete the equation for the reaction that has taken place on the Report Sheet.

E. Preparation of Methyl Salicylate

Methyl salicylate can be prepared in a manner similar to aspirin, but at a lower temperature. Since the yield of the product will not be determined for this preparation, exact quantities of starting materials will not have to be used. Place a small amount of salicylic acid (about the size of a pencil eraser) in a 20-cm test tube. To this add about 5 mL of methyl alcohol and 5 drops of concentrated sulfuric acid. (*Caution!*) Place the test tube in a 70°C water bath and hold the temperature at 70°C for 15 minutes while the reaction proceeds. Stir the contents to dissolve the salicylic acid in the alcohol. The presence of methyl salicylate can be detected by its minty aroma.

---

[1] This is the correct way to observe the odor of a reagent: Fill you lungs half full of air, then as you waft the vapors toward you, inhale slowly to detect the odor. If the odor is foul or irritating, expel the air in your lungs through your nose to sweep the substance out of your body. Having air in your lungs *before* you attempt to smell a substance is the safe thing to do.

Name ______________________________ Locker Number__________ Date___________

Please print; last name first

PRELIMINARY EXERCISES: *Experiment 18*

The Synthesis of Aspirin

1. Give the molecular formula and molecular weight of the compounds shown below. The atomic weights are C = 12.0, H = 1.00, and O = 16.0.

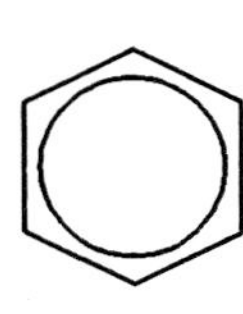

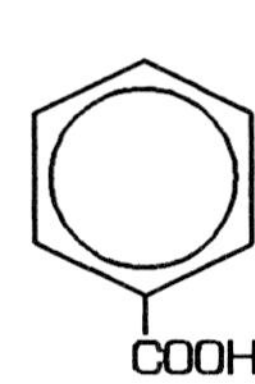

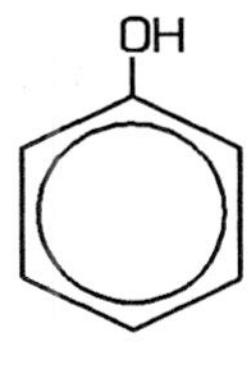

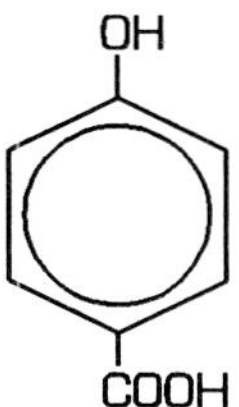

Molecular formula: __________ __________ __________ __________

Molecular weight: __________ __________ __________ __________

2. Which of the following structural formulas is acetylsalicylic acid?

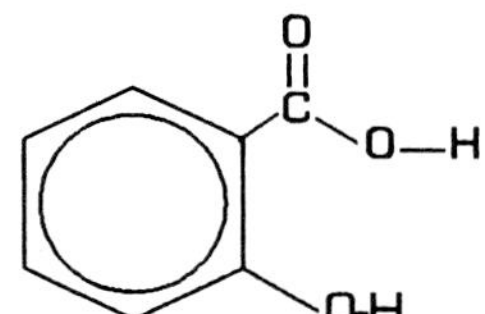

3. What mass of aspirin can be prepared by reacting 5.00 g of salicylic acid with an excess of acetic anhydride? Please show your work.

Mass = __________

(The Preliminary Exercises continue on the following page.)

4. If 1.15 g of aspirin is recovered from an aspirin synthesis reaction that began with 2.00 g of salicylic acid, what would the percent yield be for the preparation? (See Table 1 in the Discussion section for useful data.) Please show your work.

Percent yield = ___________

5. How many grams of pure aspirin are in twelve 5-grain aspirin tablets? (The mass of aspirin in one tablet is given in Section A of the Discussion section.) Please show your work.

Mass = ___________

6. A typical 5-grain aspirin tablet weighs 0.372 g, the weight of the aspirin plus the binder. What is the percent aspirin by weight in one 5-grain aspirin tablet? Please show your work.

Percent = ___________

Name ______________________ Locker Number__________ Date__________
Please print; last name first

REPORT SHEET: *Experiment 18*

The Synthesis of Aspirin

A. Preparation of Aspirin

1.	Mass of weighing paper + salicylic acid:	
2.	Mass of weighing paper:	
3.	Mass of salicylic acid:	
4.	Mass of dry filter paper + aspirin:	
5.	Mass of dry filter paper:	
6.	Mass of aspirin:	

7. Calculate the mass of aspirin that should have been obtained. Show all work, please.

8. Calculate the percent yield of your synthesis. Show all work, please.

B. Iron(III) Chloride Test

Instructor's OK: ______________________

Test Tube	Observations	Conclusions
salicylic acid		
commercial aspirin		
your aspirin		
control		

C. Starch Test

Instructor's OK: ______________________

Test Tube	Observations
Commercial aspirin	
Your aspirin	
control	

D. Optional Tests

1. Reaction with Moist pH Paper. Record your observations. What did you learn?

2. Reaction with Aqueous $NaHCO_3$. Record your observations. Complete the equation.

$$\underset{\text{aspirin}}{C_9H_8O_{4(aq)}} + \underset{\text{sodium bicarbonate}}{NaHCO_{3(aq)}} \rightarrow \underset{\text{sodium salt of aspirin}}{C_9H_7O_4Na_{(aq)}} + \underset{\text{water}}{H_2O} + \underline{\qquad\qquad}$$

3. Reaction with Water. What aroma did you detect? Complete the equation.

$$\underset{\text{aspirin}}{C_9H_8O_{4(aq)}} + \underset{\text{water}}{H_2O_{(l)}} \rightarrow \underset{\text{salicylic acid}}{C_7H_6O_{3(aq)}} + \underline{\qquad\qquad}$$

E. Preparation of Methyl Salicylate

Describe the aroma of methyl salicylate.

Experiment 19

The Properties and Preparation of Esters and Soaps

Purpose:

In this experiment you are going to prepare several esters and note their characteristic aromas. You will also prepare soap and compare its properties in solution to those of a synthetic detergent.

Discussion:

A. Organic Compounds Related to This Experiment

1. Alcohols

Alcohols can be considered derivatives of hydrocarbons in which a hydrogen atom is replaced by a hydroxyl group, –OH. The general formula of an alcohol is R–OH, where R represents the hydrocarbon portion of the molecule. Structural formulas of the first four members of the alcohol class are shown below with their systematic and common names. The systematic names are derived from rules set forth by the International Union of Pure and Applied Chemistry (IUPAC).

```
     H                 H   H
     |                 |   |
 H — C — OH        H — C — C — OH
     |                 |   |
     H                 H   H
```

	CH_3OH	CH_3CH_2OH
systematic name:	methanol	ethanol
common name:	methyl alcohol	ethyl alcohol
also called:	wood alcohol	grain alcohol

```
     H   H   H                 H   H   H
     |   |   |                 |   |   |
 H — C — C — C — OH        H — C — C — C — H
     |   |   |                 |   |   |
     H   H   H                 H   OH  H
```

	$CH_3CH_2CH_2OH$	$CH_3CH(OH)CH_3$
systematic name:	1-propanol	2-propanol
common name:	propyl alcohol	isopropyl alcohol
also called:		rubbing alcohol

2. Carboxylic Acids (the organic acids)

Carboxylic acids contain the carboxyl functional group, $-\overset{\overset{\displaystyle O}{\|}}{C}-OH$, which may also be written as —COOH. Carboxylic acids are weak acids.

The general formula for a carboxylic acid is $R-\overset{\overset{\displaystyle O}{\|}}{C}-OH$ or R—COOH, where R represents either hydrogen or a hydrocarbon group. The first two members of the carboxylic acid family are:

$$H-\overset{\overset{\displaystyle O}{\|}}{C}-OH \quad \text{or} \quad HCOOH \qquad\qquad H-\overset{\overset{\displaystyle H}{|}}{\underset{\underset{\displaystyle H}{|}}{C}}-\overset{\overset{\displaystyle O}{\|}}{C}-OH \quad \text{or} \quad CH_3COOH$$

systematic name: methanoic acid — ethanoic acid
common name: formic acid — acetic acid

Carboxylic acids that have long, straight-chain hydrocarbon groups (with 12 or more carbon atoms) are commonly classified as **fatty acids**, and several are of biological importance. Fatty acids occur in animal or vegetable tissue and nearly always have an even number of carbon atoms. Those that have only single bonds between carbon atoms are **saturated fatty acids**, while those with one or more carbon-carbon double bond are **unsaturated fatty acids**. The general formula of a saturated fatty acid is:

$$C_nH_{2n+1}COOH$$

Stearic acid is a saturated fatty acid with 17 carbon atoms in the hydrocarbon group, R, so its general formula is:

$$C_{17}H_{2(17)+1}COOH \quad = \quad C_{17}H_{35}COOH$$

The principal saturated fatty acids and their formulas are given in Table 1.

Table 1. Principal Saturated Fatty Acids

Common Name	Molecular Formula
lauric acid	$C_{11}H_{23}COOH$
myristic acid	$C_{13}H_{27}COOH$
palmitic acid	$C_{15}H_{31}COOH$
stearic acid	$C_{17}H_{35}COOH$
arachidic acid	$C_{19}H_{39}COOH$

Carboxylic acids that have an unsaturated straight-chain hydrocarbon group (one or more carbon-carbon double bonds) are **unsaturated fatty acids**. Four important unsaturated fatty acids are listed below in Table 2. The first three may be considered derivatives of stearic acid since each has 17 carbon atoms in the hydrocarbon group. But unlike stearic acid, these acids have one or more carbon-carbon double bond in the hydrocarbon chain. Oleic acid has one carbon-carbon double bond and its molecular formula has two fewer hydrogen atoms than that of stearic acid. Linoleic and linolenic acid have two and three double bonds, respectively, and their formulas show the corresponding loss of two hydrogen atoms for each double bond.

Table 2. Principle Unsaturated Fatty Acids

Common Name	Number of Double Bonds	Molecular Formula
oleic acid	1	$C_{17}H_{33}COOH$
linoleic acid	2	$C_{17}H_{31}COOH$
linolenic acid	3	$C_{17}H_{29}COOH$
arachidonic acid	4	$C_{19}H_{31}COOH$

Linoleic, linolenic and arachidonic acids are not synthesized in our bodies and for that reason are classified as **essential fatty acids**. They must be supplied in our diet.

All carboxylic acids can be neutralized with bases (KOH, NaOH, etc.) to form salts. The general neutralization equation is:

$$\underset{\text{carboxylic acid}}{R{-}COOH} + \underset{\text{base}}{NaOH} \rightarrow \underset{\text{salt}}{R{-}COO^-, Na^+} + \underset{\text{water}}{H_2O}$$

The salts have names that end in the characteristic *-ate*. A few organic salts are shown below.

CH_3COO^-, Na^+	$HCOO^-, K^+$	$C_{17}C_{33}COO^-, Na^+$	$(C_{17}H_{35}COO^-)_2, Ca^{2+}$
sodium acetate	potassium formate	sodium oleate	calcium stearate

3. Esters

The general formula for an **ester** is $R{-}\overset{\overset{\displaystyle O}{\|}}{C}{-}O{-}R'$. An ester can be prepared by reacting an alcohol with a carboxylic acid. The R group in the general formula was part of the acid. The other group, symbolized R′ was part of the alcohol. A general equation for the preparation of an ester is given below. Sulfuric acid catalyzes the reaction.

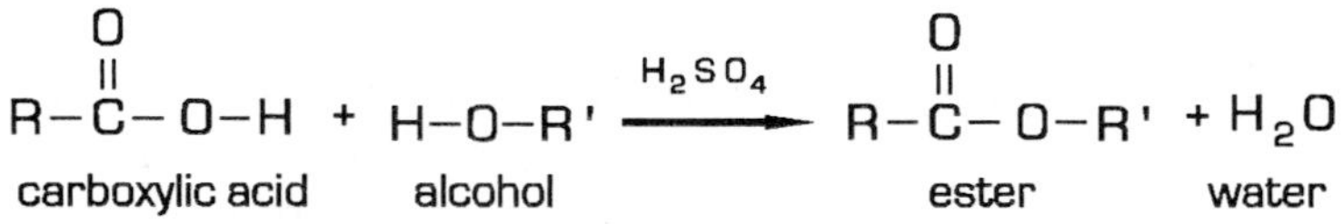

$$\underset{\text{carboxylic acid}}{R-\overset{\overset{\displaystyle O}{\|}}{C}-O-H} + \underset{\text{alcohol}}{H-O-R'} \xrightarrow{H_2SO_4} \underset{\text{ester}}{R-\overset{\overset{\displaystyle O}{\|}}{C}-O-R'} + \underset{\text{water}}{H_2O}$$

The name of an ester indicates the acid and alcohol that combined to form the compound. The first word in the name comes from the alkyl group of the alcohol, (R′), and the second word is the name of the acid with its *-ic* ending changed to *-ate*. The names of esters are similar in form to those of the salts of carboxylic acids. Formulas of four esters are shown below with their IUPAC names and common names written below the IUPAC names.

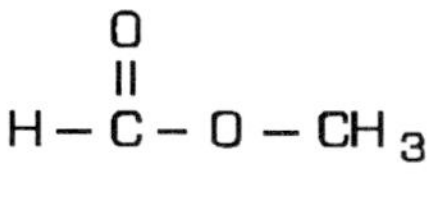

$$H-\overset{\overset{\displaystyle O}{\|}}{C}-O-CH_3$$

methyl methanoate
methyl formate

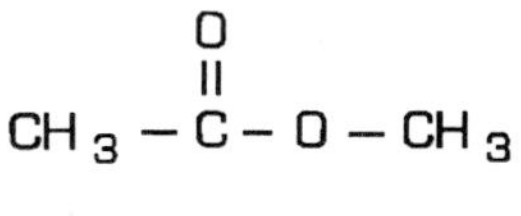

$$CH_3-\overset{\overset{\displaystyle O}{\|}}{C}-O-CH_3$$

methyl ethanozte
methyl acetate

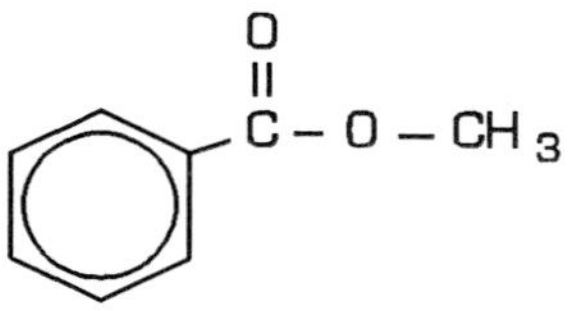

$$C_6H_5-\overset{\overset{\displaystyle O}{\|}}{C}-O-CH_3$$

methyl benzoate

$$CH_3-\overset{\overset{\displaystyle O}{\|}}{C}-O-CH_2-CH_3$$

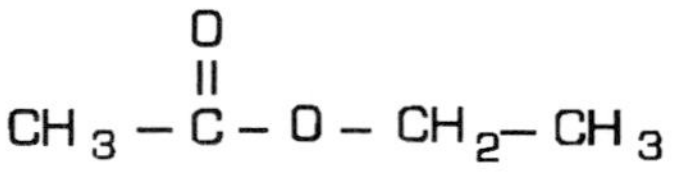

ethyl ethanoate
ethyl acetate

Many esters are colorless liquids with pleasant, fruity aromas. The aroma and flavor of many foods are due to the presence of esters.

4. Triglycerides

Triglycerides are *tri*esters, containing three ester functional groups per molecule. They are large molecules that can be considered the product of the reaction of glycerol (an alcohol containing three hydroxyl groups that is also called glycerine) with three fatty acids. Fats and oils are triglycerides. Structural formulas of glycerol and a general triglyceride are below.

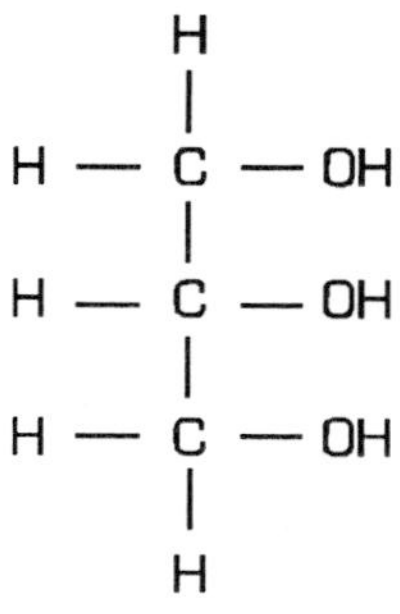

$$\begin{array}{c} H \\ | \\ H-C-OH \\ | \\ H-C-OH \\ | \\ H-C-OH \\ | \\ H \end{array}$$

glycerol (glycerine)

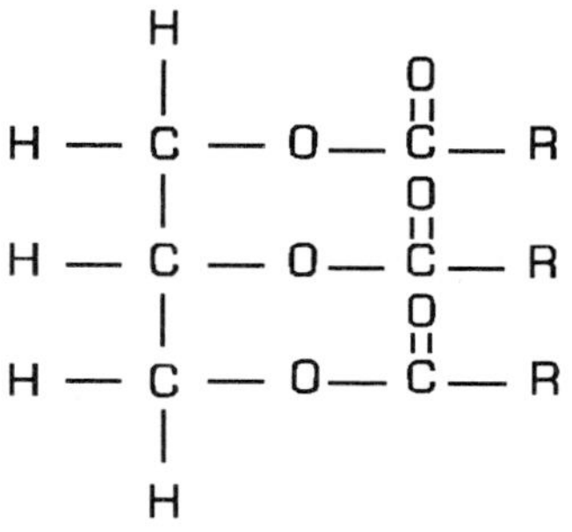

$$\begin{array}{l} \quad\;\; H \\ \quad\;\; | \qquad\quad\;\; O \\ H-C-O-\overset{\|}{C}-R \\ \quad\;\; | \qquad\quad\;\; O \\ H-C-O-\overset{\|}{C}-R \\ \quad\;\; | \qquad\quad\;\; O \\ H-C-O-\overset{\|}{C}-R \\ \quad\;\; | \\ \quad\;\; H \end{array}$$

a general triglyceride

One property that distinguishes a fat from an oil is its physical state at room temperature. Fats are solids and oils are liquids at room temperature. Generally, the long fatty acid hydrocarbon chains are saturated in fats and unsaturated in oils. Both fats and oils can hydrolyze (react with water) to form glycerol and three fatty acids as products, as shown below. In the laboratory the reaction must be heated, and a catalyst like sulfuric acid is required to carry out the reaction. In the body, lipase (an enzyme) catalyzes the reaction in the small intestines.

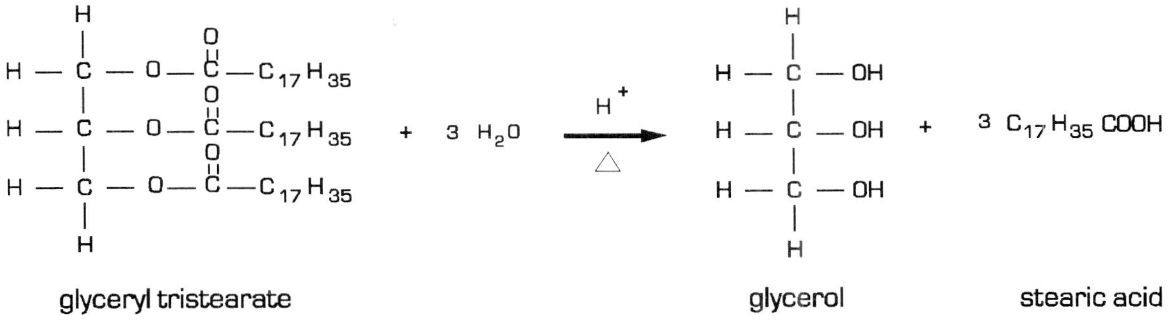

$$\begin{array}{l} \quad\quad H \\ \quad\quad | \\ H-C-O-\overset{O}{\overset{\|}{C}}-C_{17}H_{35} \\ \quad\quad | \\ H-C-O-\overset{O}{\overset{\|}{C}}-C_{17}H_{35} \\ \quad\quad | \\ H-C-O-\overset{O}{\overset{\|}{C}}-C_{17}H_{35} \\ \quad\quad | \\ \quad\quad H \end{array} + 3\ H_2O \xrightarrow[\triangle]{H^+} \begin{array}{l} \quad\quad H \\ \quad\quad | \\ H-C-OH \\ \quad\quad | \\ H-C-OH \\ \quad\quad | \\ H-C-OH \\ \quad\quad | \\ \quad\quad H \end{array} + 3\ C_{17}H_{35}COOH$$

glyceryl tristearate — glycerol — stearic acid

B. **Soap and Cleansing Action**

1. **The Preparation of Soap - Saponification**

Soap is prepared by the hydrolysis of triglycerides in the presence of a strong base like sodium hydroxide, NaOH, or potassium hydroxide, KOH. Instead of obtaining a fatty acid as a product of the hydrolysis, the sodium or potassium salt of the fatty acid is obtained. If a fatty acid did form in the hydrolysis, it would quickly be neutralized by the NaOH or KOH to form the soap.

The soluble salt of a fatty acid is a **soap**. The reaction that produces soap is called **saponification**, and it is of great industrial importance. Sodium stearate is produced when glyceryl tristearate, from animal fat, is heated in the presence of aqueous sodium hydroxide. Sodium stearate is the primary component of Ivory® soap.

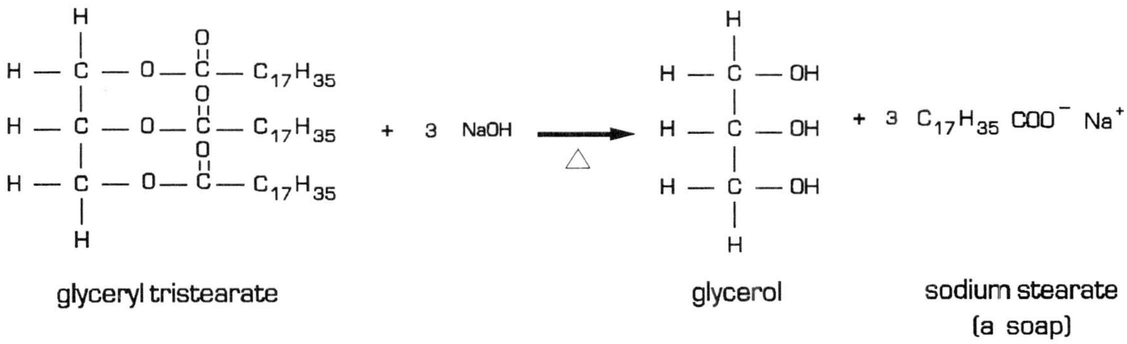

$$\begin{array}{l} \quad\quad H \\ \quad\quad | \\ H-C-O-\overset{O}{\overset{\|}{C}}-C_{17}H_{35} \\ \quad\quad | \\ H-C-O-\overset{O}{\overset{\|}{C}}-C_{17}H_{35} \\ \quad\quad | \\ H-C-O-\overset{O}{\overset{\|}{C}}-C_{17}H_{35} \\ \quad\quad | \\ \quad\quad H \end{array} + 3\ NaOH \xrightarrow[\triangle]{} \begin{array}{l} \quad\quad H \\ \quad\quad | \\ H-C-OH \\ \quad\quad | \\ H-C-OH \\ \quad\quad | \\ H-C-OH \\ \quad\quad | \\ \quad\quad H \end{array} + 3\ C_{17}H_{35}COO^-\ Na^+$$

glyceryl tristearate — glycerol — sodium stearate (a soap)

2. The Cleansing Action of Soaps

A soap molecule is composed of a long nonpolar hydrocarbon chain (which is **hydrophobic**, meaning repelled by water) with a highly polar carboxylate salt on one end (which is **hydrophilic**, meaning water soluble). It is a molecule with a split personality, so to speak, both polar and nonpolar, with one end soluble in water and the other insoluble in water. Sodium stearate, a common soap, is shown below.

CH_3 CH_2 CH_2 CH_2 CH_2 CH_2 CH_2 CH_2 CH_2 O

CH_2 CH_2 CH_2 CH_2 CH_2 CH_2 CH_2 CH_2 C O^- Na^+

hydrophobic nonpolar hydrocarbon chain hydrophylic polar end

When soap molecules are in water, the hydrophilic ends of the molecules are accommodated by the polar water molecules, but the nonpolar hydrophobic ends would be better accommodated in a nonpolar, hydrocarbon-like environment. In the presence of an oil (most dirt is held to clothes by a thin film of oil), the hydrophobic chains dissolve in the nonpolar oil while the hydrophilic ends remain dissolved in the polar aqueous phase. Scrubbing breaks up the oil into tiny droplets, and the soap dissolves in and covers the surface of the droplet. The polar, negatively charged, carboxylate ends of the soap molecules project from the surface of the droplet, giving it a coat of negative charge. These small, electrically coated units are called **micelles**. Because all the micelles are of similar charge on their surfaces, they repel one another and do not coalesce to form larger drops.

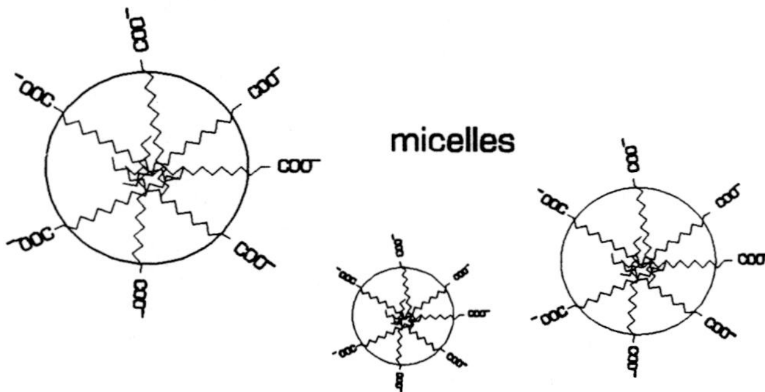

The micelles remain suspended in solution and are rinsed away from clothing and skin. Soap also lowers the surface tension of water, which increases its ability to remove oily dirt.

Soaps have one principal disadvantage. They form insoluble materials when combined with certain metal ions found in hard water, most notably calcium and magnesium ion. The insoluble salts they form with the ions of fatty acids compose what is commonly called soap scum. The formation of insoluble calcium stearate in hard water is shown below:

$$2\ C_{17}H_{35}COO^-,\ Na^+ \quad + \quad Ca^{2+} \quad \rightarrow \quad (C_{17}H_{35}COO^-)_2,\ Ca^{2+} \quad + \quad 2\ Na^+$$

sodium stearate (soluble) — calcium stearate (insoluble)

The "softening" of hard water is nothing more than the removal of these ions either by distillation or deionization processes. In recent years, the problems caused by hard water have been significantly overcome by the use of synthetic detergents in place of soap. Detergents can clean as well as soap but have a much lower tendency to form insoluble salts with calcium and magnesium ion.

3. Synthetic Detergents

Detergent is a general term used to denote any cleansing agent. However, the popular use of the term has come to mean the synthetic detergents that in recent years have largely replaced soap. Synthetic detergents have the desirable property of not forming insoluble precipitates with calcium, magnesium or iron ions. As a result, the problem of soap scum is eliminated. There are more than 1000 synthetic detergents available in the United States. A typical synthetic detergent, sodium laurylsulfate, is shown below. It is used in toothpaste and shampoo. Notice that the polar, ionic end of the molecule, the hydrophilic end, is not a carboxylate ion but a sulfate ion, which has a much lower attraction for calcium or magnesium ion.

$$CH_3\,CH_2\,CH_2\,CH_2\,CH_2\,CH_2\,CH_2\,CH_2\,CH_2\,CH_2\,CH_2\,CH_2{-}O{-}\overset{\overset{\displaystyle O}{\|}}{\underset{\underset{\displaystyle O}{\|}}{S}}{-}O^{-}\quad Na^{+}$$

sodium laurylsulfate

Experimental Procedure:

> **CAUTION: Strong acids and bases are used in several procedures. If you spill a corrosive liquid on your skin, wash the area immediately with cold water and inform the instructor. As always, eye protection must be worn at all times.**
>
> **ENVIRONMENTAL ALERT: Concentrated acids, bases and certain organic compounds should not simply be poured down the drain. Your instructor will advise you on proper waste disposal in this experiment.**

A. Preparation of Some Esters

1. Heat about 400 mL of tap water in a 600-mL beaker to 85°C. While the water is heating, label four 15-cm test tubes A, B, C and D, and obtain corks for each. Place the following reagents in the labeled test tubes:

Test tube A: 20 drops of ethyl alcohol (C_2H_5OH), 20 drops of glacial acetic acid (CH_3COOH), and 10 drops of concentrated sulfuric acid (H_2SO_4) *Caution. Handle concentrated sulfuric acid with care.*

Test tube B: a quantity of salicylic acid (HOC_6H_4COOH) equal to the size of a paper match head, 20 drops of methyl alcohol (CH_3OH), and 3 drops of concentrated sulfuric acid

Test tube C: 20 drops of pentyl alcohol ($C_5H_{11}OH$), 20 drops of glacial acetic acid, and 5 drops of concentrated sulfuric acid

Test tube D: a match-head quantity of benzoic acid (C_6H_5COOH), 20 drops of ethyl alcohol, and 10 drops of concentrated sulfuric acid

2. Stir the contents of each test tube with a glass stirring rod and loosely place a cork in each. When the water bath has reached 85°C, turn off the burner and place each stoppered test tube in the hot water.

3. After 8 to 10 minutes, remove the stopper from each tube and note the odor produced. You should be able to detect the aroma of banana, butter rum, fingernail polish remover and the minty scent of wintergreen. Record the aroma of each mixture on the Report Sheet, then complete the equation and name the ester.

B. Preparation of Soap

1. In an evaporating dish, mix 2 mL of vegetable oil and 3 mL of ethyl alcohol, which is a good solvent for both the triglyceride (the vegetable oil) and sodium hydroxide. Add 20 drops of 50% (19 M) sodium hydroxide. (*Caution, make certain you are wearing eye protection*).

2. Heat the mixture, while stirring, with a moderate, almost luminous flame (reduce the air flow into the burner to produce a moderate flame) until it becomes a thick paste. *Do not overheat.* Allow the evaporating dish to cool. Note the appearance of the product on the Report Sheet, then complete the equation and name the products of the saponification reaction.

C. Properties of Soap

1. Place the soap prepared in Section B in a 150-mL beaker with about 50 mL of deionized water. Heat the mixture, with stirring, for 5 minutes. Allow the solution to cool, then use the soap solution as needed in steps 2 through 5.

2. *Salting Out Soap*: Soap can be forced out of solution by dissolving NaCl in the solution. This technique has been used for years to purify soap. The by-products of the saponification reaction are more soluble in saltwater and remain in solution as the soap comes out of solution.

Pour 10 mL of the soap solution into a 50-mL beaker and add sodium chloride (NaCl) a little at a time, stirring, until no more NaCl dissolves and the bottom of the beaker is covered with NaCl. Record what you observe on the Report Sheet.

Remove the mass of soap floating on the surface by gathering it to one edge with a piece of filter paper and lifting it off with the paper. Rinse this piece of soap with tap water, then attempt to form suds by shaking a small piece with deionized water in a 15-cm test tube. Record your observations on the Report Sheet.

3. *Formation of Fatty Acids*: Add up to 4 drops of 6 M HCl, one drop at a time, to 5 mL of the soap solution (from step 1) in a 15-cm test tube until a precipitate forms. Pour off the aqueous layer and place the precipitate onto a piece of filter paper. Record your observations on the Report Sheet.

4. *Formation of Soap Scum*: To 5 mL of the soap solution (from step 1), add 10 drops of 1 M calcium chloride ($CaCl_2$) solution. Record your observations on the Report Sheet.

5. *pH of the Soap Solution*: Place a drop of the soap solution (from step 1) on a piece of pH paper. Record the pH on the Report Sheet.

D. Properties of a Synthetic Detergent

Obtain 25 mL of a synthetic detergent solution from the stock in the laboratory. Repeat steps 2 through 5 from Section C using the detergent. Record your observations on the Report Sheet.

Name ______________________ Locker Number__________ Date__________
Please print; last name first

PRELIMINARY EXERCISES: *Experiment 19*

The Properties and Preparation of Esters and Soaps

1. Define the following:

a. ester
b. triglyceride
c. soap
d. fatty acid

2. Write the balanced equation for the saponification of glyceryl tristearate with aqueous sodium hydroxide. The formula of glyceryl tristearate is given in the Discussion Section.

3. Write the equation describing the formation of an ester from stearic acid and ethyl alcohol. The formula of stearic acid appears in Table 1. Ethyl alcohol is C_2H_5OH.

4. Why is the hydrocarbon chain of a soap more soluble in oil than in water?

Name ______________________ Locker Number__________ Date____________
Please print; last name first

REPORT SHEET: *Experiment 19*

The Properties and Preparation of Esters and Soaps

A. Preparation of Some Esters

Test tube A: Reaction of Ethyl Alcohol with Acetic Acid

1. Describe the aroma. ______________________

2. Complete the equation for the ester formation and *name the ester.*

$$CH_3CH_2OH + H-O-C(=O)-CH_3 \xrightarrow{H_2SO_4}$$

Test tube B: Reaction of Salicylic Acid with Methyl Alcohol

1. Describe the aroma. ______________________

2. Complete the equation for the ester formation and *name the ester.*

$$CH_3OH + H-O-C(=O)-C_6H_4-OH \xrightarrow{H_2SO_4}$$

Test tube C: Reaction of Pentyl Alcohol with Acetic Acid

1. Describe the aroma ______________________

2. Complete the equation for the ester formation and *name the ester.*

$$CH_3CH_2CH_2CH_2CH_2OH + H-O-C(=O)-CH_3 \xrightarrow{H_2SO_4}$$

Test tube D: Reaction of Benzoic Acid with Ethyl Alcohol

1. Describe the aroma. ____

2. Give the equation for the ester formation and *name the ester.*

$$CH_3CH_2OH + H-O-\overset{\overset{\displaystyle O}{\|}}{C}-C_6H_5 \xrightarrow{H_2SO_4}$$

B. Preparation of Soap

1. Describe the appearance of the soap.

2. Complete the equation for the saponification of glyceryl trioleate, a vegetable oil, and name all products.

$$3\,NaOH + \begin{array}{l} \quad\;\; H \\ \quad\;\; | \\ H-C-O-\overset{\overset{\displaystyle O}{\|}}{C}-C_{17}H_{33} \\ \quad\;\; | \\ H-C-O-\overset{\overset{\displaystyle O}{\|}}{C}-C_{17}H_{33} \\ \quad\;\; | \\ H-C-O-\overset{\overset{\displaystyle O}{\|}}{C}-C_{17}H_{33} \\ \quad\;\; | \\ \quad\;\; H \end{array} \xrightarrow[\Delta]{}$$

C. Properties of Soap

Salting Out Soap

1. What are your observations? Describe the appearance of the material obtained from the salt-saturated solution.

Name ______________________ Locker Number__________ Date___________
Please print; last name first

REPORT SHEET (continued): ***Experiment 19***

The Properties and Preparation of Esters and Soaps

2. Did the collected material form suds when shaken with distilled water? What are your conclusions?

3. The "salting out" process is used commercially to obtain purified soap. What substances can be separated from soap using this process? Think about the substances in solution *before* and *after* the preparation of soap that may not be desirable in a commercial product.

Formation of Fatty Acids

1. Describe your observations and the appearance of the resulting material.

2. Complete the equation for the reaction of the soap, sodium oleate, with $HCl_{(aq)}$ and name the products.

$C_{17}H_{33}COO^-, Na^+ \ + \ HCl_{(aq)} \ \rightarrow$

Formation of Soap Scum

1. Describe the appearance of the product of the reaction:

2. Complete the equation for the reaction of $CaCl_2$ with the soap, sodium oleate. (Be certain to balance the equation and name the products.)

$CaCl_2 \ + \ C_{17}H_{33}COO^-, Na^+ \ \rightarrow$

What is the pH of the soap solution? ___________

D. Some Properties of a Synthetic Detergent

Salting Our Detergent

1. Can a detergent be "salted out" with sodium chloride?

2. Describe the appearance of the final solution.

Formation of Fatty Acids

1. Does the addition of $HCl_{(aq)}$ to a detergent cause the formation of an insoluble material?

2. Describe the appearance of the final solution.

Formation of Soap Scum

1. Does the synthetic detergent form a precipitate when combined with a solution of $CaCl_2$?

2. What are the advantages of using a synthetic detergent in place of a soap in hard water?

What is the pH of the detergent solution? ____________

Experiment 20

Alcohols

Purpose:

In this experiment you will study selected chemical and physical properties of alcohols. The solubility of a series of alcohols in water and in an organic solvent will be examined, and the similarities and differences of primary, secondary and tertiary alcohols will be seen in two chemical tests.

Discussion:

Alcohols are organic compounds that contain a hydroxyl group (−OH) bonded to a carbon atom that is *not* a carbonyl carbon (C=O). If the −OH group is joined to a carbon in an aromatic ring, the compounds are **phenols**, which have properties different from alcohols. We will only be concerned with alcohols here. Alcohols can be separated into three subclasses, **primary** (p), **secondary** (s) and **tertiary** (t), based on the number of alkyl groups (R−) joined to the carbon atom with the −OH group. The general formula for an alcohol can be abbreviated R−OH.

```
     H              R'              R'
     |              |               |
 R — C — OH     R — C — OH      R — C — OH
     |              |               |
     H              H               R"
```

primary alcohol   secondary alcohol   tertiary alcohol

The names of alcohols end in *-ol*. Several alcohols are shown below with their IUPAC names and the subclass to which each belongs (p, s or t). Their common names are also given.

CH_3OH	CH_3CH_2OH	$CH_3CH_2CH_2CH_2OH$
methanol (p) (methyl alcohol)	ethanol (p) (ethyl alcohol)	1-butanol (p)

```
                   OH                 CH3
                   |                  |                    CH3
CH3 — CH2 — C — CH3         CH3 — C — OH                |
                   |                  |             CH3 CH CH2 — OH
                   H                  CH3
```

2-butanol (s)   2-methyl-2-propanol (t) (tertiary butyl alcohol)   2-methyl-1-propanol (p) (isobutyl alcohol)

The common alcohols are colorless liquids at room temperature. Their boiling points are considerably higher than those of alkanes or ethers of similar molecular weight due the ability of alcohols to engage in hydrogen bonding with one another. Their ability to hydrogen bond is also responsible for the solubility of low molecular-weight alcohols in water. In general, alcohols with one −OH group and fewer than five carbon atoms are soluble in water, but as the nonpolar hydrocarbon chain gets larger, hydrogen bonding through the single $-OH$ group (which enhances solubility) is not sufficient to keep the entire molecule in solution. An important chemical property of alcohols is their ability to be oxidized to form other compounds. Chemical oxidation can be accomplished using a variety of reagents: acidic potassium dichromate ($K_2Cr_2O_7-H_2SO_4$), acidic potassium permanganate ($KMnO_4-H_2SO_4$) and chromic acid ($CrO_3-H_2SO_4$). Because we are more concerned with the organic compounds in these reactions, the oxidizing agent will simply be represented as [O] above the arrow in the oxidation equation. Primary alcohols are oxidized to aldehydes. In turn, aldehydes are oxidized to carboxylic acids.

```
    H   H   H   H                  H   H   H   H                  H   H   H   OH
    |   |   |   |       [O]        |   |   |   |       [O]        |   |   |   |
H - C - C - C - C - OH  ----> H - C - C - C - C = O   ----> H - C - C - C - C = O
    |   |   |   |                  |   |   |                      |   |   |
    H   H   H   H                  H   H   H                      H   H   H
```

1-butanol (p)	butanal	butanoic acid
primary alcohol	aldehyde	carboxylic acid

Secondary alcohols are oxidized to ketones, which react no further.

```
    H   H   OH  H                        H   H   O   H
    |   |   |   |          [O]           |   |   ||  |
H - C - C - C - C - H     ---->     H - C - C - C - C - H
    |   |   |   |                        |   |       |
    H   H   H   H                        H   H       H
```

2-butanol	butanone
secondary alcohol	ketone

Tertiary alcohols are resistant to chemical oxidation.

In summary:

$$\text{primary-alcohol} \xrightarrow{[O]} \text{aldehyde} \xrightarrow{[O]} \text{carboxylic acid}$$

$$\text{secondary-alcohol} \xrightarrow{[O]} \text{ketone}$$

$$\text{tertiary-alcohol} \xrightarrow{[O]} \text{no reaction}$$

Another useful chemical property of secondary and tertiary alcohols is their reaction with hydrochloric acid in the presence of zinc chloride to form alkyl chlorides. This reaction is the basis of the Lucas test, which, because of the marked difference in rates of reaction, serves as a convenient way to distinguish between primary, secondary and tertiary alcohols.

$$\underset{\text{secondary or tertiary alcohol}}{R-OH_{(aq)}} + HCl_{(aq)} \xrightarrow{ZnCl_2} \underset{\text{alkyl chloride}}{R-Cl_{(l)}} + H_2O_{(l)}$$

Tertiary alcohols react quickly to form an alkyl chloride while secondary alcohols require between 5 and 15 minutes to react at room temperature. Primary alcohols show no reaction even after 30 minutes. The visible sign of reaction is the formation of the water-insoluble alkyl chloride as a milky suspension that, in time, separates to form a layer resting on top of the aqueous solution.

Experimental Procedure:

> **CAUTION: As always, eye protection must be worn at all times.**
>
> **ENVIRONMENTAL ALERT: Discard all waste as directed by your instructor.**

A. Solubility of Alcohols in Water

Label five test tubes A through E, then place 1 mL (20 drops) of deionized water in each one. Add 10 drops of each alcohol listed below to the appropriate tube.

A. ethanol
B. 1-butanol
C. 2-butanol
D. 1-hexanol
E. 2-methyl-2-propanol

Seal each test tube with a cork or plastic film and shake the mixture thoroughly. Then place each in a test tube rack. After 5 minutes, observe each mixture. If the alcohol is *insoluble* in water, you will see either a cloudy mixture or two separate layers. If it is *soluble*, you will see a clear (transparent) solution with no layering. Record your observations on the Report Sheet. For those alcohols you judge as insoluble, rationalize your conclusion in terms of the number of carbon atoms in the alcohol. Draw the Lewis structure of each alcohol.

B. Solubility of Alcohols in Acetone

Acetone is widely used as an organic solvent because of its ability to dissolve a wide variety of compounds. Empty the contents from the test tubes used in Section A and rinse each with a small volume of acetone. Then place 1 mL (20 drops) of acetone in test tubes B, C, D and E. Add 10 drops of the following alcohols to the appropriate test tube.

B. 1-butanol
C. 2-butanol
D. 1-hexanol
E. 2-methyl-2-propanol

Cork or seal each test tube and shake the mixture vigorously, then place each in a test tube rack. After several minutes, observe each mixture and record the solubility of the alcohol in acetone on the Report Sheet. Indicate those alcohols that were not soluble in water but were in acetone.

> **CAUTION: The chromic acid reagent and the Lucas reagent are very corrosive and must be handled with care. If either is spilled on the skin, rinse immediately and thoroughly with cold water, then contact the instructor.**

C. Chromic Acid Test: the Oxidation of Alcohols

Empty the contents of the test tubes used in Section B, rinse each with a few milliliters of acetone and allow them to air-dry. Place 1 mL (20 drops) of *reagent-grade* acetone into each of five test tubes labeled A through E. Then add 2 drops of each alcohol listed below to the appropriate test tube. No alcohol will be added to test tube E. It will serve as a control so you can observe the behavior of chromic acid in the absence of alcohol. If you are analyzing an unknown alcohol, (Section E), test it at the same time you test the known samples.

A. ethanol (p)
B. 1-butanol (p)
C. 2-butanol (s)
D. 2-methyl-2-propanol (t)
E. control - no alcohol is added

Thoroughly mix the alcohol with the acetone, then add one drop of chromic acid reagent to each test tube. *Caution: Handle the chromic acid reagent with care.* A color change from orange to green or to blue-green within 5 seconds is a positive test, indicating the alcohol has been oxidized. *Ethanol will give a positive test.* Record your observations on the Report Sheet and, for those alcohols that gave a positive test, draw the Lewis structure(s) of the expected oxidation product(s).

D. Lucas Test: the Conversion of Alcohols to Alkyl Chlorides

The Lucas reagent is a mixture of $ZnCl_2$ in 12 M hydrochloric acid, HCl. It is used to determine whether an alcohol is primary (p), secondary (s) or tertiary (t). This test is limited to water-soluble alcohols, those containing fewer than six carbon atoms.

The Lucas reagent will convert a *soluble* alcohol to an *insoluble* alkyl chloride. The appearance of a milky suspension in the reaction mixture indicates the reaction has taken place. The *time* it takes for you to see evidence that the reaction is occurring depends on the structure of the alcohol. *Tertiary alcohols* react almost immediately. *Secondary alcohols* usually require 5 to 15 minutes before evidence of the reaction is observed. *Primary alcohols* will show little or no reactivity in 30 minutes.

Add 2 mL (40 drops) of Lucas reagent to five labeled test tubes. *Caution: Handle the Lucas reagent with care.* Then add 4 drops of each alcohol listed below to the appropriate test tube. Use 4 drops of deionized water for the control. Cork or seal each test tube and mix the reagents well. Then place them in a test tube rack for observation at specific times.

A. cyclohexanol (s)
B. 1-butanol (p)
C. 2-butanol (s)
D. 2-methyl-2-propanol (t)
E. control

Record your observations for each alcohol after **1** minute, **5** minutes, **15** minutes and **30** minutes on the Report Sheet. Note if your observations are consistent with the primary, secondary or tertiary structure of each alcohol.

E. Analysis of an Unknown Alcohol

In this analysis you will be given an alcohol containing fewer than six carbon atoms, and you are to determine if it is a primary, secondary or tertiary alcohol. You will use the chromic acid test and the Lucas test for this determination. It is most convenient to analyze the unknown alcohol at the same time you perform these tests on the known alcohols in Section C and D.

Record the unknown number, your observations and conclusion in on the Report Sheet. Be certain to justify your conclusion.

Name ______________________ Locker Number__________ Date____________
Please print; last name first

PRELIMINARY EXERCISES: *Experiment 20*

Alcohols

1. Define the following terms:

a. primary alcohol
b. secondary alcohol
c. tertiary alcohol

2. The fact that the hydroxyl group of an alcohol can engage in hydrogen bonding with water is given as an explanation for the water solubility of the lighter alcohols. In the space below, show how ethanol and water engage in hydrogen bonding. You may want to consult your textbook.

3. How can you tell if an alcohol is primary, secondary or tertiary using the Lucas test?

Name ______________________________ Locker Number__________ Date____________
Please print; last name first

REPORT SHEET: *Experiment 20*

Alcohols

A. Solubility of Alcohols in Water

Alcohol	Observations (Soluble or Insoluble)	Lewis Structure of the Alcohol
A. ethanol		
B. 1-butanol		
C. 2-butanol		
D. 1-hexanol		
E. 2-methyl-2-propanol		

Rationalize the trend in solubility in terms of the number of carbon atoms in the alcohol.

B. Solubility of Alcohols in Acetone

Alcohol	Observations (Soluble or Insoluble)	Soluble in Acetone—Insoluble in Water (Yes or No)
B. 1-butanol		
C. 2-butanol		
D. 1-hexanol		
E. 2-methyl-2-propanol		
Comment on the ability of acetone to dissolve organic compounds.		

C. Chromic Acid Test: the Oxidation of Alcohols

Alcohol	Observations (Positive or Negative)	Lewis Structure(s) of Reaction Product(s)
A. ethanol		
B. 1-butanol		
C. 2-butanol		
D. 2-methyl-2-butanol		

Name ______________________________ Locker Number__________ Date____________
Please print; last name first

REPORT SHEET (continued): *Experiment 20*

Alcohols

D. Lucas Test: the Conversion of Alcohols to Alkyl Chlorides

Alcohol	Observations: Milky suspension or layers after. . .				Do observations agree with the structure? (Yes or No)
	1 minute (Yes or No)	5 minutes (Yes or No)	15 minutes (Yes or No)	30 minutes (Yes or No)	
A. cyclohex-anol					
B. 1-butanol					
C. 2-butanol					
D. 2-methyl-2-propanol					
E. control					

E. Analysis of an Unknown Alcohol

Unknown number: __________

Observations	Conclusion and Rationale
Chromic acid test:	
Lucas test:	
The unknown alcohol is: **Primary** **Secondary** **Tertiary** (circle one)	

Experiment 21

Aldehydes and Ketones

Purpose:

In this experiment you will study some of the chemical and physical properties of aldehydes and ketones, two classes of organic compounds. The solubility in water and aroma of several aldehydes and ketones will observed, and the way each class responds to oxidation will show an important chemical difference between aldehydes and ketones.

Discussion:

Both aldehydes and ketones contain the carbonyl group, $-\overset{O}{\overset{\|}{C}}-$, but that is where the similarity ends. All aldehydes have at least one hydrogen atom bonded to the carbon atom of the carbonyl group while ketones have none. In ketones, two organic groups (R–) are bonded to the carbonyl carbon. The presence or absence of the carbonyl hydrogen is responsible for the chemical differences that exist between aldehydes and ketones.

$R-\overset{O}{\overset{\|}{C}}-H$	$CH_3-\overset{O}{\overset{\|}{C}}-H$	$R-\overset{O}{\overset{\|}{C}}-R$	$CH_3-\overset{O}{\overset{\|}{C}}-CH_3$
the aldehyde group	acetaldehyde	the ketone group	acetone

The condensed formula for the aldehyde group is frequently written –CHO. Do not confuse this with that used for alcohols, –COH.

Naming the Aldehydes and Ketones

Most of the common aldehydes and ketones were known long before the IUPAC system of nomenclature was developed. As a result, common names are still widely used for these compounds. The common names for aldehydes are derived from the name of the carboxylic acid with the same number and arrangement of carbon atoms. For example, the aldehyde with two carbon atoms, CH_3CHO, is called acetaldehyde, a name derived from the carboxylic acid with two carbon atoms, acetic acid, CH_3COOH. The *-ic* or *-oic* suffix that ends the name of the acid is replaced with *-aldehyde*. Form*ic* acid, HCOOH, becomes form*aldehyde*, HCHO, and so forth. In the IUPAC nomenclature system, all aldehydes end with the suffix *-al*, which is added to the name of the longest chain of carbon atoms that contains the carbonyl group. Acetaldehyde would be named ethan*al* and formaldehyde would be methan*al*. Both the common and IUPAC names of several aldehydes are given in Table I on the next page.

Common names for ketones are usually derived from the names of the organic groups bonded to the carbonyl carbon. The simplest ketone would have the common name of dimethyl ketone, but 99% of the time it is called acetone, a very old common name. In fact, it is rare to hear the compound called dimethyl ketone. The IUPAC scheme for naming ketones

Table I – Naming Aldehydes		
Structural Formula	Common Name	IUPAC Name
$H-C(=O)-H$	formaldehyde	methanal
$CH_3-C(=O)-H$	acetaldehyde	ethanal
$CH_3CH_2-C(=O)-H$	propionaldehyde	propanal
$CH_3-CH(CH_3)-C(=O)-H$	isobutryaldehyde	2-methylpropanal
$C_6H_5-C(=O)-H$	benzaldehyde	benzaldehyde

Table II – Naming Ketones		
Structural Formula	Common Name	IUPAC Name
$CH_3-C(=O)-CH_3$	acetone (dimethyl ketone)	propanone
$CH_3-C(=O)-CH_2CH_3$	methylethylketone	butanone
$CH_3-C(=O)-CH_2CH_2CH_2CH_2CH_3$	methylpentylketone	2-heptanone
$C_6H_5-C(=O)-CH_3$	methylphenylketone	acetophenone

requires that the name ends with *-one*. The name of the ketone is derived from the longest chain of carbon atoms containing the carbonyl group with its *-e* ending replaced by *-one*; propan*e* becomes propan*one*. The position of the carbonyl group in this chain is given by a number which is determined by counting carbon atoms from the nearest end of the chain. See the examples of ketone names in Table II.

The Solubility of Aldehydes and Ketones in Water

In general, aldehydes and ketones with fewer than 5 carbon atoms will be completely soluble in water, but as the number of carbon atoms increases solubility falls off markedly. The reason for this lies in the fact that only the carbonyl group can be accommodated by water through hydrogen bonding and dipole-dipole interactions. If the carbon chains are not large, the interaction of the carbonyl group with water is sufficient to keep the entire molecule in solution. But as the number of carbon atoms increases, the water-repelling character of the carbon chain overwhelms the water-loving nature of the carbonyl group and it is increasing difficult for the organic molecule to stay in solution. If an organic molecule is to be soluble in water it must be accommodated by the solvent by interacting with water in ways similar to the way water molecules interact with themselves. Since hydrogen-bonding between molecules is of critical importance in water, the solubility of an aldehyde or ketone is enhanced by its ability to hydrogen bond with water as is shown above in Figure 1.

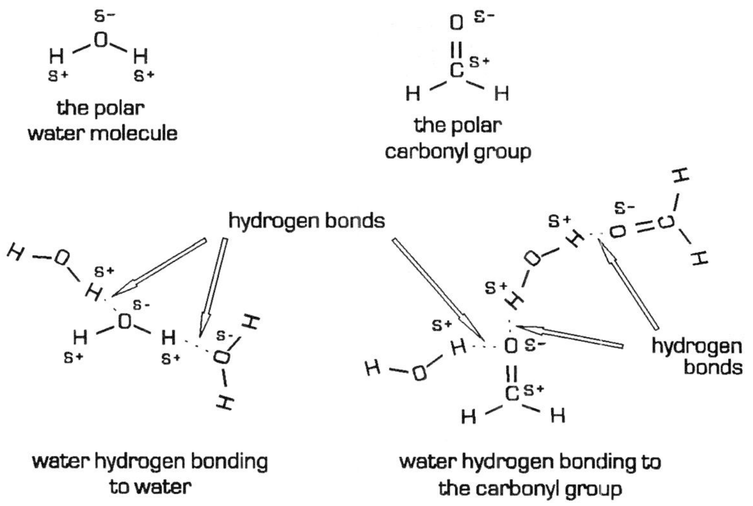

the polar water molecule

the polar carbonyl group

hydrogen bonds

hydrogen bonds

water hydrogen bonding to water

water hydrogen bonding to the carbonyl group

Figure 1 The Importance of Hydrogen Bonding

Some Important Aldehydes and Ketones

The simplest aldehyde is **formaldehyde** (common name) and its structural formula and IUPAC name can be found in Table I. Pure formaldehyde is a colorless gas at room temperature and pressure, but it is more often encountered as a 40% solution in water and as such is known as *formalin*. Formalin is used to preserve tissue and it has been used to sterilize surgical instruments. Formaldehyde is easily oxidized to formic acid by oxygen in the air. Its use to preserve tissue is because of this ease of oxidation. Oxygen cannot attack tissue if it is consumed by a faster reaction with formaldehyde. The next higher aldehyde is **acetaldehyde** (common name) and its structural formula and IUPAC name are also in Table I. Acetaldehyde is a colorless liquid at room temperature and pressure with a sharp, irritating odor. Like formaldehyde, it is readily oxidized to a carboxylic acid, acetic acid, in this case. The oxidation of aldehydes will be discussed later on in this experiment.

The structural formulas of some aldehydes that are found in nature are shown in Figure 2. The characteristic aroma of each one suggests it source.

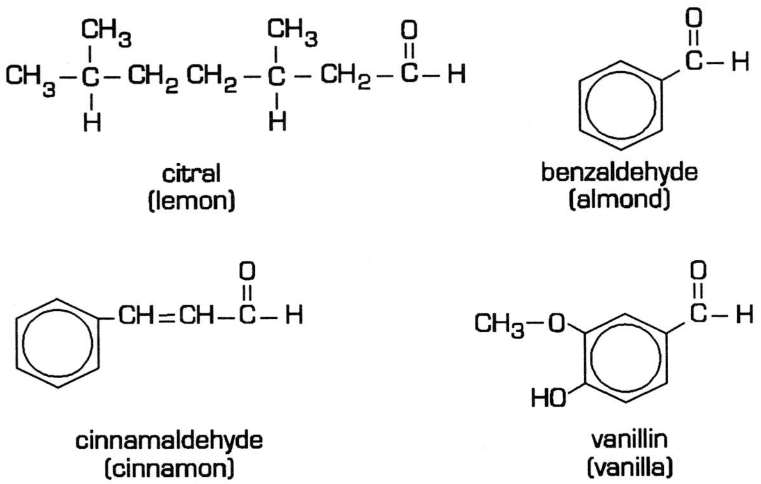

$$CH_3-\overset{\displaystyle CH_3}{\underset{\displaystyle H}{C}}-CH_2CH_2-\overset{\displaystyle CH_3}{\underset{\displaystyle H}{C}}-CH_2-\overset{\displaystyle O}{\overset{\|}{C}}-H$$

citral
(lemon)

$C_6H_5-\overset{O}{\overset{\|}{C}}-H$

benzaldehyde
(almond)

$C_6H_5-CH=CH-\overset{O}{\overset{\|}{C}}-H$

cinnamaldehyde
(cinnamon)

vanillin
(vanilla)

Figure 2 Four Aldehydes Found in Nature

The simplest and most industrially important ketone is **acetone** (common name) and its structural formula and IUPAC name is in Table II. Acetone is a colorless liquid at room temperature and pressure and is very soluble in water. Acetone is a common solvent for organic compounds, and the aroma of fingernail polish remover is largely that of acetone. Small amounts of acetone are normally produced in the body during the metabolism of fats. However, under conditions of starvation or diabetes mellitus the production of acetone can increase to life-threatening levels. The sweet, fruity order of acetone can be detected on the breath of a diabetic experiencing elevated levels of acetone in the blood.

The camphor tree produces a fragrant cyclic ketone, commonly called **camphor**, that is used in medicines and liniments. Another, rather exotic ketone is **muscone** obtained from the preputial follicles of the musk deer of northern Asia and used in perfumes.

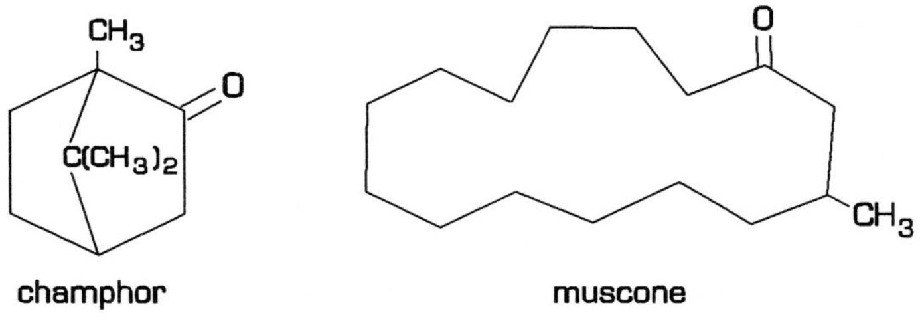

champhor

muscone

Figure 3 Two Ketones Found in Nature

The Preparation of Aldehydes and Ketones From Alcohols

Aldehydes and ketones can be synthesized by the oxidation of alcohols as was described in the previous experiment.

a. Oxidation of a *primary alcohol* produces an *aldehyde*.
b. Oxidation of a *secondary alcohol* produces a *ketone*.
c. *Tertiary alcohols* are not oxidized under normal laboratory conditions.

In the equation below the [O] symbol above the arrow indicates oxidation. Note that aldehydes can be further oxidized to form carboxylic acids. In this experiment the oxidation will be carried out using chromic acid.

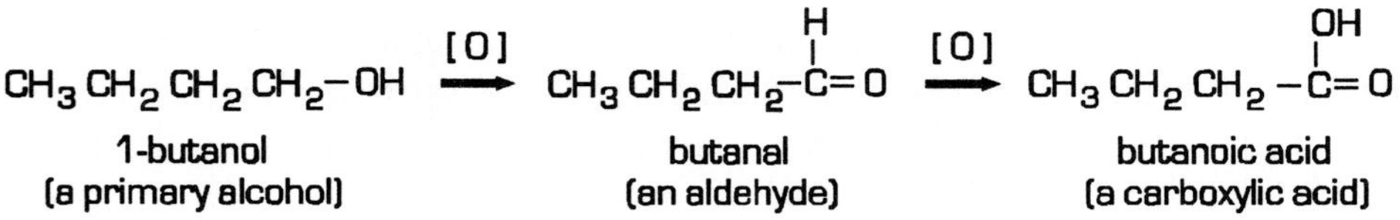

$$\underset{\text{1-butanol (a primary alcohol)}}{CH_3CH_2CH_2CH_2-OH} \xrightarrow{[O]} \underset{\text{butanal (an aldehyde)}}{CH_3CH_2CH_2-\overset{\displaystyle H}{\overset{|}{C}}=O} \xrightarrow{[O]} \underset{\text{butanoic acid (a carboxylic acid)}}{CH_3CH_2CH_2-\overset{\displaystyle OH}{\overset{|}{C}}=O}$$

$$CH_3CH_2-\underset{H}{\overset{OH}{C}}-CH_3 \xrightarrow{[O]} CH_3CH_2-\overset{O}{\overset{\|}{C}}-CH_3$$

2-butanol (a secondary alcohol) → butanone (a ketone)

$$CH_3CH_2-\underset{CH_3}{\overset{OH}{C}}-CH_3 \xrightarrow{[O]} \text{no reaction}$$

2-methyl-2-butanol (a tertiary alcohol)

The Oxidation of Aldehydes

Aldehydes are easily oxidized to carboxylic acids. Ketones, on the other hand, are not oxidized under most laboratory conditions. For many aldehydes simple exposure to oxygen in the air will result in the slow formation of acids.

Several chemical tests have been developed over the years to determine the presence of an aldehyde based on their ease of oxidation. One very sensitive test is the **Benedict test** in which the oxidation of the aldehyde is accompanied by the simultaneous reduction of copper(II) ion, Cu^{2+}, to a highly colored precipitate of copper(I) oxide, Cu_2O. The appearance of the colored product signals the oxidation of an aldehyde to the carboxylic acid. Most often the copper(I) oxide forms as a brick-red precipitate, but it can also produce a yellow or orange-yellow color too. Any of these colors indicate the presence of an aldehyde.

The Benedict reagent is a basic (NaOH) solution of copper sulfate, $CuSO_4$, that is blue. The chemical equation for the test is given below. Note that because the solution is basic, the carboxylic acid is produced as its sodium salt.

$$CH_3-\overset{H}{C}=O + 2Cu^{2+} + Na^+ + 5OH^- \longrightarrow CH_3-\overset{O^-\ Na^+}{C}=O + \underline{Cu_2O} + 3H_2O$$

acetaldehyde; Benedict Reagent (blue); sodium acetate; copper(I) oxide (brick-red)

You will use the Benedict test to identify aldehydes in this experiment.

Experimental Procedure:

> **CAUTION: As always, eye protection must be worn at all times.**
>
> **ENVIRONMENTAL ALERT: Discard all waste as directed by your instructor.**

A. Solubility of Aldehydes and Ketones in Water

Label nine small test tubes A through I, then place 1 mL (20 drops) of deionized water in each one. Add 10 drops of each liquid listed below to the appropriate teat tube. Glucose, $C_6H_{12}O_6$, is a solid, so add an amount of it to test tube F equal to the size of the head of a paper match. Glucose is a simple carbohydrate that contains an aldehyde group (−CHO) and five hydroxyl groups (−OH).

A. ethanal (acetaldehyde)
B. propanal
C. 2-methylpropanal
D. heptanal
E. benzaldehyde
F. glucose (blood sugar)
G. propanone (acetone)
H. 2-heptanone
I. acetophenone

Seal each test tube with a cork or plastic film (Parafilm® works well) and shake each mixture thoroughly, then set it aside in a test tube rack. After 5 minutes, observe each mixture. If the substance is *insoluble* in water, you will see either a cloudy mixture or two separate layers. If it is *soluble*, you will see a clear (transparent) solution with no layering. Record your observations on the Report Sheet then answer the questions.

B. The Aroma of Selected Aldehydes and Ketones

Check the aroma of the six compounds listed below. They can be found in the laboratory room in small labeled bottles. Remove the top from one bottle at a time and fan the fumes toward your nose as you inhale cautiously. It is best to fill you lungs half full of air before you test an aroma. That way you can completely exhaust the substance from your system before moving on to the next sample.

A. camphor
B. cinnamaldehyde
C. benzaldehyde
D. propanone (acetone)
E. acetaldehyde
F. vanillin

Record your observations on the Report Sheet.

C. Chromic Acid Oxidation of Alcohols, Aldehydes and Ketones

Aldehydes and ketones can be prepared by the oxidation of primary and secondary alcohols. Chromic acid is able to oxidize these alcohols. Chromic acid can also oxidize aldehydes to carboxylic acids (R−CHO → R−COOH). As oxidation occurs a color change is observed that visual sign that oxidation has taken place.

CAUTION: The chromic acid reagent is very corrosive. If spilled on the skin, rinse immediately and thoroughly with cold water, then contact the instructor.

In each of 9 small test tubes labeled A through I, place 1 mL (20 drops) of *reagent-grade* acetone. To test the quality of the chromic acid reagent, add one drop of the reagent to test tube A (acetone). The color of the mixture should remain orange. If the color changes, inform your instructor. Once you are confident the reagent is satisfactory, add 2 drops of each of the following compounds to the appropriate test tube. Mix each solution well.

A. propanone (acetone, the control)
B. ethanol (primary alcohol)
C. 2-propanol (secondary alcohol)
D. 2-methyl-2-propanol (tertiary alcohol)
E. ethanal
F. propanal
G. 2-methylpropanal
H. benzaldehyde
I. 2-heptanone

Add *one drop* of chromic acid reagent to each test tube. *Caution: Handle the chromic acid reagent with care.* Gently tap the bottom of each test tube against the palm of your hand to mix the contents. If no reaction occurs, a negative result, the solution should remain orange. Propanone will give a negative test. A color change from orange to green or to blue-green within 5 seconds is a positive test, indicating the oxidation of the compound. *Ethanol will give a positive test.* Continue to watch the samples for five minutes to see if any of the compounds react slowly to give a positive test. Record your observations on the Report Sheet. Discard all waste from the chromic acid tests in the appropriate waste container in the lab.

D. Benedict Test

The Benedict reagent can oxidize aldehydes to carboxylic acids but cannot oxidize primary or secondary alcohols. The weaker oxidizing medium of the Benedict test can be used to distinguish between aldehydes and alcohols. The Benedict test requires heating in boiling water, so before going further begin heating a 400 mL beaker half-filled with water to a boil. Label 8 six-inch test tubes A through H. In each place 2 mL (40 drops) of Benedict reagent. Then add 1 mL (20 drops) of each of the following compounds to the appropriate test tube. For glucose, a solid, use an amount equal to the size of the head of a paper match.

A. pure water (control)
B. ethanol (primary alcohol)
C. ethanal
D. propanal
E. 2-methylpropanal
F. benzaldehyde
G. glucose (a simple carbohydrate)
H. propanone (acetone, a ketone)

Seal each test tube with a clean cork or plastic film and mix thoroughly. Place each labeled test tube in a *boiling* water bath for 10 minutes. Remove, and place the test tubes in a test tube rack. As the test tubes cool, check for a red, brown or yellow precipitate. A colored precipitate indicates a positive test. Glucose will give a positive test. Record the results of the Benedict test on the Report Sheet and answer the questions. Discard all waste from the Benedict Test in the appropriate waste container.

Name ______________________________ Locker Number__________ Date____________
Please print; last name first

PRELIMINARY EXERCISES: *Experiment 21*

Aldehydes and Ketones

1. Define the following terms:

a. aldehyde
b. ketone
c. Benedict Test

2. Draw a picture showing hydrogen-bond formation between water and the carbonyl group of an aldehyde or ketone. Use a dotted line (•••) to indicate a hydrogen bond.

3. Write the balanced chemical equation for the reaction of acetaldehyde, CH_3CHO, with the Benedict Reagent.

Name ______________________ Locker Number__________ Date__________
Please print; last name first

REPORT SHEET: *Experiment 21*

Aldehydes and Ketones

A. Solubility of Aldehydes and Ketones in Water

Test Tube	Compound Tested	Soluble (S) or Insoluble (I)
A	ethanal (acetaldehyde)	
B	propanal	
C	2-methylpropanal	
D	heptanal	
E	benzaldehyde	
F	glucose (blood sugar)	
G	propanone (acetone)	
H	2-heptanone	
I	acetophenone	

1. Comparing only aldehydes (A, B, C, D and E), explain the trend in solubility that you observed.

2. Comparing only the ketones (G, H and I), explain the trend in solubility that you observed.

3. Why is glucose so soluble especially when you consider that it is a large molecule with 6 carbon atoms.

B. The Aromas of Selected Aldehydes and Ketones

Compound Tested	Description of Aroma
camphor	
cinnamaldehyde	
benzaldehyde	
propanone (acetone)	
acetaldehyde	
vanillin	

C. Chromic Acid Oxidation of Alcohols, Aldehydes and Ketones

Test Tube	Compound Tested	Observation
A	propanone (acetone)	
B	ethanol	
C	2-propanol	
D	2-methyl-2-propanol	
E	ethanal	
F	propanal	
G	2-methylpropanal	
H	benzaldehyde	
I	2-heptanone	

Name ______________________ Locker Number__________ Date__________
Please print; last name first

REPORT SHEET (continued): *Experiment 21*

Aldehydes and Ketones

C. Chromic Acid Oxidation of Alcohols, Aldehydes and Ketones (continued)

Which compounds did not undergo chromic acid oxidation?
Which compounds are alcohols? Why did all the alcohols not react the same?
Which compounds are aldehydes?
Which compounds are ketones?
Comment on the ability of aldehydes and ketones to undergo chromic acid oxidation.

Draw the structural formula for each listed compound and the product it forms as it is oxidized by chromic acid.

Starting Compound	Structural Formula—Reactant	Structural Formula—Product
ethanol		
2-propanol		
ethanal		

D. Benedict Test

Test Tube	Compound Tested	Observations
A	pure water (control)	
B	ethanol	
C	ethanal	
D	propanal	
E	2-methylpropanal	
F	benzaldehyde	
G	glucose	
H	propanone (acetone)	

1. Which compounds were oxidized by the Benedict reagent?

2. Can this test be used to tell whether a compound is an aldehyde or a ketone?

3. Why did glucose give a positive Benedict test?

Experiment 22

Organic Functional Group Tests

> **INSTRUCTIONAL NOTE: Two laboratory periods may be required to complete the entire experiment.**

Purpose:

In this experiment you will learn the importance and usefulness of the functional group concept in organic chemistry. You will carry out several qualitative tests that will allow you to identify functional groups. You will then apply what you have learned by characterizing unknown organic compounds in terms of their functional group and solubility behavior.

Discussion:

Organic chemistry is primarily concerned with the chemistry of carbon compounds. Though there are millions of organic compounds, they can be grouped together into a relatively small number of classes or families on the basis of their **functional group** or groups. Functional groups are nearly always the reactive centers of the molecules, and so it is the functional groups that are likely to undergo a structural change in a chemical reaction.

Each functional group has a specific set of chemical properties, and these become the properties of the compound. All compounds that contain the same functional group are placed in the same **class** of compounds and they share many of the same chemical properties. For example, all compounds that contain a carbon-carbon double bond (a functional group) are members of the alkene class. The chemical properties of one alkene are like those of all other alkenes.

Even those hydrocarbons that have only carbon-carbon single bonds are members of a particular class of organic compounds, the alkanes. Since many organic compounds are alkane in nature, *all other functional groups have priority over the alkane class when classifying them*. The compound shown below would be classified as an alcohol because it contains the hydroxyl group, −OH. The hydroxyl functional group takes precedence, placing this compound in the alcohol class even though most of the molecule is an alkane-like hydrocarbon chain.

```
    H   H   H   H
    |   |   |   |
H - C - C - C - C - OH
    |   |   |   |
    H   H   H   H   <- the hydroxyl functional group
     an alcohol
```

Class = alcohol
Functional group = hydroxyl

There are over twenty common functional groups in organic chemistry, and ten of the most common ones are listed in Table 1. Notice that the name of the functional group and the name of the class of compound are not always the same except in two cases, the ethers and the esters. Here the functional group and class name are the same.

Table 1. Functional Groups and Classes of Organic Compounds

Functional Group	Structure	Class of Compound
single carbon-carbon bond	$-\overset{\vert}{\underset{\vert}{C}}-\overset{\vert}{\underset{\vert}{C}}-$	alkane
double carbon-carbon bond	$-\underset{\vert}{C}=\underset{\vert}{C}-$	alkene
triple carbon-carbon bond	$-C\equiv C-$	alkyne
hydroxyl	$-\overset{\vert}{\underset{\vert}{C}}-O-H$	alcohol, phenol
ether	$-\overset{\vert}{\underset{\vert}{C}}-O-\overset{\vert}{\underset{\vert}{C}}-$	ether
carbonyl	$-\overset{O}{\overset{\Vert}{C}}-$	aldehyde $-\overset{O}{\overset{\Vert}{C}}-H$
		ketone $-\overset{\vert}{\underset{\vert}{C}}-\overset{O}{\overset{\Vert}{C}}-\overset{\vert}{\underset{\vert}{C}}-$
carboxyl	$-\overset{O}{\overset{\Vert}{C}}-O-H$	carboxylic acid
amino	$-\overset{\vert}{\underset{\vert}{C}}-\overset{\vert}{N}-$	amine
halo	$-\overset{\vert}{\underset{\vert}{C}}-X$	alkyl halide, aryl halide
ester	$-\overset{O}{\overset{\Vert}{C}}-O-\overset{\vert}{\underset{\vert}{C}}-$	ester

Many organic compounds contain two or more functional groups in the same molecule. For example, the amino acid serine contains an amino group ($-NH_2$), a carboxyl group ($-COOH$) and a hydroxyl group ($-OH$). As a result, one part of the molecule acts chemically like an amine, another like a carboxylic acid, and yet another like an alcohol.

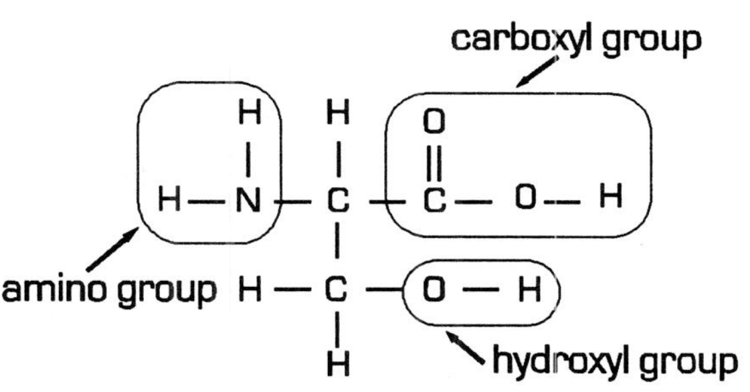

Serine

To ease your introduction to the functional groups, the compounds you will study in this experiment will be *mono*functional; that is, they will contain only *one* functional group.

A. Functional Groups

Only seven of the ten functional groups listed in Table 1 will be examined in this experiment. Esters, ethers and alkynes will be omitted. A discussion of these seven groups follows.

1. The Amino Group: $-NH_2$, $-NHR$, $-NR_2$

Class of Compound: Amine

Organic compounds containing one of the amino functional groups are members of the **amine** class. They are divided into three subclasses according to the number of alkyl or aryl (aromatic) groups (symbolized with the letter R) bonded to the nitrogen atom. **Primary amines** have a single R group bonded to nitrogen, **secondary amines** have two, and **tertiary amines**, have three R groups. R, R′ and R″ are used to indicate R groups that do not have to be identical.

R–NH–H (N bonded to H)	R–N–R′ (N bonded to H)	R–N–R′ (N bonded to R″)
primary (p) amine	secondary (s) amine	tertiary (t) amine

Some examples of amines with IUPAC names.

CH_3NH_2	$CH_3-NH-CH_3$	$CH_3CH_2CH_2-N(CH_3)-CH_2CH_3$
methyl amine (p)	dimethylamine (s)	methylethylpropylamine (t)

$CH_3CH_2-NH-CH_2CH_3$

diethylamine (s)

$C_6H_5NH_2$

aniline (p)

2. The Hydroxyl Group: −OH

Class of Compound: *Alcohol*

Compounds containing the hydroxyl group joined to a carbon atom that is *not* a carbonyl carbon (C=O) are members of the **alcohol** class. If −OH is joined to a carbon in an aromatic ring, the compounds are **phenols**. We will only be concerned with alcohols here. Like amines, alcohols can be divided into three subclasses, primary, secondary and tertiary, based on the number of alkyl groups bonded to the carbon atom to which the −OH group is joined. A general formula of an alcohol can be abbreviated ROH. Several alcohols are shown below with their IUPAC names. Their common names are given in parentheses.

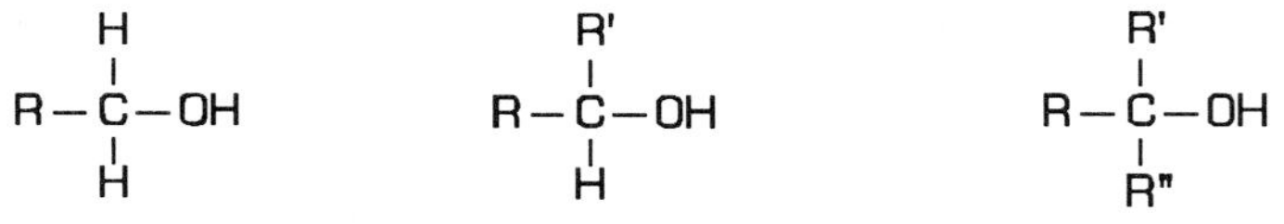

```
    H               R'              R'
    |               |               |
R — C — OH      R — C — OH      R — C — OH
    |               |               |
    H               H               R"
```

primary (p) alcohol   secondary (s) alcohol   tertiary (t) alcohol

---

CH_3OH	CH_3CH_2OH	$HOCH_2CH_2OH$
methanol (p)	ethanol (p)	1,2-ethanediol (p)
(methyl alcohol)	(ethyl alcohol)	(ethylene glycol)

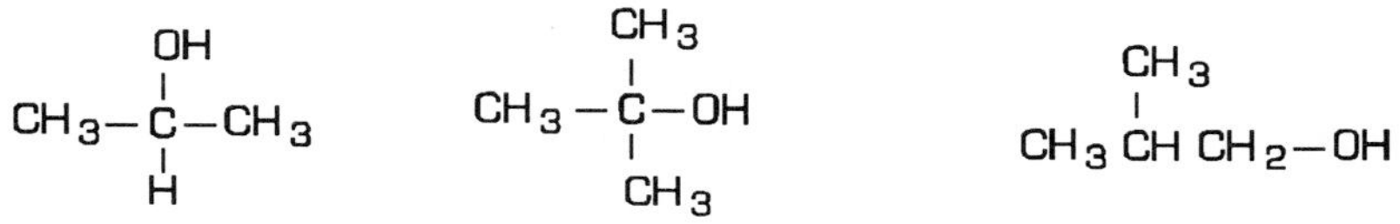

```
      OH                  CH3                  CH3
      |                   |                    |
CH3 — C — CH3       CH3 — C — OH         CH3 CH CH2 — OH
      |                   |
      H                   CH3
```

2-propanol (s)	2-methyl-2-propanol (t)	2-methyl-1-propanol (p)
(isopropyl alcohol)	(tertiary butyl alcohol)	(isobutyl alcohol)

3. The Carboxyl Group: −COOH

Class of Compound: Carboxylic Acid

Compounds containing the carboxyl functional group are members of the **carboxylic acid** class of organic compounds. They are all weak acids. The general formula for a carboxylic acid can be condensed to R-COOH or RCOOH, in which the −COOH pattern indicates the carboxyl group. Four carboxylic acids are shown below in both structural and condensed formulas. IUPAC names are given, with common names in parentheses.

```
structural          O                  H   O
formulas            ||                 |   ||
                H — C — OH         H — C — C — OH
                                       |
                                       H
```

condensed formulas	$HCOOH$	CH_3COOH
	methanoic acid (formic acid)	ethanoic acid (acetic acid)

structural formulas

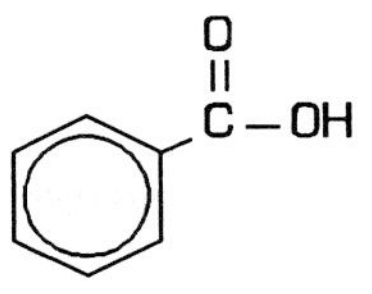

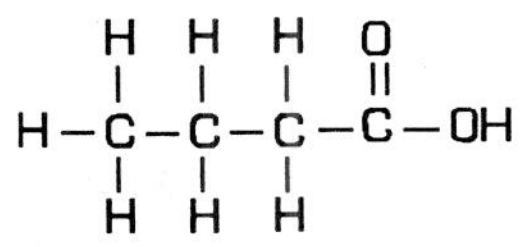

benzoic acid: C_6H_5–C(=O)–OH

butanoic acid: H–C(H)(H)–C(H)(H)–C(H)(H)–C(=O)–OH

condensed formulas

C_6H_5COOH — benzoic acid

$CH_3\,CH_2\,CH_2\,COOH$ — butanoic acid (butyric acid)

4. **The Double Bond: C=C**
Class of Compound: Alkene

If a compound contains a **double bond** between two carbon atoms, it is classified as an **alkene**. Alkenes are also called olefins or unsaturated compounds. The first two members of the alkene class are shown below. IUPAC names are given, with common names in parentheses.

structural formulas

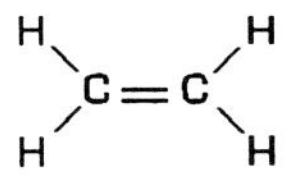

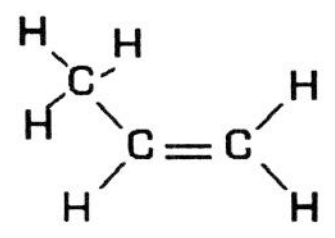

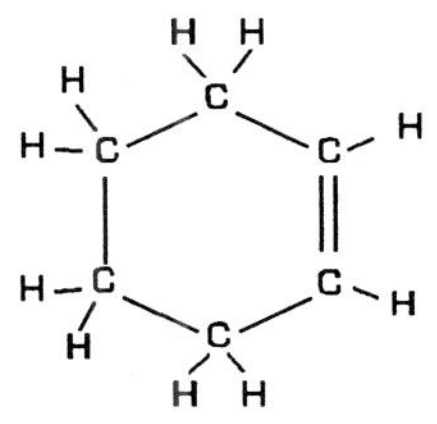

condensed formulas

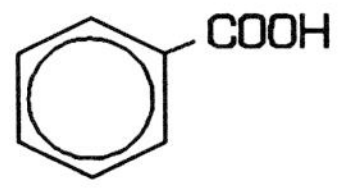

$CH_2{=}CH_2$
ethene
(ethylene)

$CH_3\,CH{=}CH_2$
propene
(propylene)

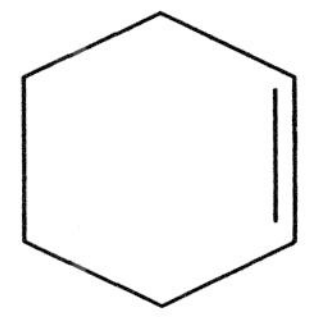

cyclohexene

5. **The Halo Group: −F, −Cl, −Br, −I**
Class of Compound: Alkyl Halide, Aryl Halide

Members of the **alkyl halide** class will be the only halogen-containing compounds considered in this experiment. The general formula for any alkyl halide is RX, in which X represents fluorine, chlorine, bromine, or iodine. Some examples of this class are shown below. IUPAC names are given, with common names in parentheses.

CH_3Cl chloromethane (methyl chloride)	Cl_3CH trichloromethane (chloroform)	CCl_4 tetrachloromethane (carbon tetrachloride)
CH_3CH_2Br bromoethane	CH_3I iodomethane	$CH_3CHBrCH_2CH_3$ 2-bromobutane

6. The Carbonyl Group: $-\overset{O}{\overset{\|}{C}}-$

Classes of Compounds: Aldehyde and Ketone

The **carbonyl** group is common to both **aldehydes** and **ketones**. In aldehydes, the carbonyl group always has a hydrogen atom bonded to it, and in the case of formaldehyde, it has a second hydrogen bonded to it. Because the aldehyde group must include at least one hydrogen atom with the carbonyl, the aldehyde function will always appear at the end of a carbon chain. Because the chemical properties of aldehydes differ from those of ketones, it is common to think of the aldehyde carbonyl unit as a functional group in its own right.

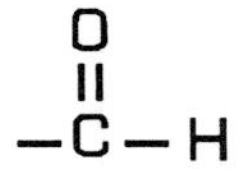

$$-\overset{O}{\overset{\|}{C}}-H$$

the aldehyde functional group

The aldehyde functional group can be condensed to $-CHO$, and the general formula for an aldehyde can be written $R-CHO$ or $RCHO$. Note that the H comes before the O in the aldehyde formula, but after the O in alcohols.

In **ketones**, the carbonyl group is always bonded to two other carbon atoms, never directly to a hydrogen atom (because then it would be an aldehyde). As with the aldehydes, it is common to think of the ketone as a functional group in its own right.

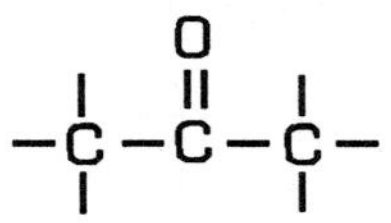

$$-\underset{|}{\overset{|}{C}}-\overset{O}{\overset{\|}{C}}-\underset{|}{\overset{|}{C}}-$$

the ketone functional group

The ketone functional group can be condensed to $R(C{=}O)R$ or $RCOR$, in which R is either an alkyl or aryl (aromatic) group. Some common aldehydes and ketones are shown below, and as before, IUPAC names are given, with common names in parentheses.

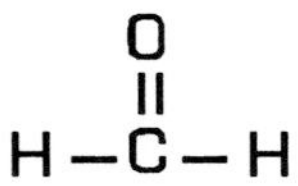

$$H-\overset{O}{\overset{\|}{C}}-H$$

methanal
(formaldehyde)

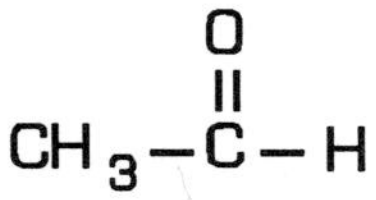

$$CH_3-\overset{O}{\overset{\|}{C}}-H$$

ethanal
(acetalcehyde)

$$CH_3-\underset{CH_3}{\underset{|}{\overset{H}{\overset{|}{C}}}}-\overset{O}{\overset{\|}{C}}-H$$

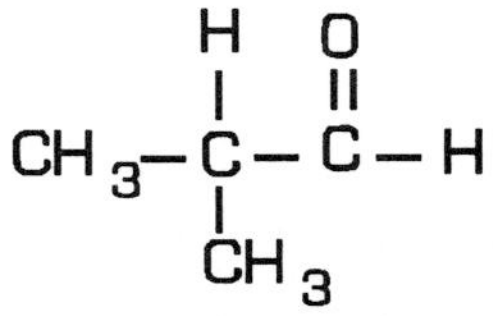

2-methylpropanal
(isobutyl aldehyde)

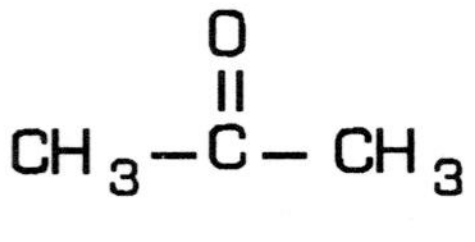

$$CH_3-\overset{O}{\overset{\|}{C}}-CH_3$$

propanone
(acetone)

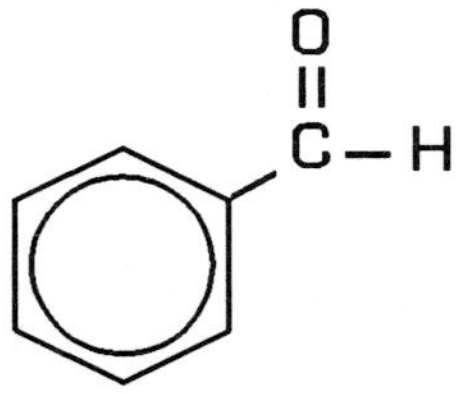

$$C_6H_5-\overset{O}{\overset{\|}{C}}-H$$

benzaldehyde

$$C_6H_5-\overset{O}{\overset{\|}{C}}-CH_3$$

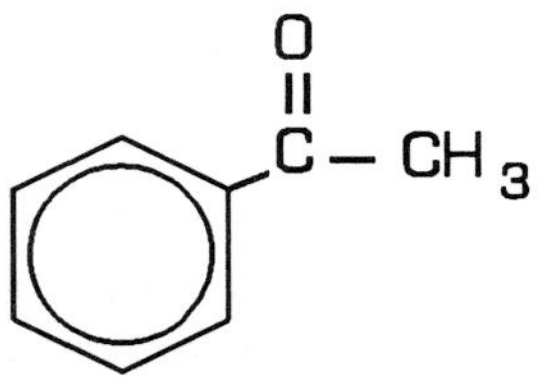

acetophenone

7. The Single Bond: C–C
Class of Compound: Alkane

The seventh class of organic compounds to be studied is the saturated hydrocarbons, the **alkanes**. Because they are saturated hydrocarbons (all carbon-carbon bonds are single bonds), they are not considered functional in the same sense as members of the six previous groups. Yet alkanes can undergo certain substitution reactions, such as that in which a bromine or chlorine atom replaces a hydrogen atom in the molecule. Because alkanes are not very reactive, these reactions require more rigorous conditions than those with compounds containing other functional groups.

As mentioned earlier, when classifying organic compounds, all other functional groups have priority over the alkane group. In the analysis scheme used in this experiment, any compound that fails to give a positive test for any of the other functional groups should be judged a member of the **alkane** class.

B. Solubility and Functional Group Tests

Each functional group has a particular set of chemical properties that allow it to be identified. Some of these properties can be demonstrated by observing solubility behavior, while others can be seen in chemical reactions that are accompanied by color changes, precipitate formation or other visible effects. When attempting to identify the functional group of an organic compound, its best to start by evaluating its solubility behavior.

1. Solubility Tests

Solubility in Water: Most organic compounds are not soluble in water, except for low-molecular-weight amines and compounds that contain oxygen (alcohols, carboxylic acids, ketones, etc). Low-molecular-weight compounds are generally limited to those with fewer than five carbon atoms. Solubility tests in water should always be done using deionized water.

Carboxylic acids with fewer than five carbon atoms are soluble in water and form solutions that give an acidic response (pH < 7) when tested with litmus or pH paper.

$$RCOOH + H_2O \rightleftarrows RCOO^- + H_3O^+$$

Amines with fewer than five carbon atoms are also soluble in water, and their solutions give a basic response (pH > 7) when tested with litmus or pH paper.

$$R_2NH + H_2O \rightleftarrows R_2NH_2^+ + OH^-$$

Ketones, aldehydes and alcohols with fewer than five carbon atoms are soluble in water and form neutral solutions (pH = 7).

Organic compounds that are not soluble in pure water should then be checked for solubility in 6 M NaOH.

Solubility in NaOH: *Solubility in 6 M NaOH is a positive identification test for acids.* A carboxylic acid that is insoluble in pure water will be soluble in 6 M NaOH, due to the formation of the soluble sodium salt of the acid as the acid is neutralized by the base:

$$\underset{\text{insoluble}}{RCOOH} + NaOH_{(aq)} \rightarrow \underset{\text{soluble salt}}{RCOO^- \, Na^+_{(aq)}} + H_2O$$

Solubility in HCl: *Solubility in 6 M HCl is a positive identification test for bases.* Amines that are insoluble in pure water will be soluble in 6 M HCl, due to the formation of a soluble organic-ammonium chloride salt. The basic amine is neutralized by the acid:

$$\underset{\text{insoluble}}{RNH_2} + HCl_{(aq)} \rightarrow \underset{\text{soluble salt}}{RNH_3^+ \, Cl^-_{(aq)}}$$

Solubility tests are done in this sequence: water, then 6 M NaOH, then 6 M HCl. If, when testing for solubility, you find that an organic substance is soluble in water, you need not check for solubility in any other solution. Once you achieve a positive solubility test, you go no further. Thus, if an organic compound is insoluble in water but soluble in 6 M NaOH, you need check no further. A great deal can be learned about an organic compound by observing its behavior in solubility tests. A flowchart showing the sequence of solubility tests, along with the appropriate conclusions, is shown in Figure 1.

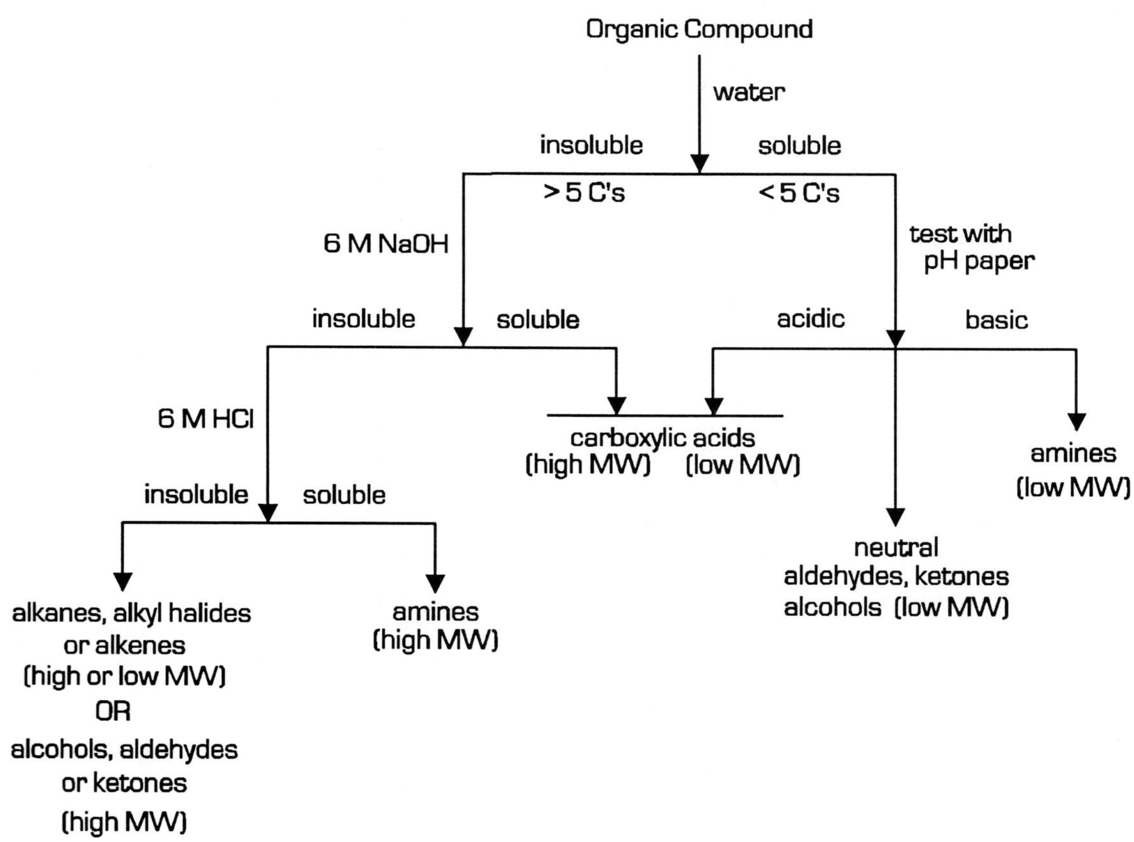

Figure 1. The Solubility Test Flowchart

2. Ceric Nitrate Test for Alcohols

The ceric nitrate test is a qualitative test for the hydroxyl functional group. If an alcohol is added to a yellow solution of ceric nitrate, a red ceric complex forms. The color change from yellow to red is a positive test for the presence of an alcohol. The equation for the general reaction that takes place is:

$$\underset{\text{(yellow)}}{[Ce(NO_3)_6]^{2-}} + ROH \rightarrow \underset{\text{(red)}}{[CeOR(NO_3)_5]^{2-}} + HNO_3$$

3. Silver Nitrate Test for Alkyl Halides

Testing alkyl halides with ethanoic silver nitrate (silver nitrate dissolved in ethyl alcohol) will result in the formation of a white or yellow silver halide precipitate that is insoluble in nitric acid. The reaction quite often proceeds slowly, and occasionally *slight* warming is necessary.

$$RX + AgNO_3 \rightarrow RONO_2 + AgX_{(s)}$$

It is important to check if the precipitate is soluble in dilute nitric acid. Carboxylic acids form insoluble silver salts that precipitate, but these dissolve in dilute nitric acid while the silver halides do not.

$$RCOOH + AgNO_3 \rightarrow RCOOAg_{(s)} + HNO_3 \quad \text{(insoluble silver salt)}$$

$$RCOOAg_{(s)} + HNO_3 \rightarrow RCOOH + AgNO_3 \quad \text{(dissolves in acid)}$$

$$AgX_{(s)} + HNO_3 \rightarrow \text{no reaction (precipitate does not dissolve)}$$

4. 2,4-Dinitrophenylhydrazine (2,4-DNPH) Test for Ketones and Aldehydes

Aldehydes and ketones are easily detected by the formation of brightly colored 2,4-dinitrophenylhydrazone precipitates. The yellow to orange precipitate forms within a few minutes and often forms as quickly as the reagents are brought together. A general equation describing the formation of a 2,4-dinitrophenylhydrazone is:

$$R\text{-}C(R')\text{=}O + H\text{-}NH\text{-}NH\text{-}C_6H_3(NO_2)_2 \longrightarrow R\text{-}C(R')\text{=}N\text{-}NH\text{-}C_6H_3(NO_2)_2 + H_2O$$

aldehyde or ketone — 2,4-DNPH — 2,4-dinitrophenylhydrazone (precipitate)

5. Tollens Test for Aldehydes

Aldehydes can be distinguished from ketones by the Tollens test, because aldehydes are easily oxidized to carboxylic acids, while ketones are not. Tollens reagent is a solution of silver nitrate in aqueous ammonia. Silver ion is the oxidizing agent and is reduced to silver metal as the aldehyde is oxidized to the carboxylic acid. The positive Tollens test is the appearance of a mirror of silver metal on the wall of the test tube, or a dark gray to black precipitate of silver metal. Note that the equation describing the Tollens test shows the formation of the ammonium salt of a carboxylic acid as one of the products. In the basic environment of this test, the carboxylic acid is quickly neutralized and exists as the soluble ammonium salt.

$$\underset{\text{aldehyde}}{RCHO} + \underbrace{2\,[Ag(NH_3)_2]^+ + 2\,OH^-}_{\text{Tollens reagent}} \rightarrow \underset{\text{soluble salt of the acid}}{RCOO^-, NH_4^+} + \underset{\text{silver mirror}}{2\,Ag_{(s)}} + H_2O + 3\,NH_3$$

6. Bromination Test for Alkenes

Alkenes can readily add bromine to the double bond, forming dihalides. An alkene added to a solution of bromine in cyclohexane will quickly remove the red color of bromine as the addition reaction takes place. The removal of the red bromine color is a positive test for an alkene. The general equation describing the test is:

$$\underset{\text{alkene (colorless)}}{>C=C<} + \underset{\text{bromine (red in cyclohexane)}}{Br_2} \longrightarrow \underset{\text{alkyl bromide (colorless)}}{-\overset{|}{\underset{Br}{\underset{|}{C}}}-\overset{Br}{\overset{|}{\underset{|}{C}}}-}$$

The tests described above are but a few of the many that are used to determine the functionality of organic compounds. More could be said about each test concerning possible interfering reactions or subtle effects that might allow a more exact qualitative judgment to be made. That will be saved for a full course in organic chemistry. There are numerous references available that give a more detailed discussion of functional group analysis and these should be consulted if you desire more information.

Experimental Procedure:

CAUTION: Strong acids and bases are used in several procedures. Wear eye protection at all times. If you spill a corrosive liquid on your skin, wash the area immediately with running cold water and contact the instructor.

ENVIRONMENTAL ALERT: Most organic compounds should not go down the drain. Also, some tests involve compounds of heavy metals. Follow the disposal instructions given to you by your instructor.

A. Characterizing Known Compounds

The following procedures are to be carried out using compounds of known identity so you can observe a positive result for each test. Record your observations carefully on the Report Sheet. You will need to consult them later when you analyze compounds of unknown identity.

1. Solubility Tests

Known Compounds: *aniline*, $C_6H_5{-}NH_2$ *benzoic acid*, $C_6H_5{-}COOH$

acetone, $CH_3(C{=}O)CH_3$ *diethylamine*, $(CH_3CH_2)_2NH$

Solubility in Water: Place 1 mL (20 drops) of deionized water in each of four small (10-cm or smaller) test tubes. To the first test tube add 3 drops of aniline; to the second add 3 drops of acetone; to the third add 3 drops of diethylamine; and to the fourth add a few crystals of benzoic acid. Stopper each with a cork or cover with a small piece of plastic film , such as Parafilm®, and thoroughly mix the contents. Set each tube aside for 1 minute. If two layers form, or if the mixture is cloudy, consider the organic compound to be insoluble in water. If a transparent solution with no boundaries results, judge the sample to be soluble in water. Record your observations on the Report Sheet. Discard all solutions as advised by your instructor.

For each of those which dissolved in water, obtain a drop of the solution on the end of a *clean* stirring rod and touch it to a fresh piece of litmus paper or pH paper. Record the pH response as *acidic*, *neutral* or *basic*. Those compounds that were soluble in water need not be tested in 6 M NaOH or 6 M HCl.

Solubility in 6 M NaOH: For those compounds that were insoluble in deionized water or nearly so, test their solubility in 6 M NaOH. Add 3 drops (or several crystals) of each organic compound to 1 mL (20 drops) of 6 M NaOH. If the compound is soluble in 6 M NaOH solution, you will not test its solubility further. Record your observations on the Report Sheet. Discard all solutions as advised by your instructor.

Solubility in 6 M HCl: Those compounds that were insoluble or only partially soluble in deionized water and in 6 M NaOH must be tested in 6 M HCl. Add 3 drops (or several crystals) of each compound to be tested to 1 mL (20 drops) of 6 M HCl. Record your observations on the Report Sheet. Discard all solutions as advised by your instructor.

Note: If you are analyzing an unknown and the solubility tests lead you to believe it is a carboxylic acid, be certain to also carry out the alcoholic silver nitrate test (test 3) on that sample. Tests 3 and 5 require heating mixtures in a hot water bath. So before you continue, begin heating about 200 mL of water in a 250-mL beaker on a tripod or ring stand. Heat with a moderate flame and do not let it come to a boil.

2. Ceric Nitrate Test for Alcohols

Known Compound: ethyl alcohol (ethanol), CH_3CH_2OH

In a small test tube, mix 1 mL (20 drops) of ceric nitrate solution with 6 drops of ethyl alcohol and swirl the mixture. Record your observations.

If you are carrying out this test on an unknown compound that is a water-soluble solid, first dissolve a match-head-size amount of the solid in 20 drops of deionized water, then add 30 drops of ceric nitrate solution. On the other hand, if the compound is a water-insoluble liquid, first dissolve 3 drops of it in 40 drops (2 mL) of dioxane, then add 20 drops of ceric nitrate solution. Record your observations on the Report Sheet. Discard all solutions as advised by your instructor.

3. Silver Nitrate Test for Halides

Known compound: ethyl bromide, CH_3CH_2Br

In a small test tube, dissolve about 10 drops of ethyl bromide in 10 drops of 100% ethyl alcohol. To this, add 10 drops of alcoholic silver nitrate solution. If no sign of a reaction is visible in 3 minutes, warm the mixture in hot water. To ensure that the precipitate is a silver halide and not the salt of a carboxylic acid, add 10 drops of 6 M nitric acid to the solution with the precipitate. Stir the mixture. If the precipitate does not dissolve, you have a positive test for an alkyl halide. Record your observations on the Report Sheet. Discard the solution and any precipitate as your instructor advises.

4. 2,4-Dinitrophenylhydrazine (2,4-DNPH) Test for Aldehydes and Ketones

Known compounds: isobutyl aldehyde, $(CH_3)_2CHCHO$ and acetone, $CH_3(C{=}O)CH_3$

In two small test tubes, place about 2 mL (40 drops) of the 2,4-dinitrophenylhydrazine solution. To one tube, add 2 drops of acetone. Swirl the mixture and record the results. To the other, add 2 drops of isobutyl aldehyde. Swirl the mixture and record the results. Discard all solutions as advised by your instructor.

5. Tollens Test for Aldehydes

Known compounds: isobutyl aldehyde, $(CH_3)_2CHCHO$ and acetone, $CH_3(C{=}O)CH_3$

The Tollens reagent must be prepared fresh each time it is used, and after each use it must be immediately discarded.

CAUTION: Tollens reagent must be destroyed immediately after use. Pour all Tollens solutions and precipitates in a beaker of 6 M nitric acid. The beaker of acid will remain under the hood.

In each of two small, *very clean* test tubes, place 5 drops of aqueous 5% silver nitrate solution and 1 mL (20 drops) of deionized water. Add 1 drop of 5% NaOH to each test tube. Then add dropwise, stirring as you add each drop, fresh 1 M aqueous ammonia until the precipitate *just dissolves*. Add no more than 20 drops. If you have problems, check with the instructor. This is the Tollens reagent.

To one of the test tubes, add 4 drops of acetone and to the other add 4 drops of isobutyl aldehyde. Thoroughly mix each solution.

Heat the test tubes in a warm (not hot) water bath for 2 to 3 minutes. If the test tube is clean, a silver mirror on the glass surface constitutes a positive test; otherwise, a gray or gray-black precipitate of silver metal is the positive test.

A positive Tollens test for aldehydes is meaningful only if it is supported by a positive 2,4-DNPH test. (Some organic compounds can give a weak positive Tollens test result and not contain a carbonyl group.) Record your observations on the Report Sheet. When you have finished this test, dispose of the mixtures by pouring them in 6 M nitric acid under the hood.

6. Bromination Test for Alkenes

Known compounds: cyclohexene (an alkene) and hexane (an alkane)

CAUTION: Take care not to get the bromine solution on your skin. If you do, immediately rinse the area with running cold water and cover the affected area with glycerine. Notify your instructor immediately.

Place about 20 drops of cyclohexene in a small test tube. To the alkene, add bromine solution (Br_2 in cyclohexane) one drop at a time. Note what happens to the red bromine color as the first 2 drops are added. Record your observations on the Report Sheet.

Place about 20 drops of hexane in a small test tube and repeat the dropwise addition of bromine solution. Record your observations. Dispose of all solutions as advised by your instructor.

B. Characterizing Unknown Compounds

When you are confident you can recognize a positive test for each procedure, obtain two unknowns from your instructor. Record their numbers on the report sheet. For each test in Section A, substitute your unknown for the known compound. The goal is to determine the functional group in each unknown compound. Your instructor may ask you to analyze the unknown compounds during the next lab period.

Figure 2 is a flowchart summarizing the decision process used to identify a compound of unknown identity. You should follow this sequence as you analyze each of your unknowns. A positive test in each procedure will look much the same as you observed with the known samples. Each unknown will be a member of one of the following classes: carboxylic acid, amine, alcohol, alkyl halide, aldehyde, ketone, alkene, or alkane.

Also, be certain to record the solubility of your unknown in water. If it is soluble in pure water, then it probably contains less than five carbon atoms and may be considered a low-molecular-weight compound. If it is insoluble in pure water, then it may be considered a high-molecular-weight compound (more than five carbons atoms) or it may be an alkyl halide, an alkene, or an alkane, which are insoluble in water even when they contain only a few carbon atoms.

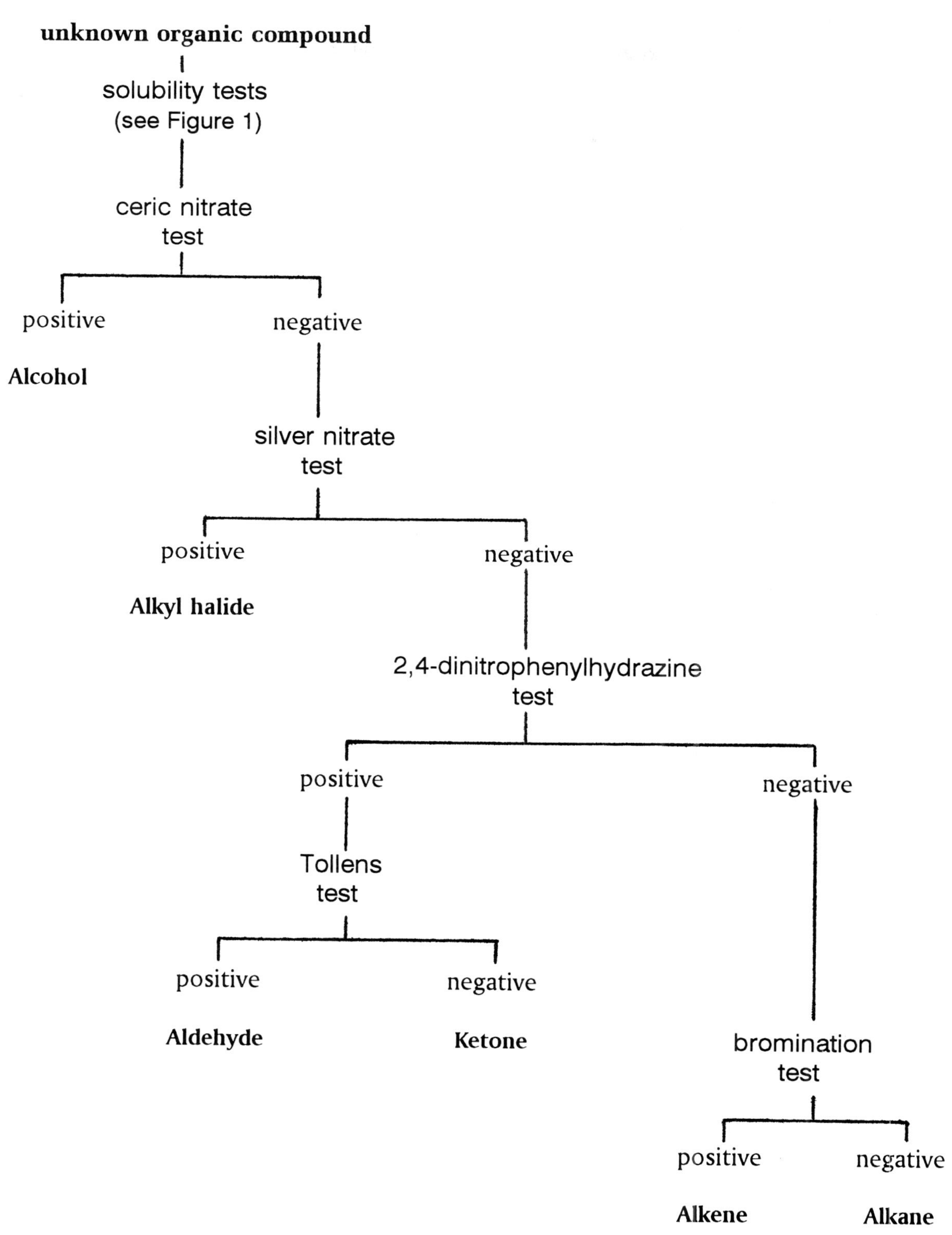

Figure 2. Test flowchart for Identifying Functional Groups in Organic Compounds

Name ______________________ Locker Number__________ Date__________
Please print; last name first

PRELIMINARY EXERCISES: *Experiment 22*

Organic Functional Group Tests

1. Name the functional group or groups that appear in each of the following organic compounds.

a. CH_3CH_2OH

____

b. CH_3-O-$CH_2CH=CHCH_2Cl$

____

c. $CH_3CH_2CH_2Br$

____

d. CH_3COOCH_3

____

e. $(CH_3CH_2)_2NH$

____

f. CH_3CH_2COOH

____

g. C_6H_5CHO

____

h. $CH_3(C=O)CH_3$

____

2. Predict the solubility for each of the following organic compounds in water, in 6 M NaOH and in 6 M HCl. Move from left to right across each row below, stopping after a positive solubility test is encountered. If soluble, mark (+). If insoluble, mark (−).

Compound	Predicted Solubility Results		
	Deionized Water	**6 M NaOH**	**6 M HCl**
a. CH_3OH			
b. $CH_3CH_2NH_3^+, Cl^-$			
c. C_6H_5-NH_2			
d. CH_3CH_2Br			
e. $CH_3CH_2CH_2CH_2CH_2CH_2OH$			
f. C_6H_5-COOH			
g. $CH_3CH_2CH_2CH_2CH_2CH_2CH_3$			
h. $CH_3CH_2CH_2COOH$			
i. $(CH_3CH_2)_2NH$			

3. Predict which test or tests will be *positive* for each of the compounds listed in the left-hand column. Mark (+) to indicate a positive test.

Compound	Predicted Positive Test Results				
	Ceric Nitrate Test	Silver Nitrate Test	Tollens Test	2,4-DNPH Test	Bromination Test
a. CH_3COOH					
b. HCHO					
c. CH_3CH_2Br					
d. $CH_3(C{=}O)CH_3$					
e. $CH_3CH_2CH_2CH_3$					
f. $(CH_3)_2NH$					
g. C_6H_5-COOH					
h. $CH_3CH{=}CH_2$					
i. CCl_3H					
j. CH_3CH_2OH					

4. What class of organic compound with less than five carbon atoms dissolves in water to form a slightly acidic solution? ____________________

5. What class of organic compound with less than five carbon atoms dissolves in water to form a slightly basic solution? ____________________

6. What class of organic compound with more than five carbon atoms is insoluble in water but readily dissolves in 6 M NaOH? ____________________

7. Suppose an unknown organic compound gives a positive test with 2,4-dinitrophenylhydrazine. Why would you want to follow up that positive test with the Tollens test?

____

8. Suppose an unknown organic compound formed a precipitate when treated with ethanoic silver nitrate, which was soluble in dilute nitric acid. In which functional class would the compound belong? ____________________

Name ______________________________ Locker Number__________ Date__________
Please print; last name first

PRELIMINARY EXERCISES (continued): *Experiment 22*

Organic Functional Group Tests

9. Complete the following equations:

a. $CH_3COOH + H_2O \rightleftarrows$

b. $CH_3CH_2NH_2 + H_2O \rightleftarrows$

c. $CH_3CH_2CH_2CH_2CH_2CH_2COOH + NaOH \rightarrow$

d. $C_6H_5{-}COOH + NaOH \rightarrow$

e. $C_6H_5{-}NH_2 + HCl \rightarrow$

f. $CH_3OH + [Ce(NO_3)_6]^{2-} \rightarrow$

g. $CH_3CH_2Br + AgNO_3 \rightarrow$

h. $CH_3COO^- Ag^+ + HNO_3 \rightarrow$

i. $CH_3CHO + [Ce(NO_3)_6]^{2-} \rightarrow$

j. $(CH_3)_2CHCHO + H_2N{-}NH{-}C_6H_3(NO_2)_2 \rightarrow$
(2,4-DNPH)

k. $(CH_3)_2CHCHO + 2\,[Ag(NH_3)_2]^+ + 2\,OH^- \rightarrow$

l. $CH_3CH_2CH{=}CHCH_2CH_3 + Br_2 \rightarrow$

m. $(CH_3)_2C{=}O + H_2N{-}NH{-}C_6H_3(NO_2)_2 \rightarrow$
(2,4-DNPH)

n. cyclohexane $+ Br_2 \rightarrow$

Name ______________________________ Locker Number__________ Date____________
Please print; last name first

REPORT SHEET: *Experiment 22*

Organic Functional Group Tests

A. Characterizing Known Compounds

1. Solubility Tests

Indicate which compounds are soluble in deionized water, and for those that are soluble, indicate whether the solution is acidic, neutral or basic.

Known Compound	Solubility (mark + or −)	Acidic, Basic or Neutral
aniline		
acetone		
diethylamine		
benzoic acid		

Which compounds are insoluble in water but soluble in 6 M NaOH?	
Which compounds are insoluble in water and 6 M NaOH, but soluble in 6 M HCl?	

Note: For tests 2 through 6, carefully record your observations. Accurately describe each positive test and include factors such as reaction time, color change, precipitate formation, and the need for heating, stirring or shaking.

2. Ceric Nitrate Test

Known Compound: ethyl alcohol (ethanol)

Observations:

3. **Silver Nitrate Test**
Known Compound: ethyl bromide

Observations:
What happened when HNO_3 was added to the precipitate?

4. **2,4-Dinitrophenylhydrazine Test**
Known compounds: acetone (a ketone) and isobutyl aldehyde (an aldehyde)

Observations with acetone:
Observations with isobutyl aldehyde:
Were any differences observed between the reaction with acetone and that with isobutyl aldehyde?

5. **Tollens Test**
Known compounds: acetone (a ketone) and isobutyl aldehyde (an aldehyde)

Observations:

6. **Bromination Test**
Known compounds: cyclohexene (an alkene) and hexane (an alkane)

Observations:
What can you determine about an organic compound with the bromination test?

Name ______________________________ Locker Number__________ Date___________
Please print; last name first

REPORT SHEET (continued): *Experiment 22*

Organic Functional Group Tests

B. Characterizing Unknown Compounds

Unknown numbers: a. __________ b. __________

Test	Observations	Conclusions
solubility in water and pH of the solution	(a)	(a)
	(b)	(b)
solubility in 6 M NaOH (perform if needed)	(a)	(a)
	(b)	(b)
solubility in 6 M HCl (perform if needed)	(a)	(a)
	(b)	(b)
ceric nitrate test	(a)	(a)
	(b)	(b)
silver nitrate test	(a)	(a)
	(b)	(b)

Test	Observations	Conclusions
2,4-Dinitrophenylhydrazine test	(a)	(a)
	(b)	(b)
Tollens test	(a)	(a)
	(b)	(b)
bromination test	(a)	(a)
	(b)	(b)

Indicate the *class* of the unknown:

a. Unknown number _______: ________________________________

b. Unknown number _______: ________________________________

From its solubility in water, is the unknown of low molecular weight? (Answer *yes*, *no*, or *test was indecisive*.)

a. Unknown number _______: ________________________________

b. Unknown number _______: ________________________________

Experiment 23

The Detection of Fats, Proteins and Carbohydrates in Foods

Purpose:

The purpose of this experiment is to introduce you to some general tests which detect fats, proteins and carbohydrates in foods.

Discussion:

A. Carbohydrates

The word **carbohydrate** entered the vocabulary of chemistry many years ago when people thought the empirical formula for most carbohydrates, CH_2O, indicated they were "hydrates" of carbon, in which water molecules were associated with carbon atoms in some way. Shortly thereafter, it was shown that these compounds were not hydrates at all, but rather aldehydes and ketones that contained several hydroxyl groups per molecule. Today, a carbohydrate is considered to be a polyhydroxyaldehyde, a polyhydroxyketone, or another compound that yields these upon acid hydrolysis (the reaction with water in the presence of acid). Carbohydrates are the principal constituents of plants, comprising 60% to 90% of their dry weight. Plants use carbohydrates as an energy source, an energy storage medium, and as structural material. The building block of starch and cellulose, two important carbohydrates, is glucose, which is produced in photosynthesis.

$$6\ CO_{2(g)} \quad + \quad 6\ H_2O_{(l)} \quad \xrightarrow[\text{chlorophyll}]{\text{sunlight}} \quad C_6H_{12}O_{6(s)} \quad + \quad 6\ O_{2(g)}$$

Carbohydrates are divided into three major classes based on their behavior when exposed to acid hydrolysis:

1. Monosaccharides cannot be broken down into simpler carbohydrates by acid hydrolysis. They are often called simple sugars.
2. Disaccharides yield two monosaccharides upon acid hydrolysis. Sucrose, the most abundant disaccharide, yields glucose and fructose.
3. Polysaccharides yield more than two monosaccharides upon acid hydrolysis.

Monosaccharides are the simplest carbohydrates. Three common monosaccharides are glucose, galactose and fructose. **Disaccharides** are composed of two monosaccharides joined by an ether (−C−O−C−) linkage. Disaccharides are formed in nature from two monosaccharides that are joined with the elimination of one water molecule. Three common disaccharides are sucrose, lactose and maltose. Their structures are shown in Experiment 24. The cyclic structures of three monosaccharides and one disaccharide (sucrose) are shown on the next page. In solution, the cyclic molecule exists in equilibrium with the open-chain structure, which contains either an aldehyde or ketone functional group. Among the three common disaccharides, the glucose unit of lactose and maltose is able to open, forming an aldehyde group. Sucrose cannot undergo such a ring opening because of the manner in which the fructose and glucose cyclic units are joined. Both the cyclic and open-chain structures are shown below for glucose.

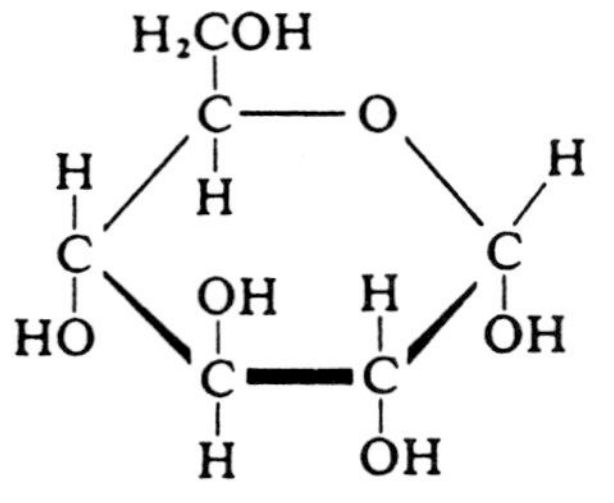

Glucose

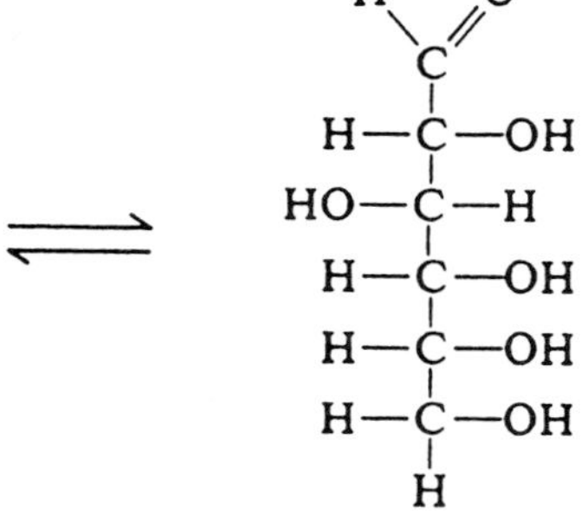

Glucose

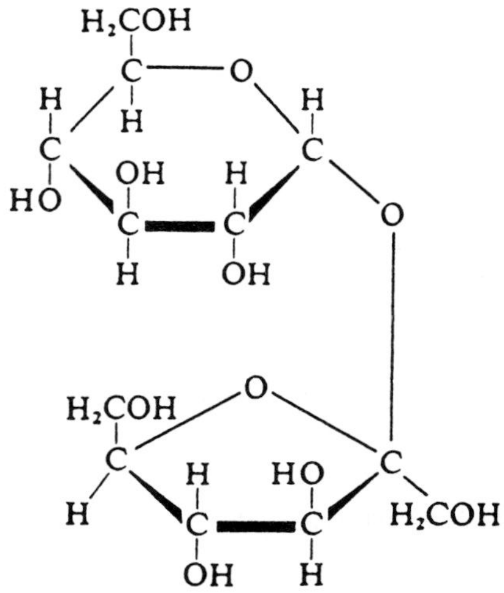

Sucrose

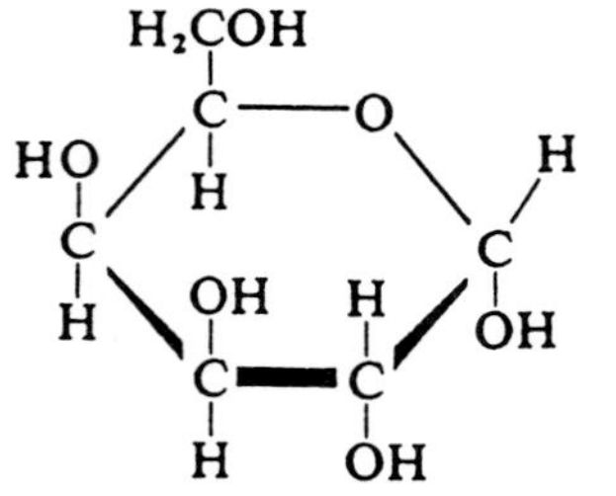

Galactose

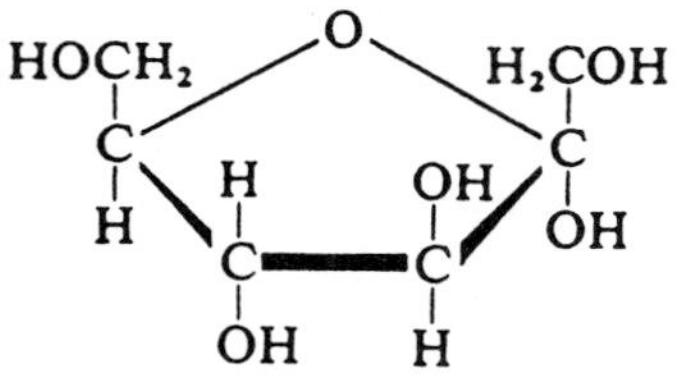

Fructose

Polysaccharides are complex carbohydrates made up of many monosaccharide units joined together by ether linkages to form a long polymeric structure. Starch, the most important polysaccharide, is a polymer of repeating glucose units. Amylose, one of the principal forms of starch, is shown below.

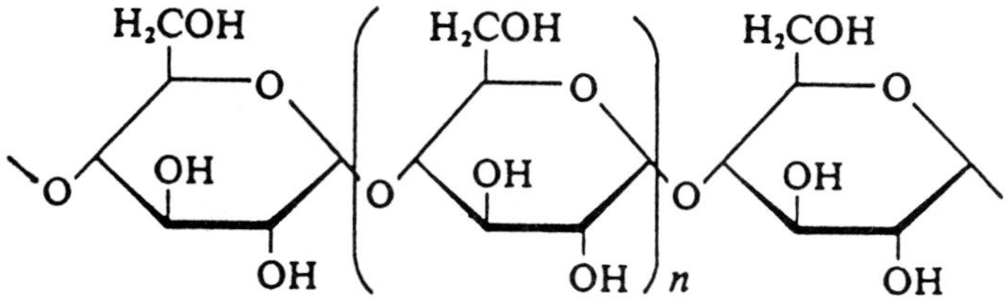

Amylose ($n = 60 - 300$)

Any carbohydrate that is able to exist in solution, to even a small extent, in an open-chain form with a free aldehyde group will be able to act as a **reducing sugar**. The **Benedict test** is used to detect the presence of reducing sugars. The Benedict test solution contains copper(II) ion, Cu^{2+}, in a soluble complex, and sodium hydroxide. This alkaline solution can oxidize the sugar's aldehyde group to a carboxylic acid group, as copper(II) ion is reduced to copper(I) oxide, a brick-red precipitate. The formation of the highly colored precipitate is visual evidence of the presence of a reducing sugar.

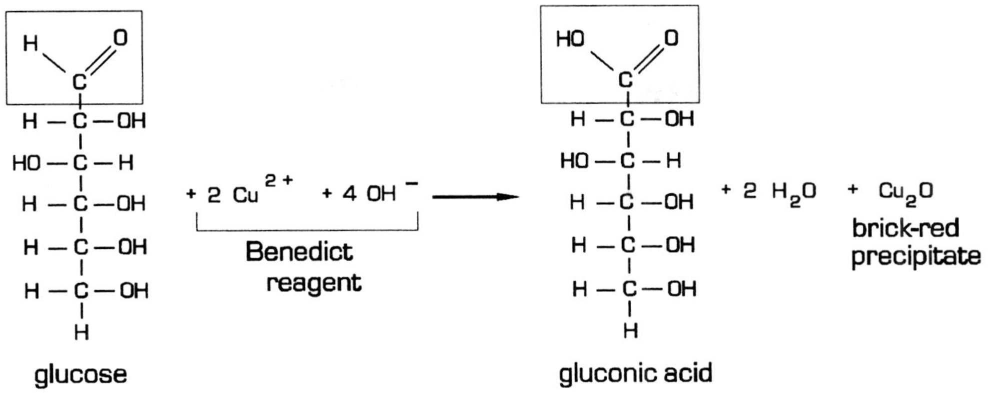

$$\text{glucose (CHO–(CHOH)}_4\text{–CH}_2\text{OH)} + 2\,Cu^{2+} + 4\,OH^- \xrightarrow{\text{Benedict reagent}} \text{gluconic acid (HOOC–(CHOH)}_4\text{–CH}_2\text{OH)} + 2\,H_2O + Cu_2O$$

brick-red precipitate

glucose — gluconic acid

Starches and other polysaccharides are not able to reduce the Benedict reagent. But if the starch is hydrolyzed, either in acid solution or by enzymes, the glucose units that are produced would give a positive Benedict test. All common monosaccharides and disaccharides, except sucrose, give a positive Benedict test.

The **iodine test** is used to detect the presence of polysaccharides. Starch will form a dark-blue complex with iodine. Other polysaccharides may respond with a complex of a different color.

B. Triglycerides (Fats and Oils)

Triglycerides are triesters of glycerol and three long-chain carboxylic acids, commonly called fatty acids, (see Experiment 21, pages 272-273). The fatty acids may be saturated (no carbon-carbon double bonds) or unsaturated (one or more double bonds). If saturated fatty acids are used to form the triglyceride, it is called a saturated triglyceride, or a **fat**. If unsaturated fatty acids are used, it is called an unsaturated triglyceride, or an **oil**. Glyceryl tristearate is a saturated triglyceride (a fat) in which the three fatty acid chains are from stearic acid, $C_{17}H_{35}COOH$.

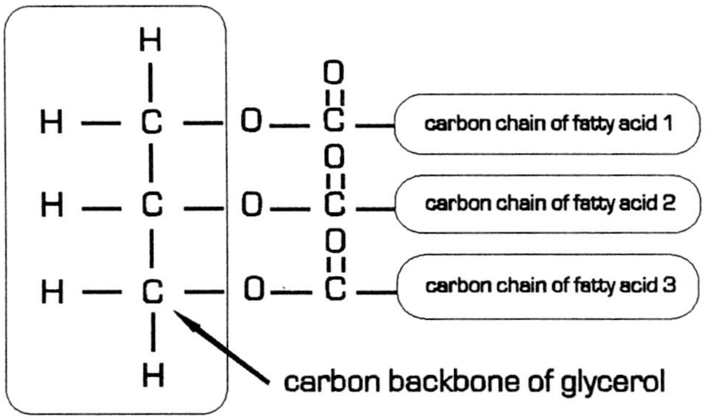

carbon chain of fatty acid 1

carbon chain of fatty acid 2

carbon chain of fatty acid 3

carbon backbone of glycerol

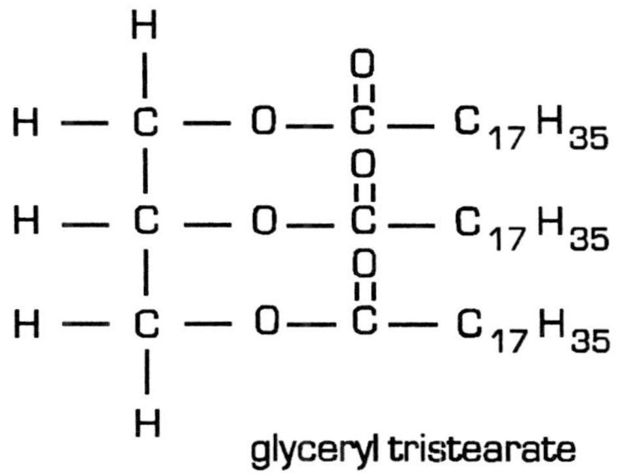

$$\begin{array}{c} CH_2-O-CO-C_{17}H_{35} \\ | \\ CH-O-CO-C_{17}H_{35} \\ | \\ CH_2-O-CO-C_{17}H_{35} \end{array}$$

glyceryl tristearate

Most triglycerides obtained from animals contain primarily saturated fatty acid chains and a small fraction of unsaturated fatty acid chains. These triglycerides are solids at room temperature and are commonly referred to as animal fat. Glyceryl tristearate is a solid animal fat. Triglycerides in plant tissue have a much higher fraction of unsaturated fatty acid chains and are liquids at room temperature. Because of this, they are often called vegetable oils or simply oils. Be aware that a vegetable oil is not the same as petroleum oil. A vegetable oil is a triester, while petroleum oil is a mixture of hydrocarbons.

Glycerol (glycerine) is the simplest trihydric (three –OH groups) alcohol. It is a colorless, oily liquid with a sweet taste. The structural formula of glycerol is shown to the right.

$$\begin{array}{c} H \\ | \\ H - C - OH \\ | \\ H - C - OH \\ | \\ H - C - OH \\ | \\ H \end{array}$$

glycerol

A simple test to show the presence of triglycerides is the **grease spot test**. An acetone solution of a triglyceride is poured onto a piece of filter paper or notebook paper. After the acetone evaporates, a translucent grease spot remains behind as evidence of the triglyceride.

C. Proteins

Proteins are complex biological substances composed of carbon, hydrogen, oxygen, nitrogen and sometimes sulfur and phosphorus. The hydrolysis of a protein produces a mixture of **amino acids**. Dietary protein provides the nitrogenous material needed to build protoplasm, enzymes and those hormones that are proteins. Amino acids have the general formula given below, where R may be a hydrogen or an aliphatic, aromatic or heterocyclic group.

$$\begin{array}{c} H \\ | \\ H_2N - C - COOH \\ | \\ R \end{array}$$

About 20 amino acids are commonly found in plant and animal tissue. Six are shown below:

$$\begin{array}{c} H \\ | \\ H_2N - C - COOH \\ | \\ H \end{array}$$

glycine

$$\begin{array}{c} H \\ | \\ H_2N - C - COOH \\ | \\ CH_3 \end{array}$$

alanine

$$\begin{array}{c} H \\ | \\ H_2N - C - COOH \\ | \\ C{-}H \\ CH_3 \quad CH_3 \end{array}$$

valine

$$\begin{array}{c} H \\ | \\ H_2N - C - COOH \\ | \\ H - C - OH \\ | \\ CH_3 \end{array}$$

threonine

$$\begin{array}{c} H \\ | \\ H_2N - C - COOH \\ | \\ CH_2 \\ | \\ C_6H_5 \end{array}$$

phenylalanine

$$\begin{array}{c} H \\ | \\ H_2N - C - COOH \\ | \\ CH_2 \\ | \\ C_6H_4 \\ | \\ OH \end{array}$$

tyrosine

Proteins are polymers of amino acids joined together by the condensation of the carboxyl group on one amino acid with the amino group on another, eliminating a water molecule. When two amino acids join, a **dipeptide** is produced.

$$H_2N-\underset{R}{\overset{H}{C}}-\overset{O}{\overset{\|}{C}}-OH + H-\overset{H}{N}-\underset{R'}{\overset{H}{C}}-COOH \longrightarrow H_2N-\underset{R}{\overset{H}{C}}-\overset{O}{\overset{\|}{C}}-\overset{H}{N}-\underset{R'}{\overset{H}{C}}-COOH + H_2O$$

dipeptide

Nutrasweet®, a commercial name for aspartame, is a dipeptide of aspartic acid and phenylalanine. The carboxylic acid group of phenylalanine on the right end of the molecule is converted to the methyl ester as shown below.

$$H_2N-\underset{CH_2-COOH}{\overset{H}{C}}-\overset{O}{\overset{\|}{C}}-\overset{H}{N}-\underset{CH_2-C_6H_5}{\overset{H}{C}}-\overset{O}{\overset{\|}{C}}-O-CH_3$$

aspartame

The linkage formed when two amino acids combine is called a **peptide link** or an **amide functional group**.

$$-\overset{O}{\overset{\|}{C}}-\overset{H}{N}-$$

peptide link

Biuret, a compound after which one of the protein tests is named, contains an amide functional group:

$$H_2N-\overset{O}{\overset{\|}{C}}-\overset{H}{N}-\overset{O}{\overset{\|}{C}}-NH_2$$

biuret

Experimental Procedure:

CAUTION: Strong acids and bases are used in the following procedures. Wear eye protection at all times. If a reagent is spilled on the skin, immediately wash the area with cold water and report to the instructor.

ENVIRONMENTAL ALERT: Dispose of all wastes properly. Strong acids and bases must not simply go down the drain. Carefully follow the directions for waste disposal given by your instructor.

A. Tests for Detecting Carbohydrates

1. Molisch Test

The Molisch test is a general test for carbohydrates, and any monosaccharide will give a positive test. Since the test solution is acidic, disaccharides and polysaccharides will hydrolyze to give monosaccharides, so they, too, will give a positive response.

Prepare and number five 15-cm test tubes as directed below:

1. 20 drops of deionized water (the control)
2. 20 drops of 1% glucose solution
3. 20 drops of 1% sucrose solution
4. 20 drops of 1% lactose solution
5. 20 drops of 1% starch solution

To each of the five test tubes, add 3 drops of Molisch reagent. Seal each test tube with a cork or film, and shake each tube to ensure mixing. Then, with the test tube inclined at a 45° angle, *slowly* add 10 drops of concentrated sulfuric acid, H_2SO_4, so that two layers form. *Use caution when handling concentrated sulfuric acid.* Keep the mouth of the test tube pointed away from yourself and others around you. Note the color produced where the two layers meet, the interface. Record your observations on the Report Sheet. A purple color is a positive test for monosaccharides, and you should see this with the glucose sample.

2. Benedict Test

Before proceeding, set up a boiling water bath in a 250-mL beaker half filled with water. Add two or three boiling chips to ensure smooth boiling.

Any carbohydrate that can form a free aldehyde group in solution will react with the Benedict reagent to form a brick-red, brown or occasionally yellow precipitate. A precipitate of any of these colors is considered a positive test. However, a color change in the solution without the precipitate is a negative test.

Prepare five clean test tubes with fresh samples of the solutions used in the Molisch test. Add 5 mL of the Benedict reagent to each test tube, and shake each to mix. Then at the same time, place all five test tubes in the boiling water bath. Remove the five test tubes after 5 minutes and record your observations on the Report Sheet. Glucose will give a positive test.

3. Iodine Test

Polysaccharides can combine with iodine to form a blue, red, violet or purple complex. Prepare five test tubes with fresh samples of the carbohydrate solutions used in the Molisch test. Add *1* drop of iodine solution to each and shake to mix. Observe the control sample. If none of the other samples has a color different than the control, add 10 drops of pure water to each test tube and 1 more drop of the iodine solution. Mix each solution well and record your observations on the Report Sheet. Starch will give a positive test.

B. Test for Detecting Fats: The Grease Spot Test

Place a small drop of vegetable oil on a piece of notebook paper or filter paper. Hold the paper up to the light. Record your observations on the Report Sheet.

C. Tests for Detecting Protein

1. Xanthoprotic Test

When treated with concentrated nitric acid, the aromatic rings on the side chains of proteins or free amino acids become substituted with nitro groups, $-NO_2$. The nitrated products are yellow, so the solution becomes yellow or a yellow precipitate forms. When the solution is made basic with sodium hydroxide, the color becomes more intense and ranges from yellow to orange to brown.

Prepare and label six 15-cm test tubes with the following solutions:

1. 20 drops of deionized water (the control)
2. 20 drops of 1% biuret solution
3. 20 drops of 2% biuret solution
4. 20 drops of 2% tyrosine solution
5. 20 drops of 2% egg-albumen solution
6. 20 drops of 2% glycine solution

To each of the six solutions, add 5 drops of concentrated nitric acid, HNO_3. *Handle concentrated nitric acid with care.* Heat the solutions in a boiling water bath for 3 minutes, then record the colors produced on the Report Sheet.

Remove the tubes, allow them to cool for 5 minutes, and *cautiously* and *slowly* add 15 drops of 6 M sodium hydroxide, NaOH, to each. Stir each with a glass stirring rod, then remove a droplet of each solution and touch it to a piece of pH paper. Add 5 more drops of NaOH to any solution that does not test strongly basic. Recheck the pH and add more NaOH if necessary until each solution tests strongly basic with pH paper. Record any color changes on the Report Sheet. Tyrosine gives a positive test.

2. Biuret Test

Obtain fresh samples of the solutions listed in the xanthoprotic test in six 15-cm test tubes. To each, add 5 drops of 6 M NaOH and 3 drops of 2% copper(II) sulfate, $CuSO_4$, solution. Seal each test tube and shake each mixture. After 3 minutes, note the colors produced. Any solution, other than the control, that does not change color should be checked to see if it is basic. If not, add NaOH until the solution tests basic, and also add 2 additional drops of the copper(II) sulfate solution. Biuret solution will give a positive test, a purple color. Record all observations on the Report Sheet.

3. Ninhydrin Test

Obtain fresh samples of the control and solutions used in the xanthoprotic test except for the 1% and 2% biuret solutions. Then, to each of the four samples, add 2 drops of 0.1% ninhydrin solution. Heat each cautiously and slowly to boiling over a burner for

1 minute. Allow the solutions to cool, and record your observations on the Report Sheet. Glycine will give a positive test.

D. Testing Foods

Your instructor may assign foods to be tested for fats, carbohydrates and proteins. Preparation of solid foods will not be the same as for liquids.

CAUTION: Acetone is a flammable liquid. Make certain there are no open flames around your work area while you are using acetone.

1. Solid Foods

Grind about 2 g of the solid in a mortar, then add acetone to cover the solid to a depth of about 2 cm. Continue to grind the food for about 2 minutes. Filter through cheesecloth and perform the *grease spot* test, Section B, using the acetone filtrate.

Spread the solid retained by the cheesecloth on a piece of filter paper and allow the acetone to evaporate. Add samples of this solid to 1 mL (20 drops) of deionized water in each of six 15-cm test tubes and perform the three carbohydrate tests (the *Molisch, Benedict* and *iodine* tests) in Section A and the three protein tests (the *xanthoprotic, biuret* and *ninhydrin* tests) in Section C. Record all results on the Report Sheet.

2. Liquid Foods

Place 20 drops (1 mL) of the liquid food in six separate 15-cm test tubes and perform the three carbohydrate tests in Section A and the three protein tests in Section C. Record your results on the Report Sheet.

Add 5 mL of acetone to 5 mL of the liquid food and mix well in a test tube. Using a medicine dropper, transfer 3 drops of the acetone layer (the top layer) to a clean sheet of notebook or filter paper for the *grease spot* test. Record your results on the Report Sheet.

E. Testing Milk (Any type of milk may be used.)

1. Place 1 mL (20 drops) of milk in each of three clean test tubes. Perform the three carbohydrate tests (the *Molisch*, *Benedict* and *iodine* tests) described in Section A. Record all observations on the Report Sheet.

2. Mix 1 mL (20 drops) of milk with 2 mL of acetone. If necessary, filter the solution through two layers of cheesecloth. Place a few drops of the filtered solution on the center of a piece of clean notebook or filter paper. Compare the results with that of the **grease spot test** obtained in Section B. Record your observation on the Report Sheet.

3. Using 1-mL (20-drop) samples of milk, carry out the three protein tests (the *xanthoprotic*, *biuret* and *ninhydrin* tests) described in Section C. Record all observations on the Report Sheet.

Name ______________________ Locker Number__________ Date__________
Please print; last name first

PRELIMINARY EXERCISES: *Experiment 23*

The Detection of Fats, Proteins and Carbohydrates in Foods

1. Define the following terms:

a. carbohydrate
b. protein
c. monosaccharide
d. fatty acid
e. acid hydrolysis
f. triglyceride

2. Give a word equation that describes the reaction of glucose with Benedict reagent.

3. State the products formed when each of the following is hydrolyzed:

a. starch
b. sucrose
c. glyceryl tristearate
d. protein (give the general products)

Name ______________________ Locker Number__________ Date__________
Please print; last name first

REPORT SHEET: *Experiment 23*

The Detection of Fats, Proteins and Carbohydrates in Foods

A. Tests for Detecting Carbohydrates

1. Molisch Test

Sample	Interface Color
deionized water (the control)	
1% glucose solution	
1% sucrose solution	
1% lactose solution	
1% starch solution	

Which carbohydrates gave a positive test? ______________________

____

2. Benedict Test

Sample	Precipitate Color
deionized water (the control)	
1% glucose solution	
1% sucrose solution	
1% lactose solution	
1% starch solution	

Which carbohydrates gave a positive test quickly? ______________________

____

3. Iodine Test

Sample	Solution Color
deionized water (the control)	
1% glucose solution	
1% sucrose solution	
1% lactose solution	
1% starch solution	

Which carbohydrates gave a positive test? ______________________________

B. Test for Detecting Fats: Grease Spot Test

Describe your observations.

C. Tests for Detecting Protein

1. Xanthoprotic Test

Sample	Color before Adding NaOH	Color after Adding NaOH
deionized water (the control)		
1% biuret solution		
2% biuret solution		
2% tyrosine solution		
2% egg-albumen solution		
2% glycine solution		

Which samples gave a positive test? ______________________________

____

Name ______________________ Locker Number__________ Date__________
Please print; last name first

REPORT SHEET (continued): *Experiment 23*

The Detection of Fats, Proteins and Carbohydrates in Foods

2. Biuret Test

Sample	Solution Color
deionized water (the control)	
1% biuret solution	
2% biuret solution	
2% tyrosine solution	
2% egg-albumen solution	
2% glycine solution	

Which samples gave a positive test? ______________________

____

Was there any difference between the two biuret samples? If so, describe.

3. Ninhydrin Test

Sample	Solution Color
deionized water (the control)	
2% tyrosine solution	
2% egg-albumen solution	
2% glycine solution	

Which samples gave a positive test? ______________________

____

D. Testing Foods

Describe the food or foods tested and the results of each test. Summarize your observations in detail and report your conclusions.

Food 1	**Food 2**
Description of the food:	Description of the food:
Carbohydrate tests:	Carbohydrate tests:
Fat test:	Fat test:
Protein tests:	Protein tests:
Conclusions:	Conclusions:

Name ______________________________ Locker Number__________ Date____________
Please print; last name first

REPORT SHEET (continued): ***Experiment 23***

The Detection of Fats, Proteins and Carbohydrates in Foods

E. Testing Milk

Source and description of the milk sample: ________________________________

1. Tests for Detecting Carbohydrates

Molisch test	Observations:	Conclusions:
Benedict test	Observations:	Conclusions:
Iodine test	Observations:	Conclusions:

2. Test for Detecting Triglycerides: Grease Spot Test

Observations:
Conclusions:

(Continued on the following page.)

3. Tests for Detecting Proteins

Xanthoprotic test	Observations:	Conclusions:
Biuret test	Observations:	Conclusions:
Ninhydrin test	Observations:	Conclusions:

4. Summarize what you have learned about the milk sample.

Experiment 24

The Characterization of Carbohydrates

INSTRUCTIONAL NOTE: Two laboratory periods (5½ hours total) are required to complete the entire experiment

Purpose:

In this experiment, you will learn how color tests can be used to characterize carbohydrates. The tests will show whether a carbohydrate is a monosaccharide, disaccharide or polysaccharide and if it is a pentose, hexose or reducing sugar.

Discussion:

A carbohydrate is a polyhydroxyaldehyde, a polyhydroxyketone or a compound that yields these upon acid hydrolysis (the reaction with water in the presence of acid). Carbohydrates are the principal constituents of plants, comprising 60% to 90% of their dry weight. Plants produce carbohydrates as a means of storing energy (starch) and as structural material (cellulose). Starch and cellulose are both polymers of glucose, which is produced in photosynthesis.

There are three classes of carbohydrates based on the products formed when they are hydrolyzed:

1. Monosaccharides cannot broken down into simpler carbohydrates by acid hydrolysis. They are called simple sugars.

2. Disaccharides yield two monosaccharides upon acid hydrolysis.

3. Polysaccharides yield more than two monosaccharides upon acid hydrolysis.

A. Monosaccharides

Monosaccharides are the simplest carbohydrates. They can be subdivided further according to the number of carbon atoms in the monosaccharide molecule. Most monosaccharides are either **trioses** (three carbon atoms), **pentoses** (five carbon atoms), or **hexoses** (six carbon atoms in the molecule). Each of these subgroups can be divided still further into aldoses and ketoses. An **aldose** has an aldehyde functional group, and a **ketose** has a ketone functional group. The *-ose* suffix is characteristic of carbohydrates.

The simplest aldose is glyceraldehyde, and the simplest ketose is dihydroxyacetone. Because there are three carbon atoms in glyceraldehyde, it could be classified as an aldotriose. Dihydroxyacetone would be a ketotriose.

```
H—C=O
H—C—OH
  CH2OH
```

glyceraldehyde
($C_3H_6O_3$)

```
CH2OH
C=O
CH2OH
```

dihydroxyacetone
($C_3H_6O_3$)

Xylose, arabinose, ribose and deoxyribose are all pentoses, and they are all aldopentoses. Ribose and deoxyribose are important components of RNA and DNA.

```
 H—C=O
 H—C—OH
HO—C—H
 H—C—OH
   CH2OH
```

xylose
($C_5H_{10}O_5$)

```
 H—C=O
HO—C—H
 H—C—OH
 H—C—OH
   CH2OH
```

arabinose
($C_5H_{10}O_5$)

```
H—C=O
H—C—OH
H—C—OH
H—C—OH
  CH2OH
```

ribose
($C_5H_{10}O_5$)

```
H—C=O
H—C—H
H—C—OH
H—C—OH
  CH2OH
```

deoxyribose
($C_5H_{10}O_4$)

Glucose, mannose, galactose and fructose are all hexoses. Fructose is a ketohexose. The rest are aldohexoses.

```
 H—C=O
 H—C—OH
HO—C—H
 H—C—OH
 H—C—OH
   CH2OH
```

glucose
($C_6H_{12}O_6$)

```
 H—C=O
HO—C—H
HO—C—H
 H—C—OH
 H—C—OH
   CH2OH
```

mannose
($C_6H_{12}O_6$)

```
 H—C=O
 H—C—OH
HO—C—H
HO—C—H
 H—C—OH
   CH2OH
```

galactose
($C_6H_{12}O_6$)

```
   CH2OH
   C=O
HO—C—H
 H—C—OH
 H—C—OH
   CH2OH
```

fructose
($C_6H_{12}O_6$)

Up to this point, only the open-chain structures have been shown for monosaccharides. In their natural environment though, only a small fraction of pentose and hexose sugars exist in the open-chain structures, and they do so in equilibrium with their cyclic structures. In these equilibria, the cyclic structures predominate. Three hexoses are shown below in their cyclic structures.

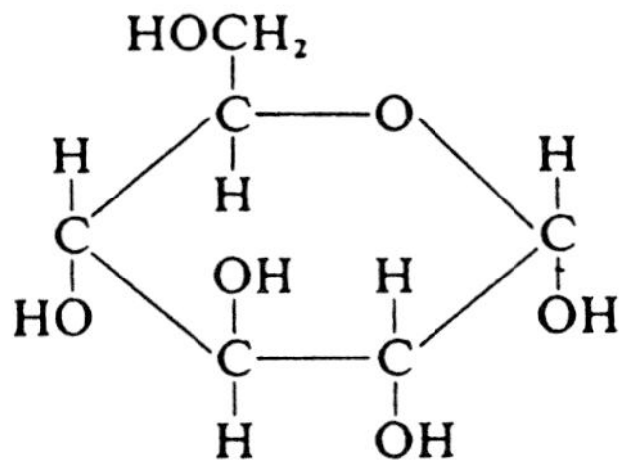

α-glucose α-galactose α-fructose

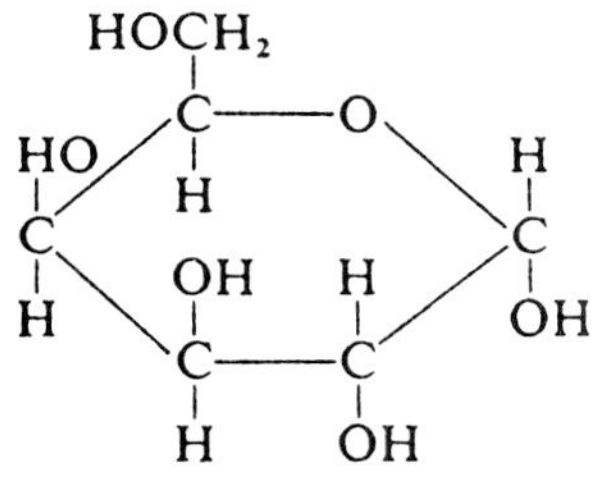

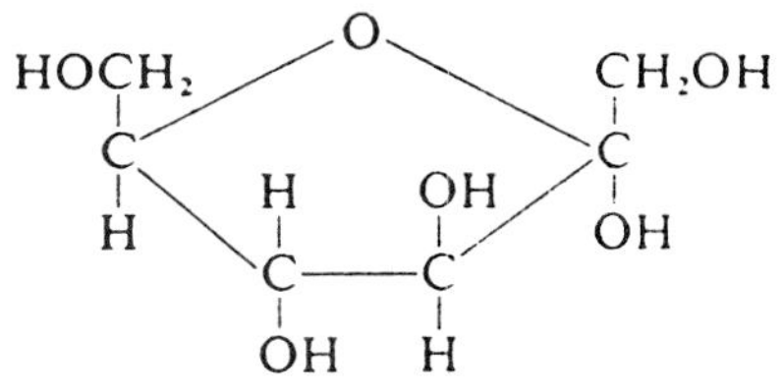

Each cyclic structure can be abbreviated by omitting most of the atomic symbols. The same monosaccharides are reproduced below in their simplified structures. In these structures, it is understood that at each corner there is a carbon atom, with a hydrogen atom attached if needed, to complete four bonds.

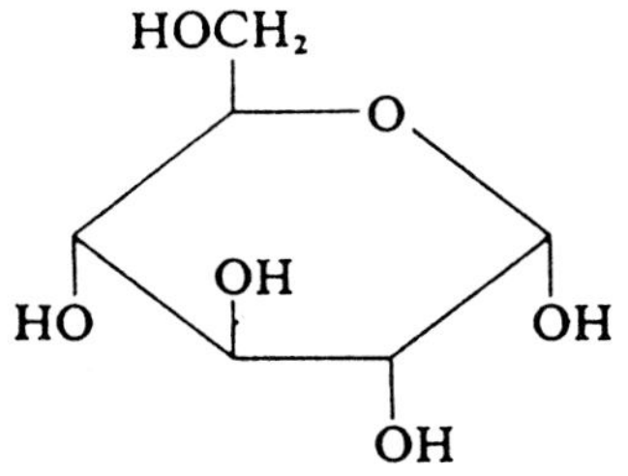

α-glucose α-galactose α-fructose

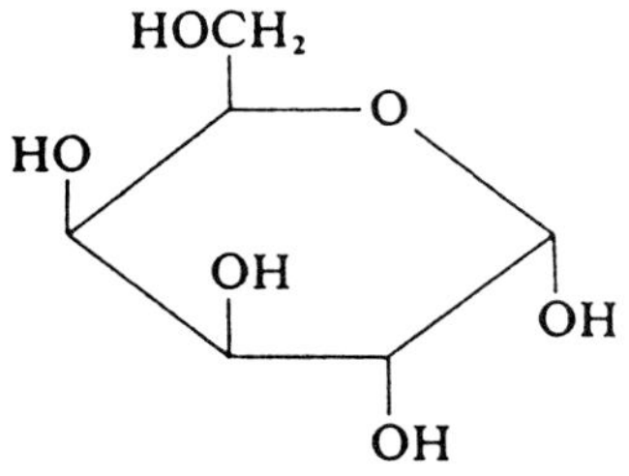

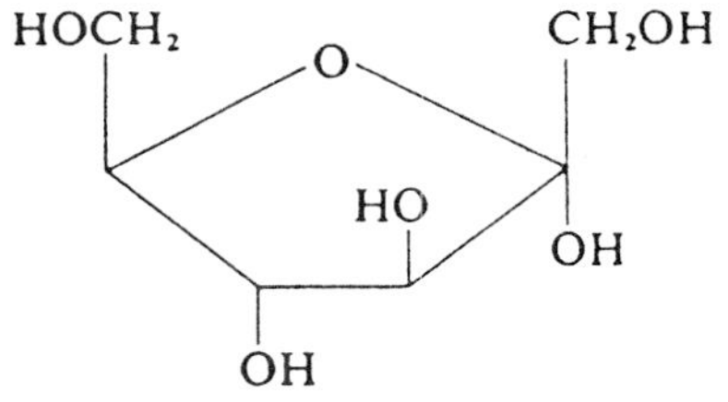

B. Disaccharides

Disaccharides are composed of two monosaccharide units connected together by an ether linkage (–C–O–C–), which forms as the two monosaccharides join and eliminate a water molecule. The three most important disaccharides are sucrose (table sugar), maltose (malt sugar) and lactose (milk sugar). The two monosaccharides that joined to form each disaccharide are reformed as the disaccharide undergoes acid hydrolysis.

$$\text{sucrose} + H_2O \xrightarrow{\text{acid}} \text{glucose} + \text{fructose}$$

$$\text{maltose} + H_2O \xrightarrow{\text{acid}} \text{glucose} + \text{glucose}$$

$$\text{lactose} + H_2O \xrightarrow{\text{acid}} \text{glucose} + \text{galactose}$$

Structural formulas of these three disaccharides are shown below.

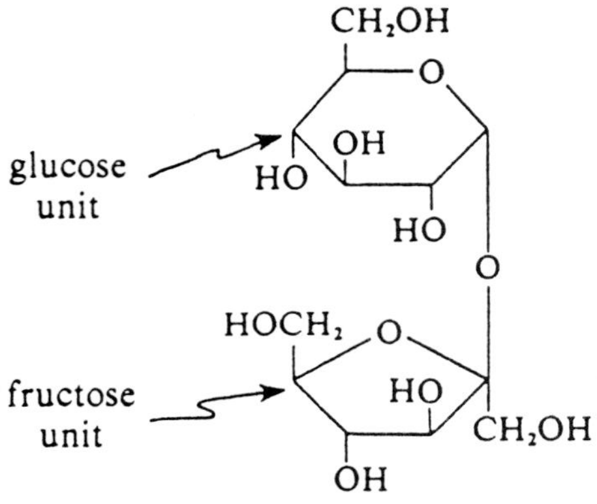

sucrose

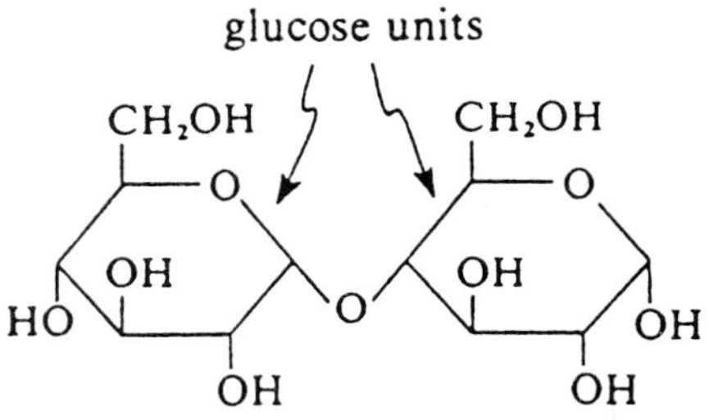

maltose

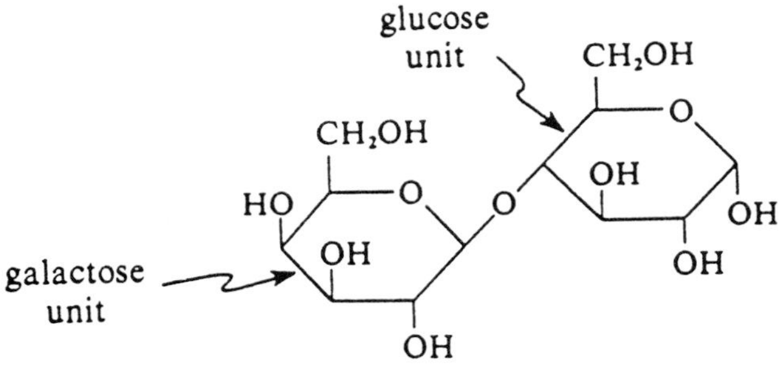

lactose

C. Polysaccharides

Polysaccharides are high-molecular-weight polymers of monosaccharides. Some, such as starch and glycogen, are used to store glucose in plants and animals, respectively. Others, most notably cellulose, are used as structural material in plants. Starch, glycogen and cellulose are the most important polysaccharides, and each is a polymer of glucose.

Starch is the most important carbohydrate in our diet. Starch composes about 25% of the food we eat, and it is primarily obtained from cereal grains (corn, wheat and rice) and vegetables (beans, potatoes and peas). Plants store glucose in the cytoplasm of cells as starch granules, which are insoluble and unable to pass through cell membranes.

Starch is a mixture of two polysaccharides, amylose and amylopectin. Amylose makes up from 10% to 20% of starch by weight and is composed of from 60 to 300 glucose units joined by α-1,4 ether linkages. The amylose polymer coils into a helical structure.

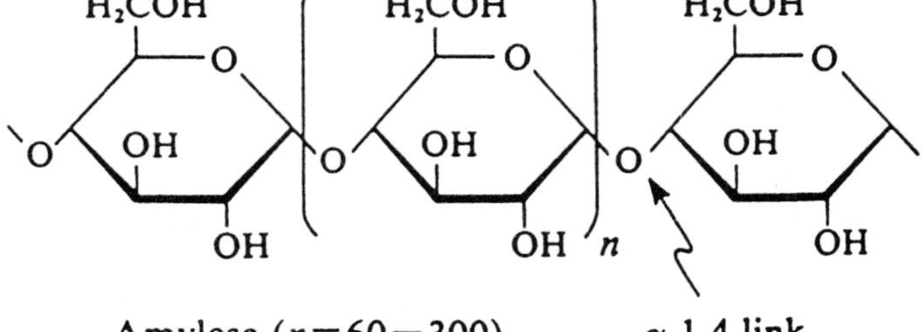

Amylose (n=60−300) α-1,4 link

The remaining 80% to 90% of starch is amylopectin, a polymer formed from 300 to 6000 glucose units joined by α-1,4 ether bonds. But unlike amylose, the amylopectin structure includes several short chains of from 25 to 30 glucose units that branch out from the longer chain. The branching points are α-1,6 ether linkages. Even with branching, each chain of polymerized glucose exists in a helical form.

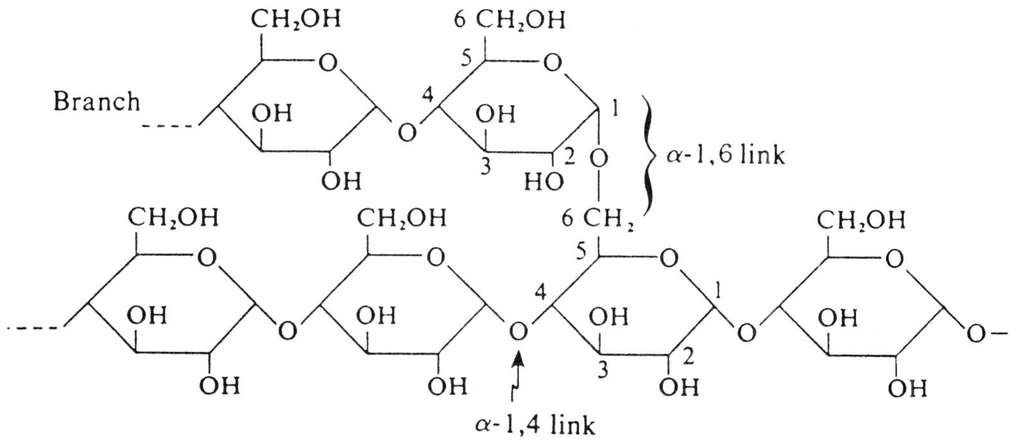

amylopectin

Glycogen is the storage form of carbohydrate in animals, and for that reason it is sometimes called animal starch. It is found to some extent in all body tissue, but primarily in muscle and liver tissue. Glycogen serves an important role in maintaining the correct glucose level in the blood and tissue. Glycogen has essentially the same branched structure as amylopectin, but the branching occurs in shorter intervals. Because of their size, glycogen polymers cannot pass through cell membranes, though glycogen is somewhat more soluble than starch.

Cellulose is a linear polymer of glucose units, but unlike starch, the glucose units are joined through β-1,4 linkages, not α-1,4 linkages. This single difference makes cellulose indigestible in humans and other carnivorous animals. Cellulose simply passes through the digestive tract unchanged. It is referred to as "fiber" in the diet.

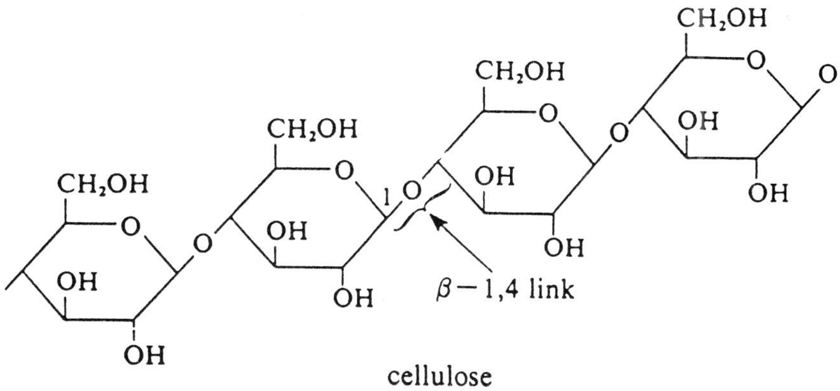

cellulose

Cellulose is the structural component of the cells of higher plants. Cotton fiber and paper are more than 90% cellulose, and wood fiber is nearly 50% cellulose. Cellulose can be hydrolyzed in acid solution or in the presence of the enzyme cellulase. The product of hydrolysis is glucose. Humans lack cellulase in our digestive tract, but ruminants, such as cows, horses and sheep, have microorganisms in their digestive systems that produce this enzyme. Termites have these microorganisms too.

Experimental Procedure:

CAUTION: Strong acids and bases are used in many of these procedures. Wear eye protection at all times. If you spill a corrosive liquid on your skin, wash the area immediately with running water, then contact the instructor.

ENVIRONMENTAL ALERT: Dispose of all wastes properly. Strong acids and bases must not simply go down the drain. Carefully follow the directions for waste disposal given by your Instructor.

To complete the experimental work in the allotted time, you may need to perform the procedures for the known and unknown carbohydrates (Sections A and B) at the same time.

A. Characterizing Known Carbohydrates

Several procedures require a boiling water bath, so first begin heating a 250-mL beaker half full of tap water on a tripod or ring stand. Be sure to add a few boiling chips.

1. Molisch Test

The Molisch test is a general test for carbohydrates, and monosaccharides will give a positive test rapidly. Disaccharides and polysaccharides will slowly hydrolyze to produce monosaccharides in the acidic medium of the test and, therefore, will also give a positive test. Since polysaccharides hydrolyze more slowly than disaccharides, you may be able to distinguish one from the other. A polysaccharide will slowly give a positive response.

Procedure: Label six 15-cm test tubes with numbers 1 to 6. Place 1 mL (20 drops) of each of the following solutions into five of the test tubes. Shake the bottle of starch solution before use. Add 20 drops of deionized water to the sixth test tube to serve as the control.

1. 1% xylose solution (an aldopentose)
2. 1% glucose solution (an aldohexose)
3. 1% fructose solution (a ketohexose)
4. 1% sucrose solution (a disaccharide)
5. 1% starch solution (a polysaccharide)
6. deionized water (the control)

To each test tube, add 3 drops of Molisch reagent. Seal each test tube with a cork or plastic film, and shake each to ensure mixing. With the test tube held at a 45° angle, *carefully* and *slowly* run 10 drops of concentrated sulfuric acid down the wall of the tube so that two layers form. *Caution: Add the acid slowly, and do not look directly into the tube since spattering may occur.* Note the color produced at the point where the two layers meet. Record the *color* you observe for each sample on the Report Sheet. A purple color constitutes a positive test, and glucose will give a positive test. Also, record the approximate *time* required for the positive response to occur. If you used corks, they should be discarded at this time.

2. **Iodine Test**

This test will detect the presence of polysaccharides. Polysaccharides will combine with iodine to form a blue, red, violet or purple color as the signal of a positive test. Iodine is adsorbed onto the surface of the polysaccharide, forming a colored complex.

Procedure: Place 1 mL (20 drops) of each of the following solutions into nine separate, numbered 15-cm test tubes:

1.	1% xylose solution	(an aldopentose)
2.	1% glucose solution	(an aldohexose)
3.	1% sucrose solution	(a disaccharide)
4.	1% starch solution	(a polysaccharide)
5.	1% glycogen solution	(a polysaccharide)
6.	1% amylose solution	(a polysaccharide)
7.	1% amylopectin solution	(a polysaccharide)
8.	1% cellulose solution	(a polysaccharide)
9.	deionized water	(the control)

To each test tube, add 1 drop of iodine solution. Seal the tubes with corks or plastic film, and shake each mixture. Record the *color* of each sample on the Report Sheet.

Add 10 drops of deionized water to each test tube and 1 more drop of iodine solution. Shake each sample, and record your observations on the Report Sheet. Starch will give a positive test. If you used corks, they should be discarded at this time.

3. **Benedict Test**

Any monosaccharide or disaccharide that can form a free aldehyde group in solution will react with the Benedict reagent to form a brick-red, brown, green or occasionally yellow precipitate. A precipitate of any of these colors is considered a positive test. A colored *solution* without a precipitate is a negative test. The reaction is carried out in basic solution. The precipitate is insoluble copper(I) oxide, Cu_2O, that forms when copper(II) ion, Cu^{2+}, is reduced by the aldehyde. Carbohydrates that give a positive Benedict test are classified as reducing sugars.

Procedure: Place 1 mL (20 drops) of each of the following into nine separate, numbered 15-cm test tubes:

1.	1% xylose solution	(an aldopentose)
2.	1% glucose solution	(an aldohexose)
3.	1% galactose solution	(an aldohexose)
4.	1% fructose solution	(a ketohexose)
5.	1% sucrose solution	(a disaccharide)
6.	1% maltose solution	(a disaccharide)
7.	1% lactose solution	(a disaccharide)
8.	1% starch solution	(a polysaccharide)
9.	deionized water	(the control)

Add 5 mL of the Benedict reagent to each solution. Seal the tubes with corks or plastic film and shake each sample. Remove the seals and place all nine test tubes in a boiling water bath *at the same time*. Remove the tubes after 5 minutes and allow them to cool. After 15 minutes, record your observations on the Report Sheet. Fructose will give a positive test; sucrose, a negative test. If you used corks, they should be discarded at this time.

4. **Barfoed Test**

The Barfoed test is used to distinguish between reducing monosaccharides and reducing disaccharides. It is similar to the Benedict test, except that it is carried out in acidic solution, and only reducing monosaccharides will form precipitates of Cu_2O within 5 minutes.

Procedure: Place 1 mL (20 drops) of each of the same carbohydrate solutions used for the Benedict test into numbered 15-cm test tubes. Again, use deionized water as the control. To each solution add 5 mL of the Barfoed reagent. Seal the tubes with corks or plastic film and shake each sample. Remove the seals and place all nine test tubes in a boiling water bath *at the same time*. Remove them after 5 minutes and let them cool. After 15 minutes, make your observations and record them on the Report Sheet. Glucose will give a positive test; maltose, a negative test. Compare your results with those obtained in the Benedict test.

5. **Hydrolysis of Disaccharides**

Disaccharides can be hydrolyzed in acidic solution to form the monosaccharides that compose them. Sucrose, a disaccharide, does not give a positive Benedict test, but after hydrolysis to form glucose and fructose, the resulting solution will give a positive Benedict test.

Procedure: Place 10 mL of 1% sucrose solution in a 15-cm test tube. *With caution*, add 2 drops of concentrated sulfuric acid to the solution. Seal the tube with a cork or plastic film, and shake the mixture. Remove the seal, and place the test tube in a boiling water bath for 3 minutes. Remove the test tube and *slowly and carefully* add 15 drops of 6 M NaOH to the mixture. Seal the tube and shake the mixture. Then, using a stirring rod, remove a drop of solution and test it with pH paper to see if it is basic. If it is not, add more NaOH, 1 drop at a time, stirring after each drop, until the solution is basic. If you used a cork, it should be discarded at this time.

Transfer 20 drops of this basic solution to a clean 15-cm test tube. Add 5 mL of Benedict reagent to this solution, mix, and place the test tube in a boiling water bath for 5 minutes. Remove the test tube, allow it to cool, then record your observations on the Report Sheet. Compare the results of the Benedict test for sucrose in Part 3 with what you observe here.

6. **Bial Test**

The Bial test will differentiate between pentoses and hexoses. Under the strongly acidic conditions of the Bial test, both pentose and hexose carbohydrates dehydrate (lose water) and react with orcinol in the presence of iron(III) chloride to produce specific colors. A blue color is a positive test for a pentose. All other colors are considered negative for pentoses.

Procedure: Place 1 mL (20 drops) of each of the following solutions in six separate, numbered 15-cm test tubes:

1. 1% xylose solution (an aldopentose)
2. 1% glucose solution (an aldohexose)
3. 1% fructose solution (a ketohexose)
4. 1% sucrose solution (a disaccharide)
5. 1% starch solution (a polysaccharide)
6. deionized water (the control)

Add 3.0 mL of Bial reagent to each. Seal the tubes with corks or plastic film and shake each sample. Remove the seals and place the six test tubes *at the same time* in a boiling water bath for 1 minute. Remove the test tubes and add 5 mL of deionized water *and* 5 mL of butanol to each. Seal and shake each mixture. If used, discard the corks at this time. Observe the color *in the butanol layer,* and record your observations on the Report Sheet. Xylose (a pentose) will give a positive test; glucose, a hexose, will not.

7. Seliwanoff Test

The Seliwanoff test will distinguish between ketohexoses and aldohexoses. Fructose, glucose, galactose and mannose are all dehydrated in the acidic environment of the Seliwanoff test, and the dehydrated products react with resorcinol to form red-colored products. The difference in the times required for the red color to appear is used to distinguish ketohexoses from aldohexoses. Fructose (a ketohexose) dehydrates quickly, and usually forms the red product within 2 minutes after addition of the test reagent. Disaccharides and polysaccharides will eventually hydrolyze to hexoses, which in time will also form red-colored solutions.

Procedure: Place 1 mL (20 drops) of each of the following solutions into eight separate, numbered 15-cm test tubes:

1. 1% xylose solution (an aldopentose)
2. 1% glucose solution (an aldohexose)
3. 1% galactose solution (an aldohexose)
4. 1% fructose solution (a ketohexose)
5. 1% sucrose solution (a disaccharide)
6. 1% lactose solution (a disaccharide)
7. 1% starch solution (a polysaccharide)
8. deionized water (the control)

To each solution, add 5 mL of the Seliwanoff reagent. Seal the tubes with corks or plastic film and shake each one. Then, if used, discard the corks. Place all eight test tubes into a boiling water bath *at the same time* for 2 minutes. After 2 minutes, remove the tubes and note the colors produced on the Report Sheet. Replace the tubes in the boiling water bath for an additional 2 minutes. Again, after 2 minutes, observe the colors produced. Heat all the samples once more for an additional 2 minutes, and observe the colors a third time. Record all observations on the Report Sheet.

B. Characterizing Unknown Carbohydrates

Your instructor will give you one or two carbohydrates to characterize using the tests described in this experiment. If the carbohydrate is a solid, dissolve about 0.4 g of the solid in 200 mL of deionized water. Use this solution for the tests. If the unknown carbohydrate is in solution as you receive it, then use that solution for the tests. Record the numbers of the unknowns on the Report Sheet.

Perform *only* the tests that are required to characterize each carbohydrate unknown. Record all observations and conclusions in brief, concise statements in the modest space provided on the Report Sheet.

1. **Solubility in Water**
   Polysaccharides and lactose, a disaccharide, are not very soluble in water. Even if your unknown is provided as a solution, you may be able to judge whether the carbohydrate is highly soluble based on the turbidity, or cloudiness, of the mixture.

2. **Iodine Test**
   If you observe a positive iodine test, you know you have a polysaccharide. Based on the color you observe, decide which polysaccharide it is: starch, glycogen or cellulose. That will be sufficient characterization for that unknown carbohydrate.

3. **Benedict and Barfoed Tests**
   Use these tests to decide if the carbohydrate is a reducing monosaccharide, a reducing disaccharide, or sucrose. If you think the unknown is sucrose, perform the hydrolysis experiment in Section A, Part 5.

4. **Bial Test**
   If you conclude the unknown is a monosaccharide, use the Bial test to determine if it is a pentose or a hexose.

5. **Seliwanoff Test**
   If you conclude the unknown is a hexose, use the Seliwanoff test to determine if it is fructose. If it is not, then it is either galactose or glucose.

Name ______________________________ Locker Number__________ Date____________
Please print; last name first

PRELIMINARY EXERCISES: *Experiment 24*

The Characterization of Carbohydrates

1. Define the following terms:

a. carbohydrate
b. aldohexose
c. reducing sugar

2. List the three major classes of carbohydrates.

   a. ____________________ b. ____________________ c. ____________________

3. Which test in this experiment is positive only for polysaccharides? ____________________

4. What can be learned about a sugar by performing both the Benedict and Barfoed tests?

5. List the three common disaccharides and their hydrolysis products.

   a. ____________________: ____________________ + ____________________

   b. ____________________: ____________________ + ____________________

   c. ____________________: ____________________ + ____________________

6. If an unknown carbohydrate gives a positive Molisch test, a negative iodine test, a negative Barfoed test, a negative Benedict test, but a positive Benedict test after hydrolysis, what is the carbohydrate?

Name ______________________ Locker Number__________ Date____________
Please print; last name first

REPORT SHEET: *Experiment 24*

The Characterization of Carbohydrates

A. Characterizing Known Carbohydrates

1. The Molisch Test

Record the *color* produced by each carbohydrate and the *time* required for a positive test:

1. xylose	3. fructose	5. starch
2. glucose	4. sucrose	6. control

Which carbohydrates gave a positive test?

Did any carbohydrate(s) require an extended time to give a positive test?

2. Iodine Test

Record the *color* produced by each carbohydrate:

1. xylose	4. starch	7. amylopectin
2. glucose	5. glycogen	8. cellulose
3. sucrose	6. amylose	9. control

Which carbohydrate(s) gave a positive iodine test?

Which of these are polysaccharides?

3. Benedict Test

Record your observation as positive (+) or negative (−) for each carbohydrate:

1. xylose	4. fructose	7. lactose
2. glucose	5. sucrose	8. starch
3. galactose	6. maltose	9. control

Which carbohydrates are reducing sugars?

Which carbohydrates are reducing monosaccharides?

Which carbohydrates are reducing disaccharides? ______________________________

Which are nonreducing disaccharides? ______________________________

4. Barfoed Test

Record your observation as positive (+) or negative (−) for each carbohydrate:

1. xylose	4. fructose	7. lactose
2. glucose	5. sucrose	8. starch
3. galactose	6. maltose	9. control

Which carbohydrates are reducing monosaccharides?

What can you learn about an unidentified carbohydrate by comparing the results of the Barfoed and Benedict tests? (Compare results in Parts 3 and 4 before answering.)

Name ______________________ Locker Number__________ Date__________
Please print; last name first

REPORT SHEET (continued): *Experiment 24*

5. Hydrolysis of a Disaccharide

Does sucrose give a positive Benedict test? Why or why not?

After hydrolysis, did sucrose give a positive Benedict test? Explain your answer.

6. Bial Test

What color indicates a positive test for a pentose sugar? ______________________

Record your observations (+ or −) for each carbohydrate in the table below:

1. xylose	3. fructose	5. starch
2. glucose	4. sucrose	6. control

Which carbohydrates are pentoses? ______________________

Which are hexoses? ______________________

Why must this test be time controlled?

7. Seliwanoff Test

Record your observations as positive (+) or negative (−) for each carbohydrate; also, indicate any *colors* observed.

1. xylose	4. fructose	7. starch
2. glucose	5. sucrose	8. control
3. galactose	6. lactose	

Which carbohydrates gave a positive test

in 2 minutes? ______________________

in 4 minutes? ______________________

in 6 minutes? ______________________

Why must this test be time controlled?

B. Characterizing Unknown Carbohydrates

Do only the tests needed to characterize the carbohydrate.

Unknown numbers:		
Describe the physical state of the carbohydrate.		
Describe the solubility of the unknown carbohydrate in water.		
Iodine Test Record your observations and conclusions.		
Benedict Test Record your observations and conclusions.		
Barfoed Test Record your observations and conclusions.		
If you believe your unknown is sucrose, do the hydrolysis experiment in Part 5. Record your observations and conclusions.		
Bial Test Record your observations and conclusions.		
Seliwanoff Test Record your observations and conclusions.		
Characterize the carbohydrate unknown.		

Experiment 25

Enzyme Action

Purpose:

This experiment is designed to introduce you to enzymes and several factors that can affect their action.

Discussion:

Thousands of chemical reactions take place within cells of living organisms. Many of these reactions can also been carried out in the laboratory, but outside the cell, they are often much slower. To speed up these reactions so they would take place as fast as they do in cells would require temperatures or pHs that are inconsistent with life, that is, conditions that simply could not exist in a living cell. How then is it possible for reactions to occur in cells at rates fast enough to meet the needs of the body? The answer is that cells produce special proteins known as **enzymes** that catalyze biological reactions, markedly increasing their rates.

The primary species acted on by an enzyme is called the **substrate** of that enzyme. Most enzymes are named after the substrate on which they act by simply adding *-ase* to the root of the name of the substrate. A lip*ase* then, would be an enzyme that acts on lipids, sucr*ase* on sucrose, and so forth. Many enzymes, such as chymotrypsin, trypsin and lysozyme, have older names that do not end in *-ase*.

Some enzymes owe their specific reactivity only to their specific protein structure. Others, however, are **conjugated proteins** that require the presence of a unique nonprotein unit to become an active enzyme. The protein portion of a conjugated enzyme is known as an **apoenzyme**, and its nonprotein portion is termed a **cofactor.** There are two kinds of cofactors. If the cofactor is an organic unit, it is commonly called a **coenzyme**. If the cofactor is a metal ion, it is called a **metal-ion activator**. Some enzymes require both types of cofactors. Some of the metal-ion activators are Na^+, K^+, Mg^{2+}, Ca^{2+}, Mn^{2+}, Co^{2+} and Zn^{2+}. Many trace metal ions found in the body are important in enzyme reactions. Some vitamins or their derivatives are coenzymes. Since vitamins and metal ions (often referred to simply as minerals) are essential for proper enzyme function, it is easy to understand why they are essential components of the diet.

A. Uses of Enzymes

Enzymes serve many critical functions in the body, catalyzing nearly every reaction that takes place. In recent years, researchers have learned to use enzymes to serve our needs in various ways. Enzymes can be powerful diagnostic tools in medicine, and it is not difficult to measure quickly and accurately the level of activity of a specific enzyme in blood serum or urine. Normally, enzymes appear in blood serum or other extracellular fluids at very low concentrations, but certain disease conditions can markedly increase the level of one or more of them. Enzymes are found principally within cells. If disease or injury damages the cell membrane, the enzymes will flow out into the extracellular fluid and eventually enter the bloodstream, where they can be detected easily. Of course, the use of enzymes is not restricted to health care. Commercially, proteolytic enzymes, such as papain, are used to tenderize meat. They catalyze the hydrolysis of connective tissue, reducing the toughness of meat. Food manufacturers use enzymes to partially digest food for infants and others with digestive problems. Certain proteolytic enzymes have been used to remove cataracts.

B. Enzyme Action

Enzymes are efficient biological catalysts. They increase the rate of a reaction by lowering its **activation energy**, the energy barrier that must be crossed over as reactants are converted to products. For example, the decomposition of hydrogen peroxide, H_2O_2, to form oxygen, O_2, and water has an activation energy in the absence of a catalyst of 18 kcal per mole of H_2O_2. This same reaction occurs in cells, but in the presence of an enzyme called **catalase**. Catalase reduces the activation energy to around 7 kcal per mole of H_2O_2, a substantial reduction, which in turn allows the reaction to take place about 100 million times faster. The effect of catalase on the activation energy of this reaction is shown in Figure 1.

$$2\ H_2O_2 \xrightarrow{\text{catalase}} 2\ H_2O + O_2$$

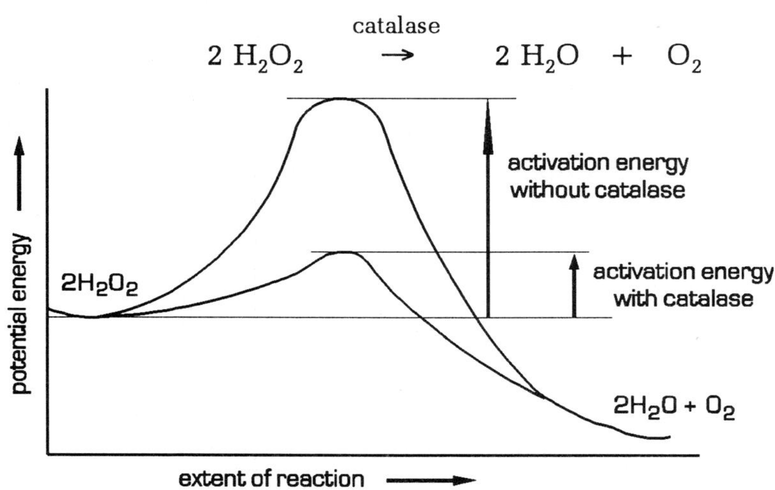

Figure 1. Effect of an enzyme on activation energy

The efficiency of an enzyme is given by its **turnover number**, the number of substrate molecules (reactant molecules) that one enzyme unit can transform in one minute. At body temperature, the turnover number for catalase is 5.6 million. This means one catalase unit can pick up and decompose 5.6 million H_2O_2 molecules each minute.

An enzyme catalyzes a reaction by changing the pathway that leads from reactants to products; that is, it changes what is called the "mechanism" of the reaction. It does this by providing a special surface for the substrate molecule (or molecules) to fit on to so that bonds can be broken and formed much more easily. The special surface is called the **active site** of the enzyme. The active site is usually a small crevice or cavity formed in the tertiary structure of the protein. The active site on an enzyme has a specific geometry or shape that will accommodate only specific substrate molecules that can fit into it. The substrate molecule, therefore, must have a structure that complements the structure of the active site, so the two fit together like a key in a lock. In fact, the idea that the enzyme and substrate molecules must fit together in a specific way is the basis for the **lock-and-key theory** of enzyme action. The enzyme is the key since it "unlocks" or changes the substrate molecule. The cofactor (coenzyme or metal-ion activator) that is required for the enzyme to function either may be part of the active site or may bond elsewhere to the surface of the enzyme and assist in a necessary but less direct way. Let us examine a step-by-step sequence of events that would be typical for an enzyme-catalyzed reaction that breaks a substrate molecule into two parts. The mechanism of the reaction can be described by three equations that represent the three steps of the process. Let E symbolize the enzyme and S the substrate. P is the product of the reaction. Note that each step is a reversible reaction.

Step 1	$E + S \rightleftarrows ES$	Formation of the enzyme-substrate complex
Step 2	$ES \rightleftarrows ES^*$	Formation of the high-energy transition-state complex, ES*
Step 3	$ES^* \rightleftarrows E + P$	Dissociation of the high-energy transition-state complex, forming the free enzyme and product or products

If the three steps are added together, the overall reaction is simply the dissociation of the substrate ($S \rightarrow P$). Since an enzyme catalyzes the reaction, "enzyme" is written over the arrow in the equation:

$$\text{substrate} \xrightarrow{\text{enzyme}} \text{product}$$

C. Factors that Influence Enzyme-Catalyzed Reactions

1. Enzyme Concentration

As the concentration of an enzyme increases, the rate at which the substrate is changed increases also. If one enzyme molecule can transform a million substrate molecules each minute, then two enzyme molecules will handle twice that number of substrate molecules in the same time. The substrate will be consumed twice as fast, which doubles the rate of reaction. The way the rate of an enzyme-catalyzed reaction is affected by the concentration of enzyme is shown in Figure 2. As long as sufficient substrate is present, the reaction rate is proportional to the enzyme concentration.

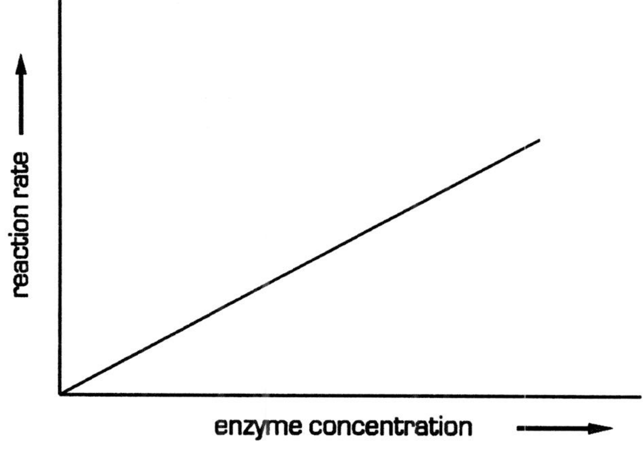

Figure 2

2. Substrate Concentration

For any given enzyme, there is some maximum number of substrate molecules that it can transform or turn over each minute. If this number is very large, the reaction will have a high maximum rate. Now let us consider a single enzyme molecule in a solution of substrate molecules. If the concentration of the substrate is very low, the number of substrate molecules that can be "caught" by the enzyme each minute will also be small. It will be lower than the number of substrate molecules the enzyme is capable of handling during that time. The rate of reaction, as measured by the number of substrate molecules changed each minute, will also be low, lower than the maximum rate. If the concentration of substrate is increased, the rate of reaction will also increase until a point is reached where the enzyme is working as fast as it can; that is, it is transforming its maximum number of substrate molecules each minute. At this point, the enzyme is said to be saturated, and further increases in the concentration of the substrate will not increase the rate of reaction. The enzyme can work no faster. The effect of substrate concentration on the rate of an enzyme-catalyzed reaction is shown in Figure 3.

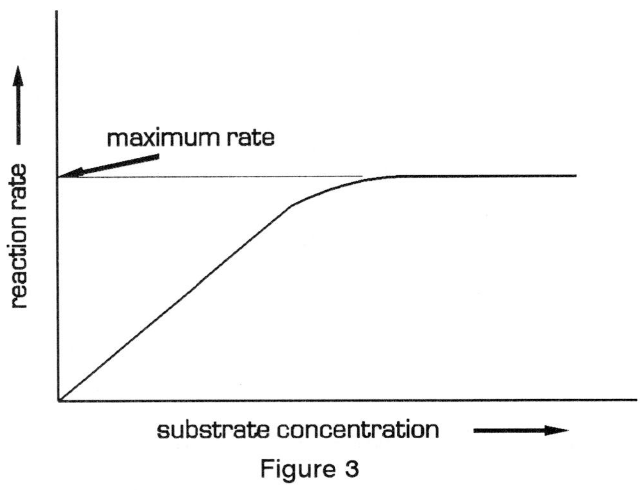

Figure 3

3. Temperature

The rate of an enzyme catalyzed reaction increases as the temperature of the reaction medium increases, but only up to a point. For every enzyme, there is one temperature at which the reaction rate will be at a maximum, and that temperature is called the **optimum temperature** for that enzyme, as shown in Figure 4. The rate of reaction at temperatures above or below the optimum temperature will be slower. The optimum temperature for most of the enzymes in the body is approximately 37°C (98.6°F), normal body temperature.

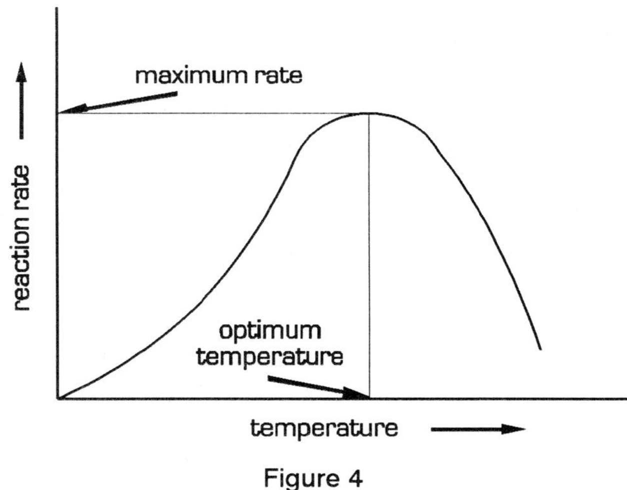

Figure 4

At temperatures above 60°C, many enzymes will denature, destroying their secondary and tertiary structures, which in turn destroys the active sites. At these temperatures, the activity of the enzyme is reduced to zero. When milk is pasteurized, or when food is canned, the high temperatures used in these processes destroy not only bacteria, but also enzymes. Low temperatures do not denature enzymes. Tissue samples, sperm, or other biological materials can be stored at subfreezing temperatures for years without markedly reducing the catalyzing power of the enzymes present in the samples.

4. pH

Just as there is an optimum temperature for enzyme activity, there is an **optimum pH** at which an enzyme's activity is greatest. At pHs above and below the optimum pH, the activity of the enzyme is reduced and reaction rates are slower, as shown in Figure 5. The optimum pH for pepsin, a proteolytic enzyme in the stomach, is around 2, close to that of the acidic environment in the stomach. Trypsin, another proteolytic enzyme, has an optimum pH of around 8, close to that of the upper intestinal tract where it is found. If trypsin were placed in the highly acidic environment of the stomach, it would likely denature and lose its catalytic activity.

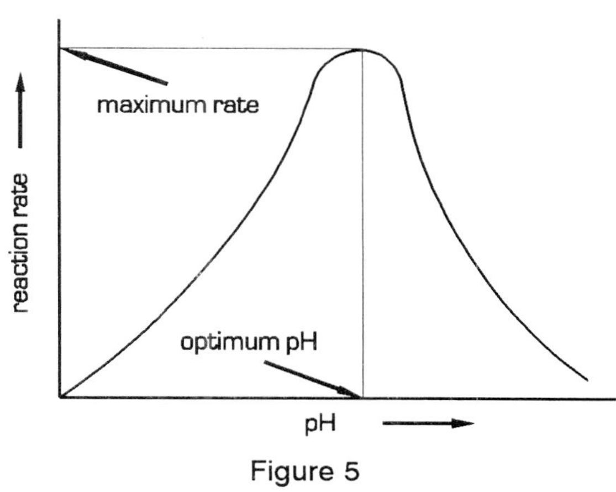

Figure 5

5. Cofactor Concentration

Cofactors are essential parts of active enzymes. If the proper cofactor is absent, an enzyme will have little or no activity. For this reason, cofactors must be present in sufficient concentration to activate enzyme molecules. If they are not, then only a fraction of the enzyme concentration will be effective in catalyzing reactions.

6. Enzyme Inhibitors

Enzyme inhibitors are substances that slow or completely stop enzyme-catalyzed reactions. This inhibition may be permanent or reversible. Natural enzyme inhibitors occur in living systems to control the rates of specific reactions; this stops the formation of a reaction product that is not needed. Enzyme inhibitors include a number of poisons, insecticides, herbicides and antibiotics. Some well-known enzyme-inhibiting poisons are cyanide ion, CN^-, arsenate ion, AsO_4^{3-}, copper (II) ion, Cu^{2+}, and heavy metal ions like Hg_2^{2+}, Hg^{2+}, Ag^+ and Pb^{2+}.

Experimental Procedure:

> **INSTRUCTIONAL NOTE: This experiment may be conveniently performed with students working in pairs.**
>
> **CAUTION: As always, eye protection must be worn at all times.**

The reaction to be studied in this experiment is the decomposition of hydrogen peroxide catalyzed by the enzyme catalase. Hydrogen peroxide is produced in many different kinds of plant and animal cells and would be toxic if allowed to accumulate. The source of catalase will be the juice from freshly peeled and ground uncooked potatoes. The equation for the decomposition of hydrogen peroxide is:

$$2\ H_2O_{2(aq)} \xrightarrow[\text{(from potato)}]{\text{catalase}} 2\ H_2O_{(l)} + O_{2(g)}$$

The rate of decomposition will be determined by measuring the amount of time required to produce 5.0 mL of $O_{2(g)}$. The apparatus shown in Figure 6 will be used to measure the volume of oxygen produced as the hydrogen peroxide decomposition takes place. The evolved oxygen will displace water from an inverted 10-mL graduated cylinder.

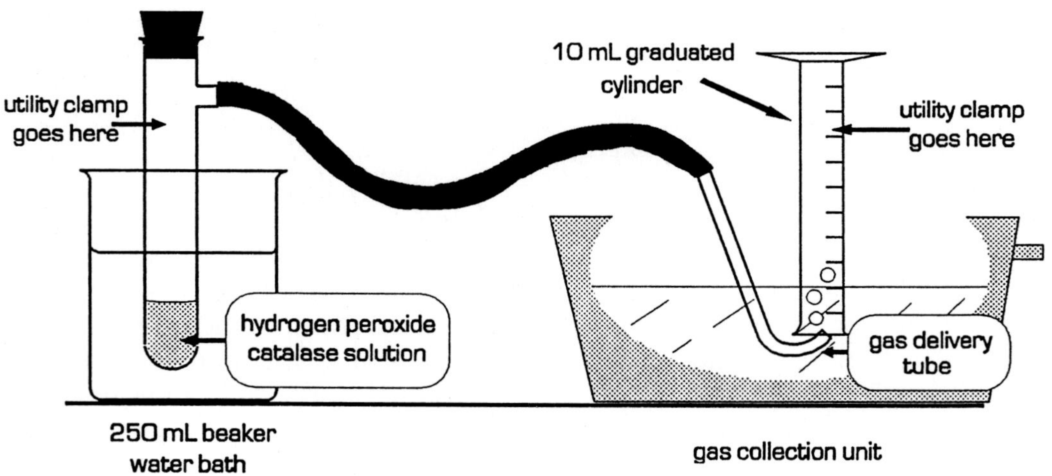

Figure 6

1. Fill a pneumatic trough two-thirds full of tap water. Then completely fill a 10-mL graduated cylinder with tap water. Place your finger securely over the mouth of the cylinder, invert it, and immerse the sealed mouth beneath the water level in the pneumatic trough. Remove your finger and secure the inverted graduate to a ring stand using a utility clamp.

2. Using a second utility clamp, attach a 15-cm sidearm test tube to another ring stand and lower it into a 250-mL beaker. Fit the sidearm test tube with a tight-fitting stopper. Later

this beaker will be filled with water to control the temperature of the catalyzed reactions. The catalyzed decomposition of H_2O_2 will take place in the sidearm test tube.

3. Fit a length of rubber tubing onto the sidearm of the test tube. It should be long enough to easily reach to the graduated cylinder. A short section of glass tubing on one end of the tubing will help direct the flow of oxygen gas into the inverted graduate.

4. Obtain two clean 2.00-mL pipets and a pipet bulb from the supply in the laboratory. One pipet will be used to measure exact volumes of the peroxide solution and the other will be used for the catalase solution. *Each must be rinsed three times with deionized water and once with the solution to be measured before use.*

> **CAUTION: It is important that all glassware that comes into direct contact with the catalase be very clean and rinsed with deionized water. Contaminants can easily affect the ability of the enzyme to function.**

A. Effect of Temperature on Enzyme Activity

The catalyzed rate of decomposition of hydrogen peroxide at four temperatures will be measured and compared. An ice bath will be used to obtain a temperature at or near 0°C. Water near the following temperatures will be available in the lab: room temperature (20° to 24°C), body temperature (37°C) and at an elevated temperature of 60°C. Be certain to record the temperature of the water bath to the nearest degree *just before combining the hydrogen peroxide and catalase solutions.*

Using a clean 2.00-mL pipet, place exactly 2.0 mL of 3% (percent by volume) hydrogen peroxide solution in the sidearm test tube. Stopper the test tube and position it inside the 250-mL beaker so its bottom is no closer than about 2 cm from the bottom of the beaker.

Fill the beaker about three-fourths full of ice, completely covering the bottom half of the test tube. In a second 15-cm test tube, place about 10 mL of fresh catalase solution obtained from the supply in the lab. Place this test tube into the ice bath also so it cools to the ice temperature. Record the temperature of the ice bath on the Report Sheet.

The next procedure must be done quickly, so if you are working with a partner, decide who will do what ahead of time. After 5 minutes, use the *freshly rinsed* pipet to add 2.0 mL of the catalase solution to the H_2O_2 in the sidearm test tube. Swirl to mix the enzyme and substrate (hydrogen peroxide) solutions, then *quickly fit the rubber stopper in the reaction tube. Note and record the time of mixing on the Report Sheet.* Continue to watch the time and record the number of seconds required to collect 5.0 mL of oxygen by water displacement in the inverted graduated cylinder.

The rate of decomposition will be expressed in milliliters of $O_{2(g)}$ evolved per second. Calculate the rate of decomposition of hydrogen peroxide in milliliters per second by dividing the volume of $O_{2(g)}$ evolved by the number of seconds required to collect the oxygen. State the rate in scientific (exponential) notation to two significant figures. The unit of the rate will be mL/sec.

Remove the rubber tubing and rinse the sidearm test tube twice with tap water and twice with deionized water, then repeat the same procedure at room temperature, body temperature and at 60°C. Be certain to record the temperature to the nearest degree just before you mix the enzyme and hydrogen peroxide solutions. Rinse the pipet with the solution to be measured before each use. Also, do not forget to place the test tubes containing the enzyme and H_2O_2 solutions in the water bath before mixing them so their temperatures will be equal when mixed.

B. Effect of pH on Enzyme Activity at Room Temperature

Fill the 250-mL beaker about two-thirds full with room temperature water. Attach the rubber tubing to the sidearm and add 2.0 mL of catalase solution to the clean sidearm test tube using a freshly rinsed pipet. Add *5 drops* of 0.1 M HCl to the catalase solution. Mix the solutions and place the clamped test tube into the water bath. Place about 10 mL of 3% H_2O_2 solution in a second test tube and hold it in the water bath alongside the reaction test tube so both solutions come to the same temperature. Fill the 10-mL graduate with water and mount it in the inverted position as before. After 5 minutes, record the temperature of the water bath. Then, using a clean and freshly rinsed pipet, add 2.0 mL of the 3% hydrogen peroxide solution to the sidearm test tube. Mix the solutions, quickly attach the rubber stopper, and collect the evolved oxygen. Note and record the time of mixing and determine the time in seconds required to evolve 5.0 mL of $O_{2(g)}$. Calculate the rate of reaction as before. The pH of the reaction solution is approximately 2. Your instructor may give you a more precise value if necessary.

The second trial will be carried out at a pH of approximately 12. Thoroughly rinse the sidearm test tube as before, reattach the rubber tubing and, using a clean and freshly rinsed pipet, add 2.0 mL of catalase solution to it. Carefully add *5 drops* of 0.1 M NaOH to the catalase solution. Swirl to mix the solutions and mount the test tube in the room-temperature water bath. Prepare the 10-mL graduate as before. After 5 minutes, record the temperature of the water bath and, using a freshly rinsed pipet, add 2.0 mL of 3% hydrogen peroxide solution to the catalase solution. Swirl to mix the solutions, record the time, and quickly attach the rubber stopper. Again, measure the number of seconds required to produce 5.0 mL of O_2, and calculate the rate of decomposition. The approximate pH of this solution is 12. If in either case a reaction is not apparent within 5 minutes, record that fact on the Report Sheet and proceed to the next experiment.

Compare the rates of reaction at pH 2 and 12 with that obtained in Section A at room temperature. The pH of the solutions in Section A was approximately 7.

C. Effect of Denaturation of an Enzyme on Enzyme Activity at Room Temperature

Add 2.0 mL of catalase solution into the clean sidearm test tube. Place the test tube into a boiling water bath for 5 to 10 minutes. The elevated temperature will denature the enzyme. Cool the reaction tube and its contents to near room temperature under running water, then place it in the room-temperature water bath. Prepare the 10-mL graduate as before. Attach the tubing to the sidearm. After 5 minutes, using a freshly rinsed pipet, add 2.0 mL of 3% hydrogen peroxide solution to the catalase solution. Swirl, record the time of mixing, and quickly attach the rubber stopper. Record the time in seconds required to obtain 5.0 mL of O_2 as before, and calculate the rate of reaction. If there is no sign of reaction in 5 minutes, record the appropriate facts and proceed to Section D.

D. Effect of Inhibitors on Enzyme Activity at Room Temperature

> **ENVIRONMENTAL ALERT: If heavy metals are used to inhibit enzyme activity in this section of the experiment, they must not go down the drain. Carefully follow your instructor's advice concerning proper waste disposal.**

The following inhibitors may be used:

1. 0.1 M $AgNO_3$
2. acetone
3. 0.1 M $Pb(NO_3)_2$
4. 2-propanol (isopropyl alcohol)
5. 0.1 M $CuSO_4$

Your instructor will tell you which of these to use.

Using a clean pipet, add 2.0 mL of catalase solution to the clean sidearm test tube, then add 4 drops of the required inhibitor solution. Swirl to mix the solutions and place the mixture in the room temperature water bath for 5 minutes. Prepare the 10-mL graduate as before. After attaching the rubber tubing to the sidearm, add 2 mL of 3% hydrogen peroxide solution to the sidearm test tube. Swirl the solutions, note the time of mixing, and quickly attach the rubber stopper. Record the number of seconds required to produce 5.0 mL of O_2 as before and calculate the rate of reaction. If no reaction occurs in 5 minutes, record the appropriate information on the Report Sheet and proceed to Section E.

E. Effect of Liver Tissue Catalase at Room Temperature (optional)

Place a few small pieces of minced liver into the clean sidearm test tube in the room-temperature water bath. Then, after attaching the tubing to the sidearm, add 2 mL of 3% hydrogen peroxide solution to the liver sample using a freshly rinsed pipet. Swirl to mix the contents of the test tube, then quickly replace the rubber stopper. Record the time of mixing on the Report Sheet. Determine the number of seconds required to produce 5.0 mL of O_2, and calculate the rate of reaction.

Name ______________________ Locker Number__________ Date____________
Please print; last name first

PRELIMINARY EXERCISES: *Experiment 25*

Enzyme Action

1. Define the following terms:

a. enzyme
b. substrate
c. cofactor
d. inhibitor
e. activation energy

2. What effect does temperature have on the rate of an enzyme-controlled reaction?

3. What effect does pH have on the rate of an enzyme-controlled reaction?

Name ______________________________ Locker Number__________ Date____________
Please print; last name first

REPORT SHEET: *Experiment 25*

Enzyme Action

A. Effect of Temperature on Enzyme Activity

Temperature (°C)	Time of Mixing	Volume of $O_{2(g)}$ Evolved (mL)	Time to Evolve $O_{2(g)}$ (sec)	Rate (mL O_2 /sec)
ice bath: ___°C				
room temp.: ___°C (Record in sections B to D too.)				
body temp.: ___°C				
elevated temp.: ___°C				

Describe how changes in temperature affect the rate of reaction.

Estimate the *optimum temperature* for catalase in this process. _____________

B. Effect of pH on Enzyme Activity at Room Temperature

pH	Time of Mixing	Volume of $O_{2(g)}$ Evolved (mL)	Time to Evolve $O_{2(g)}$ (sec)	Rate (mL O_2 /sec)
approximately 2				
approximately 7 (from Section A)				
approximately 12				

Describe how changes in pH affect the rate of reaction.

Estimate the *optimum pH* for catalase in this process. ______________

C. Effect of Denaturation of an Enzyme on Enzyme Activity at Room Temperature

Condition of Enzyme	Time of Mixing	Volume of $O_{2(g)}$ Evolved (mL)	Time to Evolve $O_{2(g)}$ (sec)	Rate (mL O_2 /sec)
not denatured (from Section A)				
denatured				

Describe the effects of denaturation on enzyme activity.

D. Effect of Inhibitors on Enzyme Activity at Room Temperature

Inhibitor	Time of Mixing	Volume of $O_{2(g)}$ Evolved (mL)	Time to Evolve $O_{2(g)}$ (sec)	Rate (mL O_2 /sec)
no inhibitor (from Section A)				
$AgNO_3$				
$Pb(NO_3)_2$				
$CuSO_4$				
acetone				
2-propanol				

Name ______________________________ Locker Number__________ Date____________
Please print; last name first

REPORT SHEET (continued): *Experiment 25*

Enzyme Action

What effect did the inhibitor have on the rate of enzyme activity compared to what you observed in Section A, both at room temperature?

Silver nitrate and 2-propanol are often used to destroy microorganisms. What do you think is the mechanism of action as they function in this capacity?

E. Effect of Liver Tissue Catalase at Room Temperature (optional)

Temperature (°C)	Time of mixing	Volume of $O_{2(g)}$ evolved (mL)	Time to evolve $O_{2(g)}$ (sec)	Rate (mL O_2 /sec)
____ °C				

Experiment 26

Analysis of Proteins and Amino Acids by Chromatography

Purpose:

In this experiment you will use paper chromatography to identify selected amino acids obtained in the hydrolysis of protein.

Discussion:

The term **protein** was first used by Gerardus Mulder in 1838 to describe the complex nitrogen-containing organic compounds that are found in all living cells. It is derived from the Greek word *proteios*, which means "of first importance." This is an appropriate description of these important compounds since proteins are involved in essentially all biochemical processes in the body.

Living cells in both plants and animals are approximately 60% water and 40% solid material, roughly half of which is protein. In plants, proteins are synthesized from carbon dioxide, water, nitrates, sulfates and smaller amounts of several other compounds. Animals, on the other hand, are not able to synthesize proteins in this way, so proteinaceous material must be obtained in the diet.

Most protein in the body is used in body building and repair. The principal component of all enzymes is protein. While lipids and, to a lesser extent, carbohydrates are stored in the body as an energy reserve, the storage of protein is almost nonexistent. Protein is little used as a source of energy. Consequently, for good health, it is necessary to have a regular intake of protein through the diet. An animal can survive for a limited time on a diet that contains only vitamins, minerals and proteins (no carbohydrates or lipids). But if the animal is fed a diet containing everything but protein, premature death will follow.

A. Amino Acids

Proteins are polymers of **amino acids**. For the most part, proteins are very large molecules composed of hundreds of amino acid units. Proteins can be broken down into the amino acids that compose them by hydrolysis (reaction with water). In the laboratory, hydrolysis reactions are usually carried out in the presence of strong acids (HCl or H_2SO_4), strong bases (NaOH or KOH) or proteolytic enzymes called proteases. Amino acids are carboxylic acids that contain an amino group, $-NH_2$, joined to the alpha-carbon of the acid. There are twenty amino acids normally found in nature. With the exception of proline, all are alpha-amino acids. The alpha position in a carboxylic acid is the carbon atom adjacent to the carboxyl group. An alpha-amino acid has an amino group attached to the alpha-carbon atom.

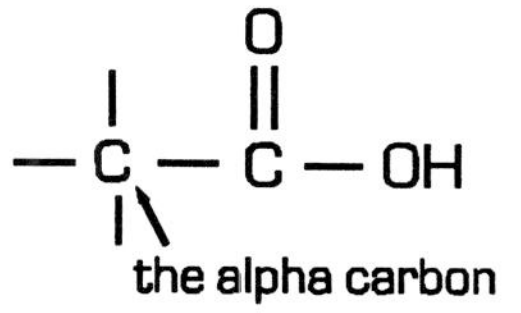

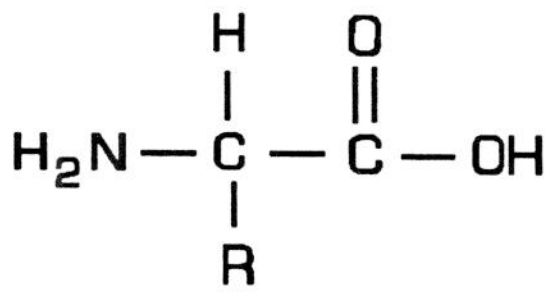

```
    |    O
    |    ||
  — C — C — OH
    |\
the alpha carbon
```

```
       H    O
       |    ||
H2N — C — C — OH
       |
       R
an alpha-amino acid
```

All alpha-amino acids can be represented by the same general formula. The only difference between one amino acid and another is the nature of the R group bonded to the alpha-carbon atom. The R group of an amino acid can be hydrogen, a simple alkyl group, a group containing an aromatic ring, a heterocyclic group or other substituted group. Each different amino acid has a unique set of properties due to the composition of its R group.

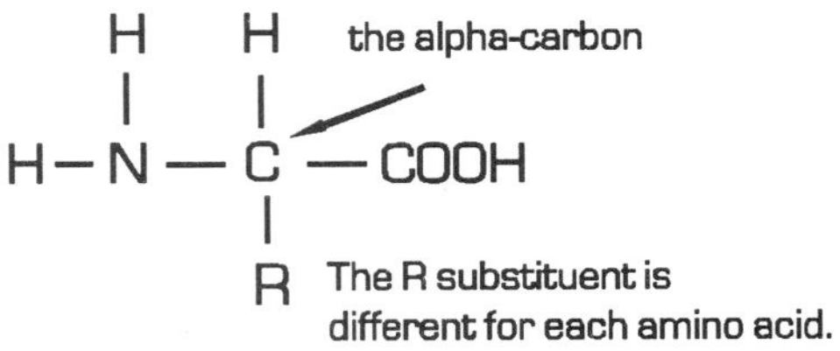

the alpha-carbon

The R substituent is different for each amino acid.

Amino acids are commonly classified in terms of the number of carboxyl, $-COOH$, and amino, $-NH_2$, groups in the molecule. **Neutral amino acids** contain one carboxyl group and one amino group. Aqueous solutions of neutral amino acids have a neutral or near neutral pH (~7). **Basic amino acids** contain one carboxyl group but more than one amino or amino-like group. Aqueous solutions of basic amino acids have a basic pH (>7). **Acidic amino acids** contain one amino group but more than one carboxylic acid group. Aqueous solutions of acidic amino acids have an acidic pH (<7).

Several of the twenty amino acids commonly found in nature are shown below. Alongside the name of each in parentheses is the three-letter abbreviation for the amino acid. Glycine, alanine and valine are neutral amino acids with aliphatic side chains.

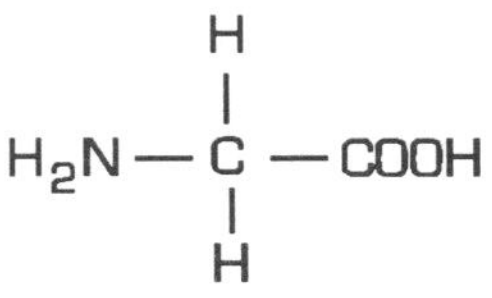

glycine (gly) 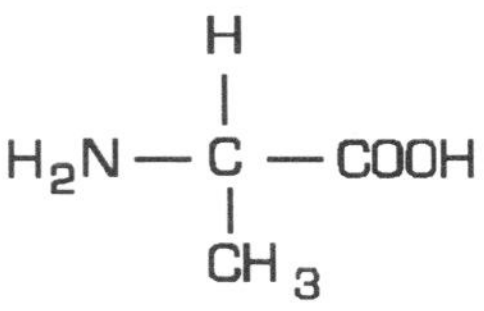 alanine (ala) 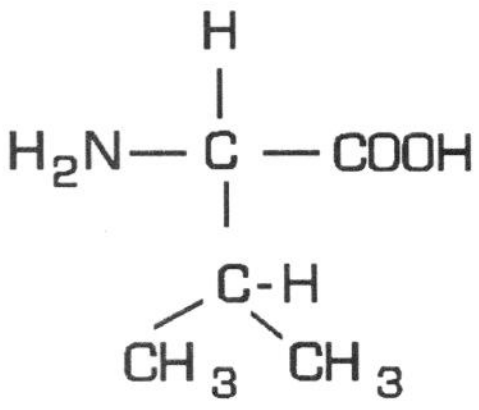 valine (val)

Threonine, phenylalanine and tyrosine are also neutral amino acids.

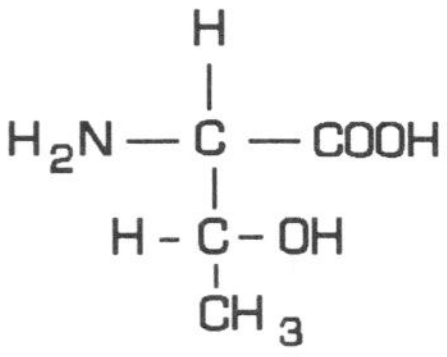

threonine (thr) 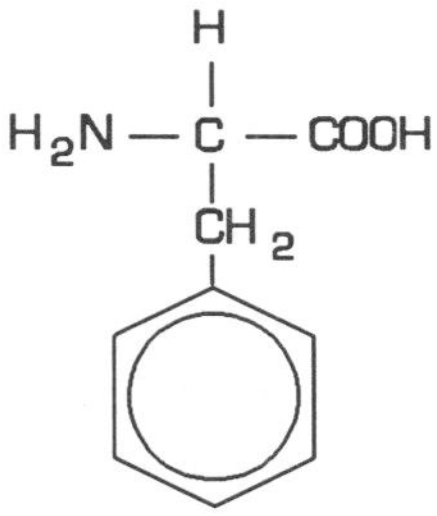 phenylalanine (phe) 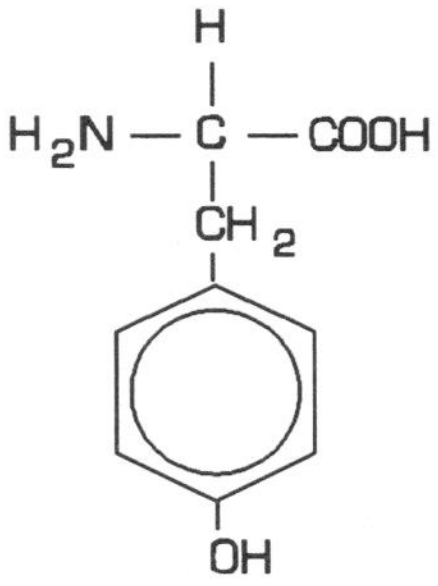 tyrosine (tyr)

Methionine and proline are also neutral amino acids. Methionine has a sulfur atom in its side chain. Proline is an alpha-imino acid with its nitrogen in a heterocyclic unit. Tryptophan is also classified as a neutral amino acid. It contains a heterocyclic ring fused to an aromatic benzene ring. The structural formulas of these amino acids follow.

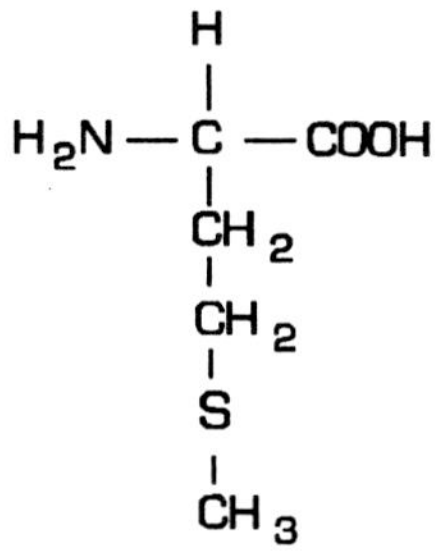

$H_2N-CH(COOH)-CH_2-CH_2-S-CH_3$

methionine (met)

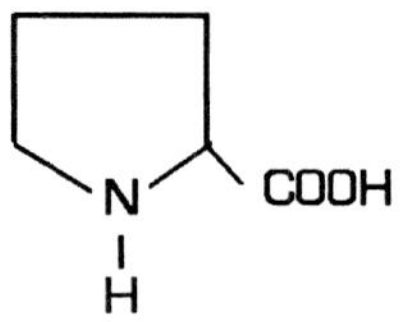

proline (pro)

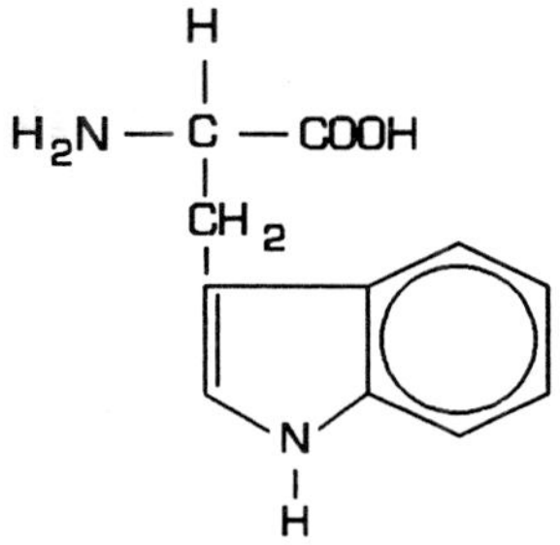

$H_2N-CH(COOH)-CH_2-$(indole ring, N–H)

tryptophan (trp)

Aspartic acid is an acidic amino acid. It has two $-COOH$ groups and one $-NH_2$ group. Lysine is a basic amino acid. It has two $-NH_2$ groups and a single $-COOH$ group.

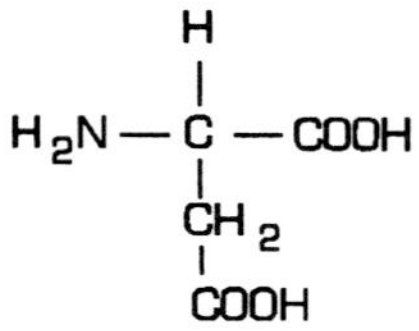

$H_2N-CH(COOH)-CH_2-COOH$

aspartic acid (asp)

$H_2N-CH(COOH)-CH_2-CH_2-CH_2-CH_2-NH_2$

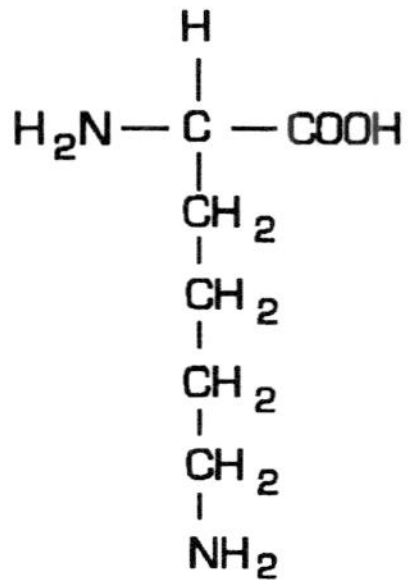

lysine (lys)

B. Peptides

As was stated earlier, the complete hydrolysis of a protein will yield a mixture of amino acids. But if the hydrolysis reaction is stopped before the protein is completely broken down, small fragments composed of only a few amino acid units can be isolated. These small fragments are called **peptides**. Each peptide will contain one or more *amide* unit commonly called a *peptide link*. A peptide link forms when amino acids join to form the protein.

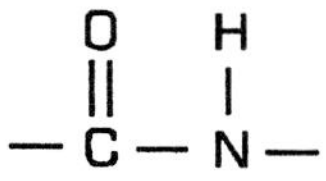

$$-\overset{O}{\overset{\|}{C}}-\overset{H}{\overset{|}{N}}-$$

peptide link

When two amino acids combine, the amino group of one amino acid reacts with the carboxyl group of the second to form water and a **dipeptide** (a peptide made up of two amino acid units). By convention, amino acids are written in an equation with the amino groups to the left and carboxyl groups to the right. A general equation for dipeptide formation is:

$$H_2N-\underset{R}{CH}-\overset{O}{\overset{\|}{C}}-OH + H-\overset{H}{\overset{|}{N}}-\underset{R'}{CH}-COOH \longrightarrow H_2N-\underset{R}{CH}-\overset{O}{\overset{\|}{C}}-\overset{H}{\overset{|}{N}}-\underset{R'}{CH}-COOH + H_2O$$

amino acid 1 | amino acid 2 | dipeptide A

If amino acid 1 and amino acid 2 react in the reverse order, a different dipeptide, an isomer of dipeptide A, results. The isomer, which we will call dipeptide B, is shown below. A third amino acid could join with either dipeptide to form a **tripeptide**, and the peptide could continue to grow in this way.

$$\mathrm{H_2N-\underset{R'}{\overset{H}{C}}-\overset{O}{\overset{\|}{C}}-\overset{H}{N}-\underset{R}{\overset{H}{C}}-COOH}$$

dipeptide B

Biochemists have developed a shorthand method for describing the composition of peptides using the abbreviations of the amino acids. The amino groups are understood to be on the left and the carboxyl groups on the right. The formation of a tripeptide from lysine (lys), valine (val) and tryptophan (trp), in that order, would be written this way:

$$\text{lys} + \text{val} + \text{trp} \xrightarrow{\text{catalyst}} \underset{\text{tripeptide}}{\text{lys–val–trp}} + 2\,H_2O$$

Experimental Procedure:

The separation technique of paper chromatography was the subject of an earlier experiment in this manual. Before starting this experiment, read again the discussion section of Experiment 5, Separation Using Chromatographic Techniques. You will need to refresh your memory about R_f values, solvent fronts and stationary versus moving phases.

This experiment concerns the separation of mixtures of amino acids obtained from the hydrolysis of proteins and peptides. Identification will be made using the R_f values of known amino acids. Since amino acids are colorless, their location on the developed chromatogram will be made visible by staining each spot with ninhydrin, a substance that reacts with amino acids to form colored products.

CAUTION: As always, eye protection must be worn at all times.

1. Obtain a clean, dry 600-mL beaker and, using a gummed label, write your name on the exterior.
2. Using a clean, dry 50-mL graduated cylinder, pour 45 mL of chromatographic development solution into the beaker. Your instructor will tell you the location of the development solution. Cover the top of the beaker tightly with Saran® wrap. Keep the plastic wrap securely in place with a rubber band, if necessary. This covered beaker will be the development chamber.
3. *Using gloves or plastic sandwich bags over your hands to minimize contamination*, obtain a sheet of chromatographic paper from the instructor. It should be 21.0 cm by 11.5 cm; if not, cut it to that size. *Amino acids in your fingerprints can contaminate the chromatographic paper, so handle the paper by the edges at all times.* Using a *pencil* and straightedge, draw a line across the long dimension of the paper 1.5 cm in from one edge, as shown in Figure 1. *Do not use a pen to draw this line.* Starting from the left end of the line, mark off ten points 2 cm apart from one another. Label each point with a letter, A

through J, writing the letters in *pencil* beneath the points. Write your initials in the upper left-hand corner of the paper.

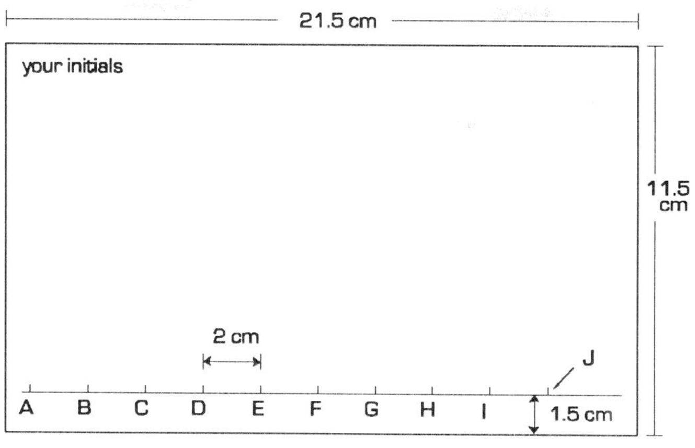

Figure 1

The protein samples you are to analyze, casein and gelatin, will have already been hydrolyzed to save laboratory time. You will also analyze a dipeptide, the artificial sweetener aspartame, commercially known as Nutrasweet®. It too will have been hydrolyzed. Finally, you will be analyzing a prepared mixture of amino acids as an unknown, along with several known amino acids. The list of ten samples you will analyze follows. Each sample should be spotted on the chromatography paper in the order in which they are listed. Read all of step 5 before spotting the samples.

A. lysine
B. aspartic acid
C. gelatin hydrolysis mixture
D. aspartame hydrolysis mixture
E. phenylalanine
F. methionine
G. unknown mixture
H. tryptophan
I. casein hydrolysis mixture
J. proline

4. Obtain an unknown mixture of amino acids from the instructor. Record the number of the unknown on the Report Sheet under step 2. The unknown mixture will contain one or more of the amino acids listed above as samples A, B, E, F, H and J.

5. A critical task in this experiment is applying the correct amount of sample to the chromatography paper. Each spot is made by transferring the *smallest possible drop* from the tip of a toothpick to the proper point on the paper. The spot must not have a diameter exceeding 3 mm, roughly the size of the letter O, and *a new toothpick must be used for each sample.* Separation is difficult if the sample is too large. Before you attempt to spot samples for analysis, practice making spots on scrap pieces of chromatographic paper until you can make them the correct size. Apply the ten samples to the appropriate points on the paper. After the spots have dried, repeat the application *two* more times. After all samples, A through J, have been applied to the paper in triplicate, set the paper aside for several minutes to allow the spots to dry.

6. Once all the spots are dry, roll the paper into a cylinder and carefully staple the ends together about 2 cm from the top and 1.5 cm from the bottom, as shown in Figure 2. The edges of the paper should not touch. Handling the paper cylinder by the edges, place it in the development chamber. *The spots should be just above the surface of the*

development solution, and the paper should not touch the walls of the beaker as shown in Figure 3. Tightly seal the development chamber with Saran® wrap and let it stand under the hood undisturbed as the solvent ascends the paper.

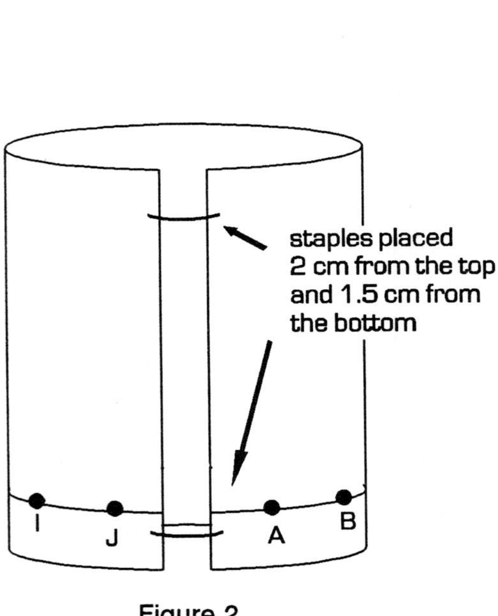

Figure 2

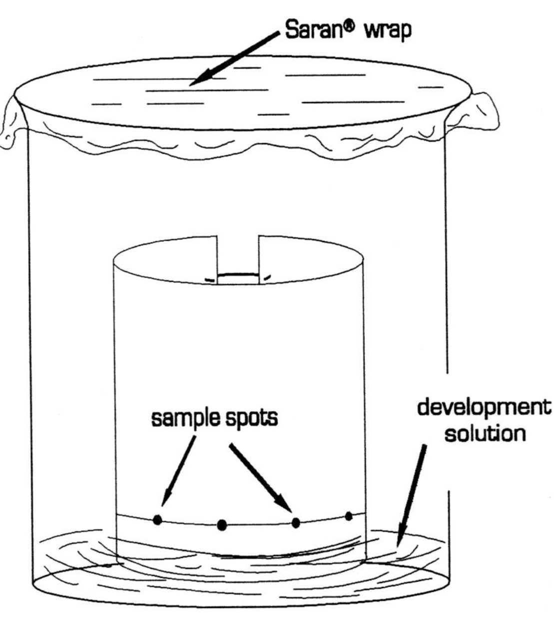

Figure 3

7. After the development solution has moved at least two-thirds of the way to the top of the paper (or after one hour has passed), remove the paper cylinder, undo the staples, and lay the paper flat on a paper towel under the hood to dry. Quickly mark the position of the solvent front with a pencil. A hair dryer may be used to speed up the drying process.

Caution: The next step must be performed under the hood.

8. When the chromatogram is *dry*, tape it against a board in the hood or lay it on a clean surface in the hood and spray the paper very *lightly and evenly* with ninhydrin. If you have any questions about this, ask the instructor.

9. Place the chromatogram in a drying oven at 90°C to 100°C for 3 to 5 minutes. Ninhydrin will react with amino acids upon heating to produce a characteristic purple, red-brown or yellow color. Once you have the developed and dried chromatogram, draw a circle around each spot that appears above the starting line and note the color of each. Make a dot in what you judge is the *center* of each spot. Some spots elongate, showing a definite head and an extended tail; the dot should be in the center of the head.

10. Measure the distance in millimeters from the starting line to the dot in the center of each separated component, and also measure the distance from the starting line to the solvent front. Write these distances with a sharp pencil on the chromatogram. If a sample produced a single spot, then it is simply a single amino acid. Those that produced two or more spots are mixtures of amino acids. Calculate the R_f values for every component spot that appears on your chromatogram. Show *all* calculations and answer the questions on the Report Sheet.

Name ______________________ Locker Number__________ Date__________
Please print; last name first

PRELIMINARY EXERCISES: *Experiment 26*

Analysis of Proteins and Amino Acids by Chromatography

1. Define the following terms that are associated with proteins and amino acids.

a. peptide link
b. amino acid
c. dipeptide
d. proline

2. Draw the structural formula for the tripeptide described as lys–val–trp.

3. What is the R_f value in paper chromatography? (Consult Experiment 5 if you need to review the meaning of the R_f value.)

4. Of what importance are R_f values in the analysis of amino acids obtained from the hydrolysis of proteins?

Name ______________________ Locker Number__________ Date__________
Please print; last name first

REPORT SHEET: *Experiment 26*

Analysis of Proteins and Amino Acids by Chromatography

After completing the Report Sheet, tape the chromatogram you obtained here.

METRIC 1 2 3 4 5 6 7 8 9 10 11 12 13 14 15

Show all R_f calculations below.

1. Calculate the R_f value and note the color of the spot for each single amino acid analyzed:

Spot A:	lysine	R_f = ________	____________
Spot B:	aspartic acid	R_f = ________	____________
Spot E:	phenylalanine	R_f = ________	____________
Spot F:	methionine	R_f = ________	____________
Spot H:	tryptophan	R_f = ________	____________
Spot J:	proline	R_f = ________	____________

2. Calculate the R_f value(s) for the component(s) of the unknown sample, spot G. Compare R_f values with those of the amino acids listed above and identify the amino acid composition of the unknown.

   Unknown number: ________

   Composition of your unknown sample: ________________________________

3. Calculate the R_f values for the two components of aspartame, spot D.

   Spot 1: ________ Spot 2: ________

   By comparing R_f values with those of the tested amino acids, identify the components of aspartame.

   Spot 1: ________________ Spot 2: ________________

4. Phenylalanine, tryptophan and lysine are three of the eight amino acids that are considered essential for good health in adults.[1] They must be obtained in foods since the body is unable to synthesize them either in sufficient quantity or at all. A *complete protein* contains all of the essential amino acids, whereas an *incomplete protein* lacks at least one of the essential amino acids.

   Examine the chromatogram of hydrolyzed gelatin, spot C, and hydrolyzed casein, spot I. Gelatin is a protein obtained from the collagen of animals and casein is the principal protein in milk. Does either contain the essential amino acid *lysine*? Justify your answer with your experimental data.

---

1 The eight essential amino acids for adults are valine, leucine, isoleucine, threonine, phenylalanine, tryptophan, methionine and lysine.

Experiment 28

Digestion

Purpose:

In this experiment you will digest various fats, proteins and carbohydrates and then use some general tests to detect the products formed in the digestion process.

Discussion:

In the digestive tract, large dietary molecules are converted into smaller molecules that can be absorbed and used by the cells of the body. This process is called **digestion.**

A. Digestion and Absorption of Carbohydrates

Starch is one of the most important carbohydrates. The digestion of starch begins in the mouth. As food is chewed, it is mixed with saliva containing the enzyme *salivary amylase*, which catalyzes the hydrolysis of starch to form smaller polysaccharides (dextrins), maltose and a small amount of free glucose. Since food is in the mouth for a brief period, only limited digestion can occur. As carbohydrates pass through the stomach, digestion ceases, but begins again in the small intestine. *Pancreatic amylase* is secreted into the small intestine to catalyze the complete hydrolysis of starch to form maltose. The intestinal enzyme, *maltase*, completes the process, converting maltose into glucose. The digestion of starch is ultimately its complete hydrolysis to glucose. The digestion of starch occurs in a stepwise fashion, as shown below:

Step 1: starch + H_2O $\xrightarrow[\text{in the mouth}]{\text{salivary amylase}}$ minor breakdown into dextrins + maltose + glucose

Step 2: starch + some digestion products from the mouth $\xrightarrow[\text{small intestines}]{\text{pancreatic amylase}}$ many maltose molecules

Step 3: maltose $\xrightarrow[\text{small intestines}]{\text{maltase}}$ many glucose molecules

The polysaccharide cellulose, which, like starch, is built from many glucose units, is not digested because humans (unlike termites) lack the enzyme *cellulase*. Cellulose passes through the digestive system unchanged and is called dietary fiber. The disaccharides lactose (from milk and milk products), maltose (found in some vegetables) and sucrose (table sugar and a minor constituent in honey and vegetables) are hydrolyzed by the respective enzymes, *lactase*, *maltase* and *sucrase*, as shown in the following equations:

$$\text{lactose} + H_2O \xrightarrow[\text{small intestine}]{\text{lactase}} \text{glucose} + \text{galactose}$$

$$\text{maltose} + H_2O \xrightarrow[\text{small intestine}]{\text{maltase}} \text{two glucose molecules}$$

$$\text{sucrose} + H_2O \xrightarrow[\text{small intestine}]{\text{sucrase}} \text{glucose} + \text{fructose}$$

The final product of carbohydrate digestion (starch along with the disaccharides) is a mixture of monosaccharides: glucose, galactose and fructose. Since our diet contains many sources of starch (grains, bread and vegetables), we normally have a large amount of glucose in the small intestine after digestion. Glucose, galactose and fructose are absorbed by the mucosal cells of the small intestine. The absorbed monosaccharides are then carried by the bloodstream to the liver, where galactose and fructose are converted into glucose. Most of the glucose is sent to cells throughout the body and used for energy, while excess glucose is converted into glycogen and stored in liver and muscle tissue. If a large excess of carbohydrate is digested, it will ultimately be converted in a multistep process into triglycerides (fats) and stored in adipose tissue. Table 1 summarizes the digestion of several carbohydrates.

Table 1. Summary of Carbohydrate Digestion

Type of Carbohydrate	Typical Sources	Site of Digestion	Required Enzymes	Digestion Products
cellulose	fruits, vegetables	none	none	none
starch	potatoes, beans, peas, breads	mouth	salivary amylase	dextrins + maltose
		small intestines	pancreatic amylase	maltose
		small intestines	maltase	glucose
sucrose	table sugar	small intestines	sucrase	fructose + glucose
lactose	milk, dairy products	small intestines	lactase	galactose + glucose
maltose	small amounts in grain	small intestines	maltase	glucose
glucose (free)	small amounts in many foods	none required	none required	glucose
fructose	equal amounts of glucose and fructose in honey, small amounts in fruits and vegetables	none required	none required	fructose

B. Digestion and Absorption of Proteins

Proteins make up about 25% of our diet. The digestion of proteins begins in the acidic environment of the stomach, where they are denatured, partially or totally unfolding the tertiary protein structure. The denatured protein exposes several peptide bonds that are then hydrolyzed in the presence of *pepsin*, a protein-splitting enzyme. About 10% of the peptide bonds are broken, forming smaller, more soluble peptides. The digested protein is now a mixture of peptides, polypeptides and free amino acids. This mixture enters the small intestines, where the hydrolysis of the remaining peptide bonds is catalyzed by the enzymes *trypsin*, *chymotrypsin* and *peptidases* (secreted by the pancreas and intestinal cells) into a complex mixture of free amino acids. The amino acid mixture is absorbed through the intestinal wall directly into the bloodstream. The amino acids are then carried by the portal circulation to the liver, the principal organ responsible for the breakdown and synthesis of amino acids. From there they go to all parts of the body, where they are used to synthesize protein for repair of old tissue and generation of new tissue. Some amino acids are also used sources of carbon and nitrogen for the synthesis of nonprotein biomolecules such as heme.

Unlike carbohydrates and fats, excess amino acids cannot be stored by the body. However, the amino groups ($-NH_2$) can be removed, and the carbon skeletons can be oxidized to carbon dioxide and water for energy or converted to carbohydrates and fats for storage. The amino groups are used to form urea and excreted. Table 2 summarizes protein digestion.

Table 2. Summary of Protein Digestion

Typical Sources	Site of Digestion	Required Enzymes	Digestion Products
meats and vegetables	Stomach: partial hydrolysis by HCl and pepsin	pepsin	peptides, polypeptides and free amino acids
	Small intestines: complete hydrolysis of products from the stomach	trypsin chymotrypsin, and stomach peptidases	free amino acids

C. Digestion and Absorption of Lipids

Dietary lipids include triglycerides (fats and vegetable oils), cholesterol and some polar lipids. Cholesterol requires no digestion and is absorbed directly into the lymph system when it reaches the small intestine. The digestion of triglycerides does not begin until they reach the small intestine, where they trigger the gallbladder to pass bile into the small intestine. Bile, being slightly basic, neutralizes the acidic mixture of partially digested food from the stomach, but its main function is to supply bile salts, which emulsify the triglycerides into small droplets. Bile salts also aid the absorption of fat-soluble vitamins from the small intestine.

The emulsified triglycerides are acted on by *lipases*, enzymes secreted by the pancreas. These hydrolytic enzymes catalyze the hydrolysis of triglycerides into glycerol and fatty acids, though some of the triglycerides may be reduced only to diglycerides or monoglycerides. The hydrolysis products are absorbed through the intestinal membrane and then resynthesized into triglycerides or phospholipids.

$$\begin{array}{l} H-\overset{\displaystyle H}{\overset{|}{C}}-O-\overset{\displaystyle O}{\overset{\|}{C}}-C_{17}H_{35} \\ H-\overset{|}{C}-O-\overset{\displaystyle O}{\overset{\|}{C}}-C_{17}H_{35} \\ H-\underset{\displaystyle H}{\underset{|}{\overset{|}{C}}}-O-\overset{\displaystyle O}{\overset{\|}{C}}-C_{17}H_{35} \end{array} + 3\ H_2O \xrightarrow[\text{bile salts}]{\text{pancreatic lipase}} \begin{array}{l} H-\overset{\displaystyle H}{\overset{|}{C}}-OH \\ H-\overset{|}{C}-OH \\ H-\underset{\displaystyle H}{\underset{|}{\overset{|}{C}}}-OH \end{array} + 3\ C_{17}H_{35}COOH$$

triglyceride — glycerol — fatty acids

The resynthesized lipids are combined with protein into a lipoprotein complex and are transported in the lymph system until they enter the venous blood. The lipoprotein is then carried to the liver, where a portion of the triglycerides may be used for energy. The remaining lipoprotein is transported to cells, where the protein is removed and the triglyceride is used for energy. If more triglycerides are present in the diet than are needed, they are stored in the body as adipose tissue.

Experimental Procedure:

This experiment will allow you to observe several reactions that occur during the digestion of food in the mouth, stomach and small intestine. In an optional exercise, your own salivary amylase will be used to catalyze the digestion of starch, paralleling its digestion in the mouth. You will observe the digestion of starch as it occurs in the small intestine, catalyzed by pancreatic amylase. This enzyme is secreted by the pancreas into the small intestine to catalyze the hydrolysis of starch into maltose and a small amount of glucose. The pancreatic amylase you will use is obtained from pork and sold under the general name pancreatin. It is a mixture of enzymes that can catalyze the digestion of starches, proteins and triglycerides (fats and oils). The digestion of protein as it occurs in the stomach will employ the stomach enzyme pepsin.

In each set of tests, control samples will be used. *The control samples contain everything but the enzyme that catalyzes the digestive process.* A control sample allows you to observe the importance of the enzyme. Some tests require longer heating periods than others. Start those that require longer heating times first.

A. Digestion of Carbohydrates

The digestion (hydrolysis) of starch will be followed using the iodine-starch color test and the Benedict test. Undigested starch will form characteristic blue-black product in the presence of iodine, I_2. But as starch is digested, it is converted into maltose and a small amount of free glucose. Maltose and glucose do not produce the blue-black color in the presence of I_2. The presence of maltose and glucose can be verified though, using the Benedict test, which was described in detail in Experiments 23 and 24. Reducing sugars produce a positive Benedict test with the formation of the a copper(I) oxide, Cu_2O, precipitate.

1. Hydrolysis of Starch by Salivary Amylase (optional)

If you have eaten within the last two hours, rinse your mouth thoroughly with a cup of water. Then chew a half-stick of dietetic gum to stimulate salivation. In a 50-mL beaker, collect a volume of saliva equal to one-third the volume of a 10-cm test tube. This will be the source of salivary amylase. Fill a 250-mL beaker about two-thirds full of tap water and begin warming the water to about 37°C. This will be used to incubate samples during the experiment. A boiling water bath (100°C) will also be needed in part 2. It is most convenient if two people share the two water baths, one at about 37°C and the other at 100°C.

Label four 15-cm test tubes A, B, C and D. Add 2.00 mL of 2% starch solution to each one. Add 4.00 mL of deionized water to test tubes A and B; these will be the controls. Add 4.00 mL of the saliva sample to test tubes C and D. Place the four test tubes into a 250-mL water bath set near 37°C (± 3°C). Then, using *very clean* glass stirring rods, stir the contents of tubes C and D for 3 minutes. Allow the four test tubes to remain in the water bath for a total of 30 minutes. During this time, begin part 2 below, which uses pancreatic amylase.

After 30 minutes, remove the test tubes and place them in a test tube rack. Add 2 drops of iodine solution to test tubes B and D. Record your observations on the Report Sheet. The solutions in test tubes A and C will be used in the following section for the Benedict test.

2. Hydrolysis of Starch by Pancreatic Amylase

Label four 15-cm test tubes E, F, G and H. Prepare each test tube as described below:

Test tube E:	4 mL of deionized water (the control)
Test tube F:	2 mL of 5% pancreatin solution + 2 mL of deionized water
Test tube G:	2 mL of 5% pancreatin solution + 2 mL of 0.5% Na_2CO_3 solution (The pH of this solution is near that of the small intestine.)
Test tube H:	2 mL of 5% pancreatin solution + 2 mL of 0.2 M HCl solution (The pH of this solution is near that of the stomach.)

To each test tube, add 2 mL of 2% starch solution. Seal each test tube with a cork or plastic film and gently shake each one to mix the contents. Remove the corks or film, place all four test tubes in the 37°C water bath, and hold them at that temperature for 30 minutes. After 30 minutes, pour half the contents of each test tube into another test tube of the same size (labeled E-2, F-2, G-2 and H-2) and add 2 drops of iodine solution to each of these second tubes and mix. Save the remaining half of each solution (tubes E, F, G and H) for the Benedict test that follows. Record your observations on the Report Sheet.

Now test for the presence of reducing sugars (glucose and/or maltose). Add 5 mL of Benedict reagent to test tubes A and C (from Section A, part 1) and to the remaining solutions in test tubes E, F, G and H. Place the six test tubes into a *boiling* water bath for 7 to 10 minutes. Record the colors you observe on the Report Sheet and indicate whether a precipitate forms. A positive test, indicating the presence of maltose and/or glucose, is a red precipitate. Your observations will allow you to answer the questions on the Report Sheet.

B. Digestion of Protein

Label and prepare three test tubes as follows:

Test tube P-1: 5 mL of deionized water + 1 mL of 0.1 M HCl (the control)

Test tube P-2: 5 mL of 2% pepsin (an enzyme) + 1 mL 0.2 M HCl (near stomach pH)

Test tube P-3: 5 mL of 2% pepsin + 1 mL deionized water (near neutral pH)

Add a small piece of boiled egg white (protein) to each test tube and place each one in a water bath at 37°C for one hour. Record any changes observed in the egg white in each test tube. Any differences observed between the control (P-1) and the other two samples can be attributed to the enzyme, pepsin, in acidic and near-neutral environments.

C. Digestion of Triglycerides

Triglycerides (fats and oils) are the principal lipids in the diet. The hydrolysis of a single triglyceride molecule produces one molecule of glycerol and three fatty acids. Fatty acids are weak acids and will change the pH of the hydrolysis solution toward the acidic side. A mixture of triglycerides will be treated with pancreatin, a source of pancreatic lipase, and sodium choleate, a bile salt that serves as an emulsifying agent.

Label three 15-cm test tubes 1, 2 and 3. Add the following materials to the test tubes, using a 2-mL pipet (*rinsed before each use*) as necessary.

Test tube 1: 2 mL of 5% sodium choleate solution (a bile salt), 2 mL of 5% pancreatin solution (an enzyme) and 10 drops of universal indicator

Test tube 2: 2 mL of 5% sodium choleate solution and 10 drops of universal indicator

Test tube 3: 2 mL of 5% pancreatin solution and 10 drops of universal indicator

To tubes 2 and 3 add 0.1% NaOH solution drop by drop until the indicator turns the color that indicates a pH of 7. Then add deionized water dropwise until the volumes of test tubes 2 and 3 have the same volume as test tube 1.

Add 5 drops of corn oil (or other triglyceride) to each test tube.

Place the test tubes in a 37°C water bath. After 10 minutes, observe the color of the indicator to determine which samples have become *more* acidic, showing that digestion of the triglyceride has taken place. Use your observations to answer the questions on the Report Sheet.

Name ______________________________ Locker Number__________ Date____________
Please print; last name first

PRELIMINARY EXERCISES: *Experiment 27*

Digestion

1. Define the following terms:

a. fats and oils
b. enzyme
c. carbohydrate
d. digestion

2. List the products formed in the complete digestion of each of the following substances:

   a. glucose: ____

   b. sucrose: ____

   c. maltose: ____

   d. a triglyceride: ______________________________________

   e. a protein: ____

Name ______________________ Locker Number__________ Date____________
Please print; last name first

REPORT SHEET: *Experiment 27*

Digestion

A. Digestion of Carbohydrates

1. Hydrolysis of Starch by Salivary Amylase (optional)

Which sample(s) (B or D) gave a positive starch-iodine test? ____________

Did digestion occur in either of these starch samples? ____________

Which sample(s) (A or C) gave a positive Benedict test? ____________

In terms of digestion, what is indicated by a positive Benedict test?

What substance catalyzed the digestion of starch?

2. Hydrolysis of Starch by Pancreatic Amylase

Which sample(s) (E-2, F-2, G-2 or H-2) gave a positive starch-iodine test? __________

Did digestion occur in any of these samples?

Which sample(s) (E, F, G or H) gave a positive Benedict test? ____________
Indicate whether a precipitate formed and its color:

E ________________ G ________________

F ________________ H ________________

At which pH was the digestion of starch the most complete? Why?

What substance catalyzed the digestion of starch in these tests?

B. Digestion of Protein

In which of the samples (P-1, P-2 or P-3) did a visible change occur? ________________
What effect does 0.2 M HCl have on the digestive process of protein?
What substance catalyzed the digestion of protein?

C. Digestion of Triglycerides

What is the color of universal indicator at pH 7? ________________ What colors were observed in each sample after heating at 37°C for 10 minutes? 1 ________________ 2 ________________ 3 ________________ Which samples became more acidic during the heating? ________________ In which samples did digestion occur? ________________________________
What substance catalyzed the digestion of triglyceride?
What is the role of sodium choleate in the digestion of triglycerides?

Appendixes:

Appendix A: Vapor Pressure of Water at Various Temperatures

Temperature (°C)	VP (torr)	Temperature (°C)	VP (torr)
15	12.8	23	21.1
16	13.6	24	22.4
17	14.5	25	23.8
18	15.5	26	25.2
19	16.5	27	26.7
20	17.5	28	28.3
21	18.6	29	30.0
22	19.8	30	31.8

Appendix B: Concentration of Common Acids and Bases

Reagent, Formula		Molarity (M)	Density (g/mL)	Percent solute (w/w)
acetic acid, $HC_2H_3O_2$	glacial	17.6	1.05	99.5
	dilute	6	1.04	34
hydrochloric acid, HCl	conc.	11.7	1.18	36
	dilute	6	1.10	20
nitric acid, HNO_3	conc.	15.6	1.42	72
	dilute	6	1.19	32
phosphoric acid, H_3PO_4	conc.	14.7	1.69	85
sulfuric acid, H_2SO_4	conc.	18.0	1.84	96
	dilute	6	1.34	44
	dilute	3	1.18	25
	dilute	2	1.13	18
aqueous ammonia, $NH_{3(aq)}$	conc.	15.1	0.90	58
	dilute	6	0.96	23
sodium hydroxide, $NaOH$	conc.	19.1	1.52	50
	dilute	6	1.22	20

Appendix C: Values of Physical Constants

Avogadro's Number, N_o	6.0221×10^{23} units mol^{-1}
Ideal Gas Law Constant, R	0.08205 L-atm/mol-K 8.314 Joule/mol-K
Volume of 1 mol of an ideal gas:	
at 1 atm and 0°C	22.414 L
at 1 atm and 25°C	24.46 L

Appendix D: Metric Prefixes

Fractions of the Metric Unit

prefix	abbreviation	meaning
atto	a	10^{-18}
femto	f	10^{-15}
pico	p	10^{-12}
nano	n	10^{-9}
micro	μ (mu)	10^{-6}
milli	m	10^{-3}
centi	cm	10^{-2}
deci	d	10^{-1}

Multiples of the Metric Unit

prefix	abbreviation	meaning
deca	da	10^{1}
hecto	h	10^{2}
kilo	k	10^{3}
mega	M	10^{6}
giga	G	10^{9}
tera	T	10^{12}

Appendix E: Conversion Factors

length	1 m = 39.36 in 2.54 cm = 1 in
mass	453.4 g = 1 lb 1 kg = 2.20 lb 28.3 g = 1 oz (avoir.) 1 lb = 16 oz (avoir.)
volume	1 L = 1.0566 qt 946.4 mL = 1 qt 28.58 mL = 1 fluid oz 1 quart = 32 fluid ounces
temperature	K = °C + 273.15 °F = 1.8(°C) + 32 °C = (°F − 32)/1.8
pressure	1 atm = 760 mmHg 1 mmHg = 1 torr 1 atm = 14.7 lb/in^2
energy	1 cal = 4.184 Joule 1 Joule = 1×10^{7} erg 1 ev = 23.06 kcal/mol

Appendix F: Introduction to Chemical Nomenclature

Chemical nomenclature concerns the assignment of names to chemical substances. At the very least the name must provide all the information necessary to write the correct formula of the substance. Names are derived from sets of nomenclature rules that are known and accepted by everyone who transmits chemical information. The International Union of Pure and Applied Chemistry (IUPAC) is responsible for creating and maintaining the rules of chemical nomenclature. The current rules of "Nomenclature of Inorganic Chemistry" appear in the *Journal of the American Chemical Society*, volume 82, page 5527, published in 1960.

Subtle differences in names can denote large differences in chemical behavior. For example, sodium cyanate and sodium cyanide are both white, crystalline compounds that are soluble in water. But, sodium cyan*ide* is extremely toxic while sodium cyan*ate* is much less so. Pay close attention to details when writing and pronouncing names.

Greek numerical prefixes are used in many names to indicate the number of atoms of an element in a substance. In practice, the Latin "nona" is preferred over the Greek "ennea" for 9. The prefixes are:

1 = mono (usually omitted)
2 = di
3 = tri
4 = tetra
5 = penta
6 = hexa
7 = hepta
8 = octa
9 = nona
10 = deca

A. Elements

Several elements in their pure state exist as molecules. To avoid confusion, nomenclature must distinguish a molecule of an element from an atom of that element. Numerical prefixes are used to do this:

Element Name and Symbol		Molecule Name and Formula	
hydrogen	H	dihydrogen	H_2
oxygen	O	dioxygen	O_2
fluorine	F	difluorine	F_2
phosphorus	P	tetraphosphorus	P_4
sulfur	S	octasulfur	S_8

Name the following molecular species:

Cl_2 ______________________ N_2 ______________________

Br_2 ______________________ O_3 ______________________

B. Binary Compounds

Binary compounds are composed of two elements. There are two distinct nomenclature schemes used for binary compounds. The first is used for compounds composed of only nonmetals, such as HI and CO; the second is for those composed of metals with nonmetals, such as NaI and FeO. In both schemes, the names end with the suffix *-ide*.

1. Binary Compounds of Nonmetals: Molecular Compounds

In general, the name of the least electronegative element is written first without any modification followed by the name of the second, more electronegative element with its ending changed to *-ide*. The *-ide* ending is added to the stem of the name of the second element, as shown below.

Element	Stem	Name in Compound
fluorine	fluor-	fluoride
chlorine	chlor-	chloride
bromine	brom-	bromide
iodine	iod-	iodide
oxygen	ox-	oxide
sulfur	sulf-	sulfide
nitrogen	nitr-	nitride
phosphorus	phosph-	phosphide
boron	bor-	boride
carbon	carb-	carbide

Specifically, the element that appears further to the right in the following series will appear to the right in a binary formula and is the one that will have its name modified with the *-ide* ending:

B, Si, C, Sb, As, P, N, H, Te, Se, S, I, Br, Cl, O, F

The number of atoms of each element in the compound is indicated with the numeric prefixes given previously and named in the order given in the formula.

CO_2	carbon dioxide	N_2O_5	dinitrogen pentaoxide
H_2O	dihydrogen oxide	PF_5	phosphorus pentafluoride
NO	nitrogen oxide	N_2O	dinitrogen oxide
CS_2	carbon disulfide	CO	carbon monoxide

Note: The vowel at the end of a numerical prefix may be dropped *if* the name of the element it precedes begins with a vowel. For example, N_2O_5, dinitrogen pent*a*oxide, can be written as dinitrogen pentoxide.

Name the following binary compounds that are composed of only nonmetalic elements:

PBr_3	____________	CCl_4	____________
SO_2	____________	P_2O_3	____________
BrF_3	____________	N_2O_4	____________
B_2H_6	____________	NCl_3	____________
IF_7	____________	Cl_2O	____________
OF_2	____________	N_2O_5	____________

2. Binary Compounds of Metals with Nonmetals: Ionic Compounds

Since several variations can occur when naming compounds composed of metals and nonmetals, a series of rules will be followed.

Rule 1. The name of the metal appears first in the name of the compound, just as the symbol of the metal is first in its formula.

Rule 2. The name of the nonmetal follows that of the metal. As with binary compounds, the name of the nonmetal ends in *-ide*.

Rule 3. If the metal commonly appears in compounds in only one charge state, *such as the metals in groups IA (1+) and IIA (2+), and Al^{3+}, Zn^{2+}, Cd^{2+} and Ag^{+}*, the name of the compound is the name of the metal followed by the name of the nonmetal with its *-ide* ending as described in rule 2. The ammonium ion (NH_4^+) is treated like a metal ion of fixed charge. The following examples contain positive ions of fixed charge.

$NaCl$	sodium chloride	NH_4Cl	ammonium chloride
NaF	sodium fluoride	$AlCl_3$	aluminum chloride
$LiBr$	lithium bromide	ZnO	zinc oxide
Na_2O	sodium oxide	$(NH_4)_2S$	ammonium sulfide
K_2S	potassium sulfide	CaO	calcium oxide

Rule 4. If the metal commonly forms simple ions of more than one charge, such as iron (Fe^{2+} and Fe^{3+}), then the size of the positive charge is written as a Roman numeral in parentheses after the name of the metal, with no space between the name and parenthesis. Nearly all *transition metals* fall into this category as do the metals at and near the bottom of groups IIIA, IVA, VA and VIA (the metals of the representative elements). Fe^{2+} is written iron(II), Fe^{3+} is iron(III), Co^{3+} is cobalt(III), and V^{2+} is written vanadium(II).

For the ions of iron, copper, mercury and certain other metals that have been known since antiquity, there also exists an older nomenclature scheme you should know about. It uses the stem of the *Latin* name for the metal and adds an *-ous* or *-ic* ending to indicate whether it carries the *lower* or *higher* positive charge, respectively. Both schemes are shown below, though the systematic name with Roman numerals is preferred. Tin is included in this list. Though tin is not a transition metal, it exists in two charge states, 2+ and 4+, and its Latin name has been used in compounds for years. The Latin names for the elements listed below are Fe, *ferrum*, Cu, *cuprum*, Hg, *hydrargyrum*, and Sn, *stannum*.

Ion	Systematic Name	Older Name
Fe^{2+}	iron(II)	ferrous
Fe^{3+}	iron(III)	ferric
Cu^{+}	copper(I)	cuprous
Cu^{2+}	copper(II)	cupric
Hg_2^{2+}	mercury(I)	mercurous
Hg^{2+}	mercury(II)	mercuric
Sn^{2+}	tin(II)	stannous
Sn^{4+}	tin(IV)	stannic

The following examples show why it is important to indicate the size of the positive charge on the metal in compounds containing these elements. If you simply called a compound "iron chloride", you would be unable to clearly identify it as $FeCl_2$ or $FeCl_3$.

Formula	Systematic Name	Older Name
$FeCl_2$	iron(II) chloride*	ferrous chloride
$FeCl_3$	iron(III) chloride	ferric chloride
Cu_2O	copper(I) oxide	cuprous oxide
CuO	copper(II) oxide	cupric oxide
Hg_2Cl_2	mercury(I) chloride	mercurous chloride
$HgCl_2$	mercury(II) chloride	mercuric chloride
$SnCl_2$	tin(II) chloride	stannous chloride
$SnCl_4$	tin(IV) chloride	stannic chloride

* Pronounced "iron-two chloride," "iron-three chloride," and so forth

The systematic names of six other transition-metal binary compounds are:

CoI_2	cobalt(II) iodide	Cr_2O_3	chromium(III) oxide
MnO_2	manganese(IV) oxide	$AuCl_3$	gold(III) chloride
$CuCl_2$	copper(II) chloride	SnF_4	tin(IV) fluoride

Name the following ionic compounds composed of a metal and a nonmetal. If the metal is a transition metal, be certain to indicate its charge.

$CaCl_2$	______________	KF	______________
BaO	______________	$FeBr_3$	______________
HgO	______________	CuS	______________
Na_2S	______________	Al_2S_3	______________
Cr_2O_3	______________	CuI	______________
TiF_4	______________	CuI_2	______________
$CoBr_2$	______________	$CoBr_3$	______________

C. Compounds Containing Three or More Elements: Ionic Compounds

There are many ionic compounds that contain polyatomic ions, and therefore, three or more elements. Most polyatomic ions are *anions*; that is, they are ions with negative charge. Ions with positive charge, ions of metals, are called *cations*. You should learn the formulas, charges and names of the common polyatomic ions.

Formula	Ion Name	Formula	Ion Name
OH^-	hydroxide ion	$C_2H_3O_2^-$	acetate ion
NO_3^-	nitrate ion	NO_2^-	nitrite ion
CN^-	cyanide ion	OCN^-	cyanate ion
HCO_3^-	hydrogencarbonate ion (also called bicarbonate ion)	CO_3^{2-}	carbonate ion
SO_4^{2-}	sulfate ion	SO_3^{2-}	sulfite ion
ClO^-	hypochlorite ion	ClO_2^-	chlorite ion
ClO_3^-	chlorate ion	ClO_4^-	perchlorate ion
MnO_4^-	permanganate ion	$H_2PO_4^-$	dihydrogenphosphate
HPO_4^{2-}	hydrogenphosphate ion	PO_4^{3-}	phosphate ion
SCN^-	thiocyanate ion	PO_3^{3-}	phosphite ion
CrO_4^{2-}	chromate ion	$Cr_2O_7^{2-}$	dichromate ion
$C_2O_4^{2-}$	oxalate ion	O_2^{2-}	peroxide ion
AsO_4^{3-}	arsenate ion	AsO_3^{3-}	arsenite ion
HSO_4^-	hydrogensulfate ion	HSO_3^-	hydrogensulfite ion

Rule 5. Names of ionic compounds that contain polyatomic anions are written with the name of the metal first followed by the name of the anion.

$Ca(OH)_2$	calcium hydroxide	$CaCO_3$	calcium carbonate
$NaOH$	sodium hydroxide	$Mg(C_2H_3O_2)_2$	magnesium acetate
$NaHCO_3$	sodium hydrogencarbonate (or sodium bicarbonate)	$Fe_2(SO_4)_3$	iron(III) sulfate
$FePO_4$	iron(III) phosphate	CaC_2O_4	calcium oxalate
$Mg(OH)_2$	magnesium hydroxide	Hg_2SO_4	mercury(I) sulfate

Names of ionic compounds containing the ammonium ion (NH_4^+), hydroxide ion (OH^-) or cyanide ion (CN^-) often have the *-ide* endings of binary compounds even though they are composed of more than two elements.

Name the following ionic compounds which contain polyatomic ions:

$Cd(OH)_2$	______________	Na_3PO_4	______________
$CaSO_4$	______________	$NaNO_3$	______________
$KC_2H_3O_2$	______________	$NaHCO_3$	______________
$Al_2(SO_4)_3$	______________	$Sr(OH)_2$	______________
NH_4Cl	______________	$KClO_3$	______________

D. Acids and Bases

The names of several common acids and bases are given below. For acids, the names apply to the substance in aqueous solution. For example, pure $HCl_{(g)}$ is named as a binary compound, hydrogen chloride, but when this gas is dissolved in water, the solution is called hydrochloric acid and symbolized $HCl_{(aq)}$. For hydroxide bases, the name in or out of solution remains the same.

Formula	Acid Name	Formula	Base Name
$HCl_{(aq)}$	hydrochloric acid	$NaOH_{(aq)}$	sodium hydroxide
$H_2SO_{4(aq)}$	sulfuric acid	$KOH_{(aq)}$	potassium hydroxide
$H_2SO_{3(aq)}$	sulfurous acid	$Ca(OH)_{2(aq)}$	calcium hydroxide
$HNO_{3(aq)}$	nitric acid	$Ba(OH)_{2(aq)}$	barium hydroxide
$HClO_{4(aq)}$	perchloric acid	$NH_{3(aq)}$*	aqueous ammonia*
$H_3PO_{4(aq)}$	phosphoric acid		
$HC_2H_3O_{2(aq)}$	acetic acid		

* This base is sometimes written as NH_4OH and called ammonium hydroxide, though this is an incorrect name.

Name ______________________ Locker Number__________ Date__________

Please print; last name first

NOMENCLATURE EXERCISES: *Part 1*

Binary Nomenclature

Correctly name the following compounds:

1. CaF_2 ____________
2. $AlCl_3$ ____________
3. FeI_2 ____________
4. NaH ____________
5. $MgCl_2$ ____________
6. $ZnBr_2$ ____________
7. $MnCl_2$ ____________
8. NH_4Cl ____________
9. FeP ____________
10. CO_2 ____________
11. PBr_3 ____________
12. BrCl ____________
13. NCl_3 ____________
14. P_2O_3 ____________
15. SeO ____________
16. CCl_4 ____________
17. PI_5 ____________
18. PbO_2 ____________
19. SiC ____________
20. As_2O_5 ____________
21. SeO_3 ____________
22. N_2O ____________
23. NO ____________
24. Al_2O_3 ____________
25. MgO ____________
26. CrO_3 ____________
27. $Fe(OH)_3$ ____________
28. Ag_2O ____________
29. HgO ____________
30. NH_4I ____________
31. Cu_2O ____________
32. CuS ____________
33. LiBr ____________
34. Cl_2O_7 ____________
35. SO_2 ____________
36. N_2O_3 ____________
37. SCl_2 ____________
38. $TiCl_4$ ____________
39. N_2O_4 ____________
40. Sb_2O_3 ____________
41. GeI_4 ____________
42. CF_4 ____________
43. AlP ____________
44. BN ____________
45. BaF_2 ____________
46. BaO ____________
47. Bi_2O_3 ____________
48. SiO_2 ____________

Write the correct chemical formula for each of the following names:

1. ferric oxide ______________
2. sodium sulfide ______________
3. vanadium(V) oxide ______________
4. silver oxide ______________
5. aluminum sulfide ______________
6. nitrogen dioxide ______________
7. cesium bromide ______________
8. calcium phosphide ______________
9. cuprous chloride ______________
10. sodium iodide ______________
11. potassium nitride ______________
12. hydrogen iodide ______________
13. copper(II) bromide ______________
14. mercury(I) oxide ______________
15. tin(IV) oxide ______________
16. cadmium chloride ______________
17. chromium(VI) oxide ______________
18. vanadium(II) oxide ______________
19. ferric chloride ______________
20. stannous fluoride ______________
21. stannic fluoride ______________
22. titanium(II) oxide ______________
23. magnesium nitride ______________
24. ferrous oxide ______________
25. chromium(III) sulfide ______________
26. titanium(IV) oxide ______________
27. cobalt(II) hydroxide ______________
28. cupric boride ______________

Name ______________________ Locker Number__________ Date____________
Please print; last name first

NOMENCLATURE EXERCISES: *Part 2*

Acids, Bases and Ternary Compounds

Write the correct chemical formula for each compound:

1. phosphoric acid ____________
2. perchloric acid ____________
3. aluminum hydroxide ____________
4. copper(I) acetate ____________
5. lithium dihydrogenphosphate ____________
6. ammonium acetate ____________
7. ammonium carbonate ____________
8. ferric chloride ____________
9. diphosphorus trioxide ____________
10. calcium nitrite ____________
11. potassium sulfate ____________
12. hydrogen peroxide ____________
13. sodium chlorate ____________
14. cadmium iodide ____________
15. magnesium bicarbonate ____________
16. sodium dichromate ____________
17. vanadium(V) sulfide ____________
18. iron(III) sulfate ____________
19. sodium nitride ____________
20. selenium hexachloride ____________
21. calcium bisulfate ____________
22. nitric acid ____________
23. lithium cyanide ____________
24. potassium phosphate ____________
25. strontium hydride ____________

Give the name, with correct spelling, for each of the following:

1. $H_2SO_{4(aq)}$ ____________________
2. $H_2SO_{3(aq)}$ ____________________
3. $H_3PO_{4(aq)}$ ____________________
4. $Fe(OH)_3$ ____________________
5. MnO_2 ____________________
6. AlP ____________________
7. $KMnO_4$ ____________________
8. $Mg(ClO_3)_2$ ____________________
9. P_2O_5 ____________________
10. NF_3 ____________________
11. PF_5 ____________________
12. $NaHSO_4$ ____________________
13. $HC_2H_3O_{2(aq)}$ ____________________
14. $CsOH$ ____________________
15. Na_2O_2 ____________________
16. $Mg(HCO_3)_2$ ____________________
17. K_3PO_4 ____________________
18. KH_2PO_4 ____________________
19. H_2CO_3 ____________________
20. $HBr_{(aq)}$ ____________________
21. HgO ____________________
22. $Na_2C_2O_4$ ____________________
23. $NaClO$ ____________________
24. $BeBr_2$ ____________________
25. $MgSO_3$ ____________________

Appendix G: Significant Figures and Rounding Numbers

It is not possible to measure anything exactly. There will always be some degree of uncertainty in any measurement that will reflect the limitations of our measuring tools. For example, suppose the mass of a gold coin is measured on two balances. The first balance is capable of measuring to the tenth of a gram, 0.1 g, while the other is able to measure to the thousandth of a gram, 0.001 g. The second balance is capable of greater precision, and this would be shown in the number of digits (significant figures) appearing in the two mass values.

The first balance:	12.5 g	This result shows that the coin has a mass between 12.4 and 12.6 g. The mass has an uncertainty of ±0.1 g. This mass is stated to *3 significant figures*, the number of figures (digits) that are certain plus the first uncertain figure, that in the tenths place.
The second balance:	12.482 g	The second balance shows the mass of the coin to be between 12.481 and 12.483 g. This mass has an uncertainty of ±0.001 g and is stated to *5 significant figures*. There is only uncertainty in the last digit.

The precision of a measured value is shown by the number of significant figures it contains (all the digits known with certainty plus one additional digit that is somewhat uncertain). The better our measuring tools, the greater the precision of our measurements and the larger the number of significant figures.

Some numbers, either by definition or by an exact count, have an infinite number of significant figures. For example, there are exactly 100 cm in 1 m and 12 in in 1 ft. These are exact definitions. You may have exactly 12 dollars in your wallet. Exactly $12, because you can count the number of dollars exactly. But most numbers you will encounter are not exact.

A. Counting Significant Figures

How can you know the number of significant figures in a number? You simply count the digits, except there is one problem. Zeros may or may not be significant, and you have to determine if they are or not. To help you do this, the following rules will be applied to resolve the "zero problem."

1. *All nonzero digits* (1, 2, 3, 4, 5, 6, 7, 8, 9) *are always significant* and must be counted.
2. *A zero standing alone to the left of a decimal point is not significant* (**0**.55 g, **0**.0055 g). These zeros simply help you notice the decimal point.
3. *In a number that is less than 1, a zero to the right of the decimal point and before a nonzero digit is not significant.* It is simply occupying a space and not a measured digit (0.**0**55 g, 0.**0000**43 g). Both examples have only 2 significant figures.
4. *A zero between two nonzero digits is significant* (2**00**85 g, 15**0**1.4 g). These examples each have 5 significant figures.

5. *A zero at the end of a number and to the right of the decimal is significant.* It tells that the measuring tool was capable if measuring a 0 for that space (0.45**0** g, 0.0376**00** g). The first number has 3 significant figures; the second has 5.

6. *A zero at the end of a number and to the left of the decimal may or may not be significant.* You cannot tell by looking at the number if zeros of this kind were actually measured or if they are serving as spacers. For example, the zero in 528**0** ft = 1 mile is significant by definition, but if you were told there were 5000 chickens on a farm, it is likely the number is not exactly 5000–it's an estimate.

   The best way to avoid confusion with this kind of zero is to either express the number in scientific notation or put a decimal point at the end of the number if the zero is significant. If there are *exactly* 5000 chickens, then state the number as 5000. chickens or as 5.000×10^3 chickens. If there are 5000 chickens ± 100, that is, between 4900 and 5100 chickens, state the number as 5.0×10^3, showing the true precision of the value. (Keep in mind, it is harder to count chickens than dollars. Unlike dollars, chickens keep moving around and they are mortal.)

The number of significant figures in the following values is given in parentheses. Examine each number and see if you agree.

1000.0 mL	(5)	2.045 g	(4)	0.00998 m	(3)	0.005060	(4)
1000. mL	(4)	2.04 g	(3)	0.99800 m	(5)	1.005060	(7)
1000 mL	(?)	2.0 g	(2)	0.9980 m	(4)	5.20×10^3	(3)

B. Using Significant Figures in Calculations

1. Multiplication and Division

In multiplication or division, the product or quotient will be given with the same number of significant figures as in the *least* significant number used in the calculation. Significant figures is abbreviated s.f. below, and the typical answer that would be seen on a calculator is shown. It always needs to be rounded to the allowed number of significant figures. Rounding will be discussed in Part 3.

$$\underset{\text{(5 s.f.)}}{3.6614\ \text{g/mL}} \times \underset{\text{(3 s.f.)}}{25.0\ \text{mL}} = \underset{\text{(shown on the calculator)}}{91.535\ \text{g}} = \underset{\text{(rounded to 3 s.f.)}}{91.5\ \text{g}}$$

$$\frac{4.55 \times 10^{-2}\ \text{m (3 s.f.)}}{9.2\ \text{sec (2 s.f.)}} = \underset{\text{(on the calculator)}}{4.9456521 \times 10^{-3}\ \text{m/sec}} = \underset{\text{(rounded to 2 s.f.)}}{4.9 \times 10^{-3}\ \text{m/sec}}$$

The following calculation includes an exact conversion factor (infinite precision), but the result can only be stated to 3 significant figures, limited by the lack of precision in the kilogram value. Again, the calculator value is shown, which demonstrates that calculators know nothing about significant figures.

$$\frac{0.0342\ \text{kg} \times 1000\ \text{g/kg}}{4.195\ \text{L}} = 8.152562575\ \text{g/L} = 8.15\ \text{g/L (3 s.f.)}$$

2. Addition and Substraction

In addition or substraction, the answer can only be as precise as the least precise number in the calculation. Usually this only concerns the number of digits to the right of the decimal point. The sum of 4.55882 + 0.042 can only be stated to the third place after the decimal. This is because all places after the third place in 0.042 are unknown.

```
   4.55882
 + 0.042??
   ---------
   4.601     (rounded from 4.60082)
```

In addition and subtraction, it is not so much a problem of significant figures as uncertainty in the tenths, hundredths or thousandths place. The following examples show this.

```
   9.43                0.00426        19.35          5
 - 6.9342            + 0.01         +  6.284       - 0.213
   ------              -------        ------         -----
   2.50                0.01           25.63          5
(rounded from 2.4958)                               (This doesn't seem
                                                     right, but it is.)
```

3. Rounding

Several numbers have been rounded in the previous calculations so that they show the degree of certainty (or uncertainty) in a measurement. Because calculators do not take precision into account, they display a window full of digits. You have to decide how many of these digits are significant and round the displayed value accordingly. All digits after the last significant figure are dropped using the following procedure:

1. If the first digit dropped (after the last significant figure) is less than 5, do nothing to the last significant figure. Rounding 5.3442 to 3 s.f. (5.34|42) gives 5.34 .
2. If the first digit dropped is greater than 5, increase the last significant figure by one. Rounding 7.3621 to 2 s.f. (7.3|621) gives 7.4 .
3. If the first digit to be dropped is 5, followed by zeros or by no other digits, then:
   a. if the last significant figure is odd, round it *up* to an even number.
   b. if the last significant figure is even, drop the 5 and keep it as an even number. 67.3$\underline{4}$5 rounded to 4 s.f. becomes 67.34; 67.3$\underline{5}$5 to 4 s.f. becomes 67.36.

These rules are applied as the following numbers are rounded to 4, 3 and 2 significant figures.

		(4 s.f.)	(3 s.f.)	(2 s.f.)
0.034567	becomes	0.03457	0.0346	0.035
132,593	becomes	1.326×10^5	1.33×10^5	1.3×10^5
72.3462	becomes	72.35	72.3 (not 72.4)	72

In multiplication and division problems, it is common practice to calculate the answer then round it off to the proper number of significant figures. Do the rounding as the last step.

Name ______________________ Locker Number__________ Date____________
Please print; last name first

CALCULATION EXERCISES:

Significant Figures and Rounding Numbers

1. Indicate the number of significant figures in each of the following:

0.034	______	9.23×10^{-6}	______	12500.	______	400.0	______
5	______	0.00230	______	0.004	______	2.001	______
4×10^3	______	0.0100	______	5001	______	98.6	______

2. In terms of significant figures, comment on the ambiguity of a measurement that is stated simply as 5000 mL. How could this ambiguity be removed?

3. Round each of the following numbers to 3 significant figures. State the rounded numbers in scientific notation.

12,345	____________	67,890	____________	0.1235	____________
1,650,953	____________	0.0003555	____________	95,000.	____________
0.082055	____________	7955	____________	40,050,000	____________

4. Carry out the following calculations, stating the answers in the correct number of significant figures. Consider each number given in each calculation to be a measured value expressed to its allowed significance. Scientific notation will be useful.

$1.334 \times 0.23 =$ ____________	$(45.25 \times 60.) \div 1215 =$ ____________
$1.2153 + 0.012 =$ ____________	$5341. - 90. =$ ____________
$1.30 \div 0.005233 =$ ____________	$(0.00450/4.2) + 1.00042 =$ ____________
$(5.62)^3 =$ ____________	$9755. + 0.05 =$ ____________
$1.00 \div 760 =$ ____________	$0.011 + 0.0111 + 0.1 =$ ____________
$(99/11) \times 9.0 =$ ____________	$(5.5 \times 10^1)^2 =$ ____________
$0.4521 - 0.0003 + 2.00 =$ ____________	$3 \times 5 \times 7 \times 9 =$ ____________

Appendix H: Solubility Rules

The factors that affect the solubility of inorganic compounds in water are complex, and the prediction of solubility requires knowing a great deal about the compound and the way it interacts with water. Nearly every substance is soluble to some small degree in water, and a line needs to be drawn to separate those considered soluble from those considered insoluble. In this laboratory manual, *a compound will be considered soluble if it can form a 0.05 M solution in water at room temperature.*

To help you predict whether a salt[1] is soluble in water, a collection of general statements has been assembled as a guide. These statements compose the *solubility rules*. As with most general rules, there are exceptions to each one, but they guide you to the correct prediction most of the time.

The solubility rules are organized in a hierarchal structure; that is, the first rule takes precedence over the second, and the second takes precedence over the third, and so forth.

1. All **sodium** (Na^+), **potassium** (K^+) and **ammonium** (NH_4^+) salts are *soluble.*
2. All **nitrate** (NO_3^-), **acetate** ($C_2H_3O_2^-$) and **perchlorate** (ClO_4^-) salts are *soluble.*
3. All **silver** (Ag^+), **lead** (Pb^{2+}) and **mercury(I)** (Hg_2^{2+}) salts are *insoluble.*
4. All **chloride** (Cl^-), **bromide** (Br^-) and **iodide** (I^-) salts are *soluble* in water (except those of silver, lead and mercury(I) as stated in the previous rule).
5. All **carbonate** (CO_3^{2-}), **phosphate** (PO_4^{3-}), **sulfide** (S^{2-}) and **oxalate** ($C_2O_4^{2-}$) salts are *insoluble* (except for those exempted by rule 1: Na^+, K^+ and NH_4^+).
6. All **sulfates** (SO_4^{2-}) are *soluble* in water *except* those of Ca^{2+}, Ba^{2+}, Ag^+, Pb^{2+} and Hg_2^{2+}.
7. Though not salts, all **oxides** (O^{2-}) and **hydroxides** (OH^-) are *insoluble* in water (except those of Na^+, K^+, Ca^{2+} and Ba^{2+}).

Nearly all insoluble hydroxides, carbonates, phosphates, sulfides and oxides are soluble in strong acids, such as hydrochloric acid, $HCl_{(aq)}$ or nitric acid, $HNO_{3(aq)}$.

There are many compounds that are not addressed in the statements given above. For compounds not considered above, consult the *Handbook of Chemistry and Physics* or Lange's *Handbook of Chemistry*. Either should be available in the library or perhaps your instructor may have them.

---

1 Many inorganic compounds are collectively classified as salts. A salt is the ionic product of an acid-base neutralization reaction. The cation of the salt comes from the base, the anion from the acid. KNO_3 is the salt in this neutralization: $H\boldsymbol{NO}_{3(aq)} + \boldsymbol{K}OH_{(aq)} \rightarrow \boldsymbol{KNO}_{3(aq)} + H_2O_{(l)}$.

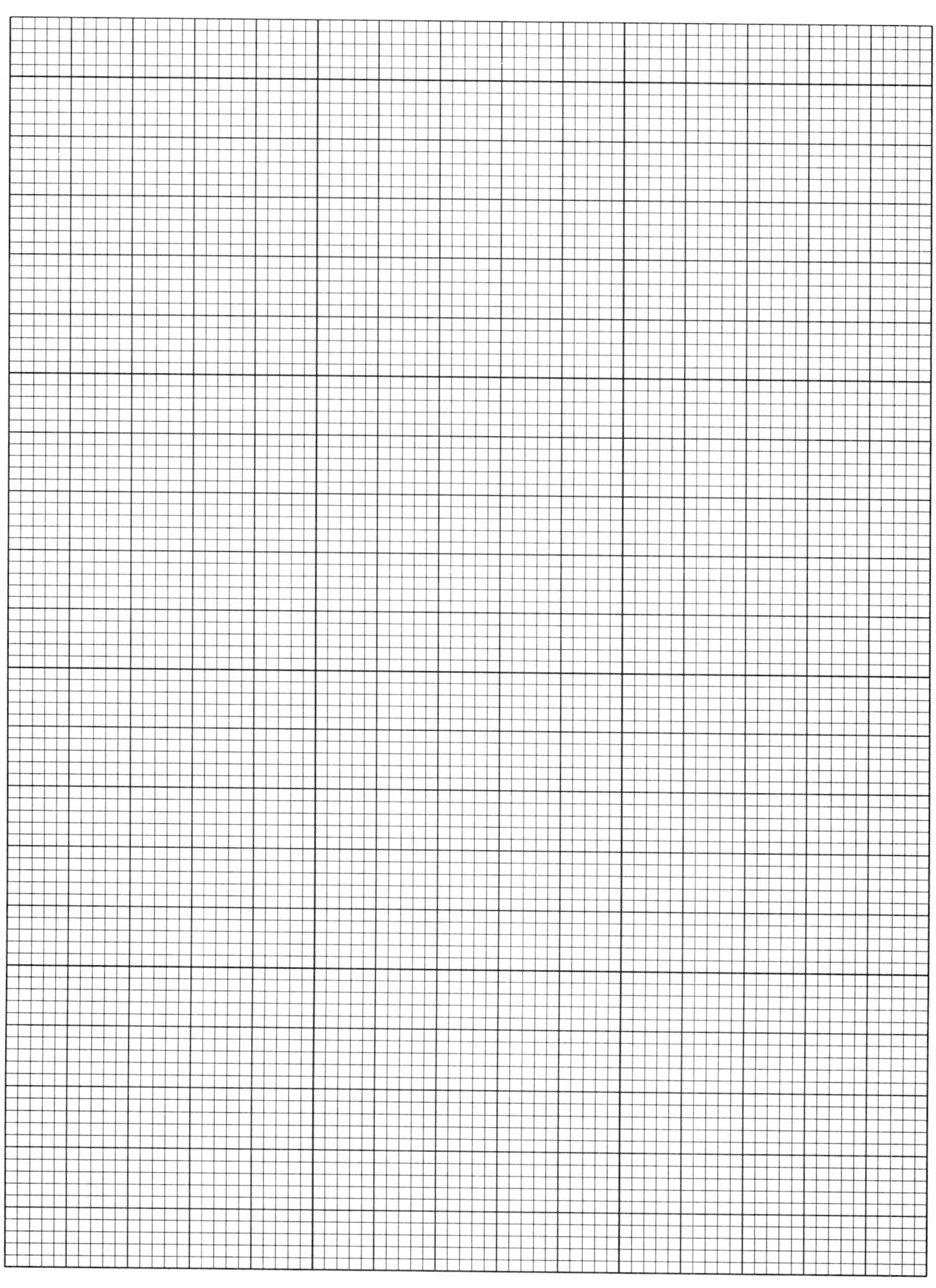

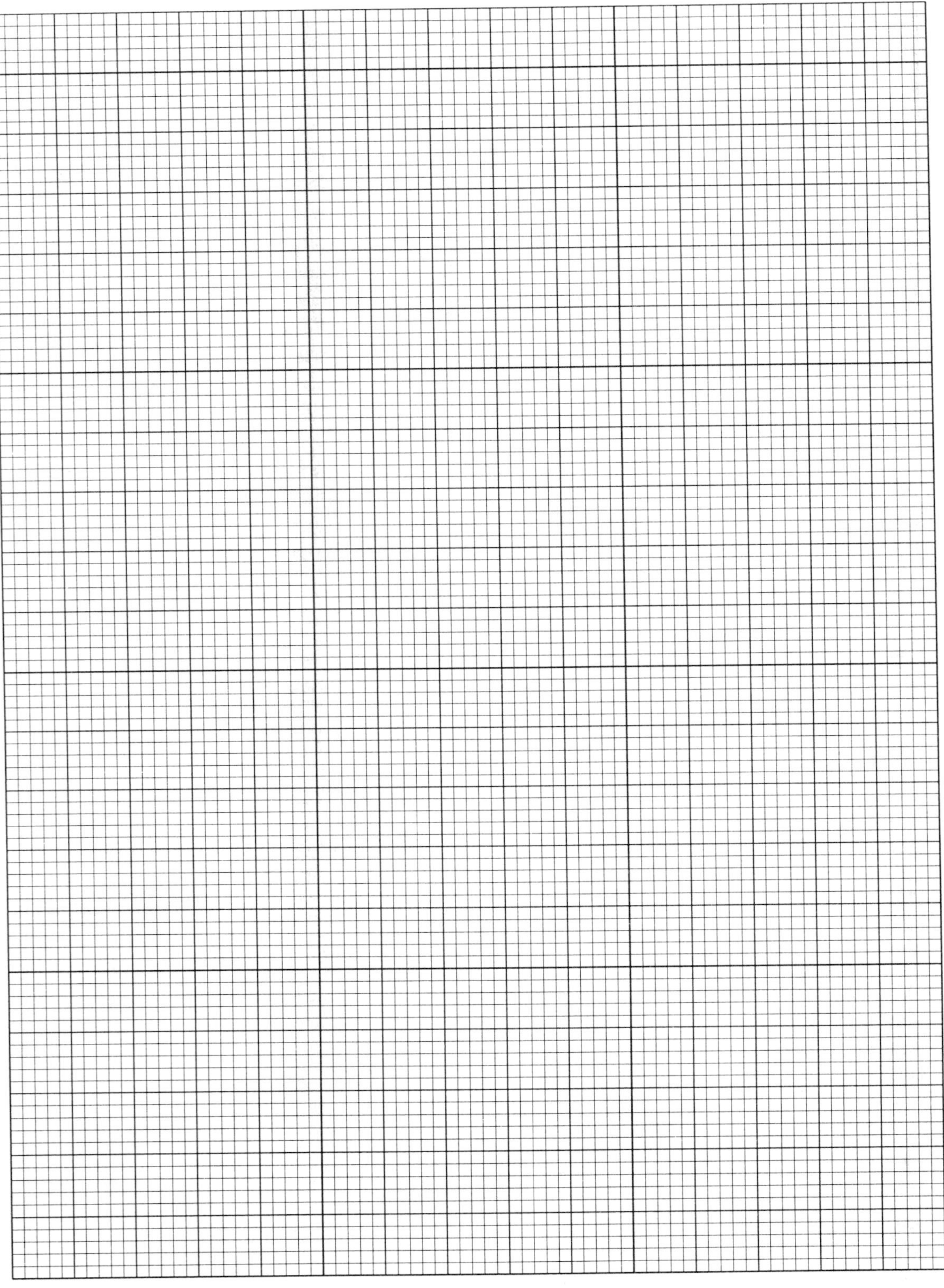